Pomeron Physics and QCD

This book describes the underlying ideas and modern developments of Regge theory. It confronts the theory with a huge variety of experimental data and with quantum chromodynamics. The book covers forty years of research and provides a unique insight into the theory and its phenomenological development. It provides comprehensive coverage of the various different theoretical approaches and considers the key issues for future theory and experiment.

The authors review experiments that suggest the existence of a soft pomeron and give a detailed description of attempts to describe this through nonperturbative quantum chromodynamics. They suggest that a second, hard pomeron is responsible for the dramatic rise in energy observed in deep inelastic lepton–nucleon scattering. The two-pomeron hypothesis is applied to a variety of interactions and is compared with perturbative quantum chromodynamics, as well as with the dipole approach.

This book will be a valuable reference for theoretical and experimental particle physicists all over the world. It is also suitable as a textbook for graduate courses in particle physics, high-energy scattering, QCD and the standard model.

SANDY DONNACHIE obtained his PhD from the University of Glasgow in 1961. He is currently Professor of Physics at the University of Manchester, a post he has held since 1969. He has been a visiting professor at departments and institutes throughout the world and is the author of around 170 publications, including one previous book.

GÜNTER DOSCH was awarded his PhD in 1963, from the University of Heidelberg. Since then he has held positions at the University of Heidelberg, CERN, the University of Karlsruhe and Massachusetts Institute of Technology. He has published numerous articles in journals and also in the proceedings of conferences and schools. He is currently a professor at the Institute for Theoretical Physics at the University of Heidelberg.

PETER LANDSHOFF qualified for his PhD from the University of Cambridge in 1962. He is Professor of Mathematical Physics there and is Vice-Master of Christ's College. He has led new research ventures in quantum information and in e-science and played an active role in the creation of the Isaac Newton Institute for Mathematical Sciences and of the Millennium Mathematics Project, whose object is to help people of all ages share in the excitement of mathematics. Professor Landshoff has spent extended periods at CERN and is an editor of *Physics Letters B*. He is the author of about 130 publications, including two previous books.

OTTO NACHTMANN obtained his PhD in 1967, from the University of Vienna. Since 1975 he has been a professor at the Institute for Theoretical Physics at the University of Heidelberg. He has spent extended periods at institutes and universities throughout the world and is the author of about 140 publications, including one previous book.

Pomeron Physics and QCD

Sandy Donnachie
University of Manchester

Günter Dosch
Universität Heidelberg

Peter Landshoff
University of Cambridge

Otto Nachtmann
Universität Heidelberg

CAMBRIDGE
UNIVERSITY PRESS

CAMBRIDGE UNIVERSITY PRESS
Cambridge, New York, Melbourne, Madrid, Cape Town, Singapore, São Paulo

Cambridge University Press
The Edinburgh Building, Cambridge CB2 2RU, UK

Published in the United States of America by Cambridge University Press, New York

www.cambridge.org
Information on this title: www.cambridge.org/9780521780391

First published 2002
This digitally printed first paperback version 2005

A catalogue record for this publication is available from the British Library

Library of Congress Cataloguing in Publication data
Pomeron physics and QCD / Sandy Donnachie . . . [et al.].
p.   cm. – (Cambridge monographs on particle physics, nuclear physics,
and cosmology; v. 19)
Includes bibliographical references and index.
ISBN 0 521 78039 X
1. Regge theory.   2. Pomerons.   3. Quantum chromodynamics.
I. Donnachie, Sandy, 1936-   II. Cambridge monographs on particle physics,
nuclear physics, and cosmology; 19
QC793.3.R4 P66 2002
539.7′21–dc21   2002023376

ISBN-13 978-0-521-78039-1 hardback
ISBN-10 0-521-78039-X hardback

ISBN-13 978-0-521-67570-3 paperback
ISBN-10 0-521-67570-7 paperback

Contents

Preface

In 1935 the Japanese physicist Hideki Yukawa predicted that there must be a particle, now known as the pion, which would transmit the strong interaction. The pion was duly discovered more than ten years later. However, we now know that although pion exchange is an important component of the static force, when the force acts between a pair of particles with high energy a very large number of particles collaborate in transmitting it. Regge theory provides a simple quantitative description of the combined effect of all these particle exchanges.

It was soon realised that the exchanges of the known particles, even though several hundred are listed in the data tables, are not sufficient to describe a striking feature of the strong force: that it retains its strength as the energy increases and even becomes yet stronger. To explain this, it must be that something else is exchanged. This new object was named after the Russian physicist Isaac Pomeranchuk. It was originally called the pomeranchukon, but this was later abbreviated to pomeron. Events in which a pomeron is exchanged are often called diffractive events. The reason for this is that pomeron exchange dominates in high-energy elastic scattering and, as we describe in chapter 3, when plotted against scattering angle the differential cross section has a striking dip, reminiscent of the intensity distribution in optical diffraction. However, we explain that actually the mechanism for dip generation in high-energy scattering is more complicated than in optical diffraction.

During the 1960s it was found that, with the inclusion of the soft pomeron, Regge theory provides a very successful description of a huge quantity of experimental data. This was summarised by Collins in his classic book[1], which was published in 1977. However, the phenomenology appeared to be complicated. It was not until the 1980s that it became apparent that the reason for this was that the early data were at comparatively low energies. When the rather-higher-energy data became available from the CERN ISR, and later on from the CERN $\bar{p}p$ collider, the phenomenology became much simpler[2], indeed considerably simpler than the known theory was able to explain.

Meanwhile, quantum chromodynamics (QCD) had been discovered in the early 1970s. It was natural to try to explain the pomeron in terms of QCD, and first attempts to do so were made by Low[3] and by Nussinov[4]. These attempts were refined over the years within the framework of perturbative QCD, notably by Cheng and Wu[5] and by Lipatov and his collaborators[6]. However, it is rather clear that the pomeron that controls the high-energy behaviour of soft hadronic reactions cannot be described by perturbation theory and work began[7] in the late 1980s on the very difficult task of modelling it through nonperturbative QCD. Even now, we still cannot claim that we have more than a rough description of the pomeron in terms of QCD.

Towards the end of the 1960s experiments had begun at the Stanford Linear Accelerator. These scattered electrons on protons and studied the rare events in which the electron momentum transfer was large. Although such events were comparatively rare, they were sufficiently copious to show that the proton contains a number of small scattering centres, which we know now to be quarks. In a real sense, these experiments marked the beginning of modern high energy physics. At the beginning of the 1990s similar experiments, but at a very much higher energy, began at the electron-proton collider HERA in Hamburg. These experiments made the quite dramatic discovery that the probability of the occurrence of large-electron-momentum-transfer events grows very rapidly with energy. At first, this was believed to be a triumphant confirmation of the perturbative-QCD calculations. That is, it suggested that actually there are two pomerons, the "soft" nonperturbative one which is responsible for the fairly gentle rise with energy of soft hadronic reactions, and a "hard" perturbative one responsible for the dramatic rise with energy of the large-electron-momentum-transfer scattering probability. Unfortunately, it was then found that the perturbative-QCD calculations receive very large nonleading-order corrections, so that the agreement between theory and experiment was lost. At present, there is no generally-accepted explanation of the dramatic HERA behaviour.

The electron-scattering experiments explore the structure of the proton. Ingelman and Schlein suggested[8] that a special class of electron-scattering events might also study the structure of the pomeron (or pomerons), thereby creating the topic of hard diffraction. This has been an active area of study at HERA; it actually began at the CERN $\bar{p}p$ collider and is continuing at the Tevatron at Fermilab.

All these things are the subject of our book, which draws together a huge amount of knowledge gathered at existing and past accelerators, as a preparation for the beginning of the operation of RHIC at Brookhaven and the Large Hadron Collider at CERN.

We have set up a web page in connection with this book:

http://www.damtp.cam.ac.uk/user/pvl/QCD/

We will use this to record corrections to the book, and perhaps some updates. It also makes available all the figures, which we are happy for others to use with due acknowledgment.

We wish to pay tribute to the Durham database,

http://www-spires.dur.ac.uk/HEPDATA/,

which we have used extensively to create our figures, particularly those for which the sources of the data have not been explicitly referenced.

We record with gratitude also that the research of AD and PVL has been supported in part by the UK Particle Physics and Astronomy Research Council, and the research of HGD and ON by the German Bundeministerium für Bildung und Forschung, which also has largely funded our meetings to prepare this book.

Sandy Donnachie
Günter Dosch
Peter Landshoff
Otto Nachtmann
December 2001

1

Properties of the S-matrix

In this chapter we specify the kinematics, define the normalisation of amplitudes and cross sections and establish the basic formalism used throughout. All mathematical functions used, and their properties, can be found in [9].

1.1 Kinematics

We consider first the two-body scattering process $1 + 2 \rightarrow 3 + 4$ of figure 1.1, where the particles have masses m_i and four-momenta P_i, $i = 1, \ldots, 4$. Our notation is that the four-momentum of a particle is $P = (E, \mathbf{p})$, where E is its energy and $\mathbf{p}$ its three-momentum, and we write

$$P_1.P_2 = E_1 E_2 - \mathbf{p}_1.\mathbf{p}_2. \tag{1.1}$$

The Lorentz-invariant variables s, t and u, called Mandelstam variables, are defined by

$$
\begin{aligned}
s &= (P_1 + P_2)^2 \\
t &= (P_1 - P_3)^2 \\
u &= (P_1 - P_4)^2
\end{aligned}
\tag{1.2}
$$

with the relation

$$s + t + u = \sum_{i=1}^{4} m_i^2. \tag{1.3}$$

Equation (1.3) means that a two-body amplitude is a function of only two independent variables. We shall normally take these to be s and t, with u defined via (1.3), and write the amplitude as $A(s, t)$. However, sometimes

1

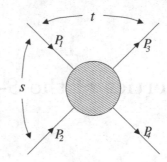

Figure 1.1.   Two-body scattering process $1 + 2 \to 3 + 4$

it will be more appropriate to use s and u, or t and u, as the independent variables, and then write the amplitude as $A(s, u)$ or $A(t, u)$.

Figure 1.1 not only describes the scattering process $1 + 2 \to 3 + 4$ in the s-channel but, by reversing the signs of some of the four-momenta, it can also represent the t-channel process $1 + \bar{3} \to \bar{2} + 4$ and the u-channel process $1 + \bar{4} \to 3 + \bar{2}$, where the bar denotes the antiparticle.

In the s-channel centre-of-mass frame of the initial particles 1 and 2, the four-momenta are given explicitly by

$$P_1 = (E_1, \mathbf{p}_1) \qquad P_2 = (E_2, -\mathbf{p}_1)$$
$$P_3 = (E_3, \mathbf{p}_3) \qquad P_4 = (E_4, -\mathbf{p}_3) \tag{1.4}$$

where E_i is the energy of particle i, $\mathbf{p}_1$ is the three-momentum of particle 1 and $\mathbf{p}_3$ the three-momentum of particle 3 in this frame. Then

$$s = (E_1 + E_2)^2 = (E_3 + E_4)^2 \tag{1.5}$$

and

$$E_1 = \frac{1}{2\sqrt{s}}(s + m_1^2 - m_2^2) \quad E_2 = \frac{1}{2\sqrt{s}}(s + m_2^2 - m_1^2)$$
$$E_3 = \frac{1}{2\sqrt{s}}(s + m_3^2 - m_4^2) \quad E_4 = \frac{1}{2\sqrt{s}}(s + m_4^2 - m_3^2) \tag{1.6}$$

and

$$\mathbf{p}_1^2 = \frac{1}{4s}[s - (m_1 + m_2)^2][s - (m_1 - m_2)^2]$$
$$\mathbf{p}_3^2 = \frac{1}{4s}[s - (m_3 + m_4)^2][s - (m_3 - m_4)^2]. \tag{1.7}$$

From (1.2) and (1.4),

$$t = m_1^2 + m_3^2 - 2(E_1 E_3 - \mathbf{p}_1 \cdot \mathbf{p}_3)$$
$$= m_1^2 + m_3^2 - 2(E_1 E_3 - |\mathbf{p}_1||\mathbf{p}_3| \cos \theta_s)$$
$$u = m_1^2 + m_4^2 - 2(E_1 E_4 + \mathbf{p}_1 \cdot \mathbf{p}_3)$$
$$= m_1^2 + m_4^2 - 2(E_1 E_4 + |\mathbf{p}_1||\mathbf{p}_3| \cos \theta_s) \qquad (1.8)$$

where θ_s is the angle between the three-momenta of particles 1 and 3 in the s-channel centre-of-mass frame, that is it is the centre-of-mass-frame scattering angle.

The physical region for the s-channel is given by

$$s \geq (m_1 + m_2)^2 \qquad \text{and} \qquad -1 \leq \cos \theta_s \leq 1. \qquad (1.9)$$

For arbitrary masses the boundary of the physical region as a function of s and t is rather complicated. It is simpler for equal masses $m_i = m$, $i = 1, \ldots, 4$, so that $\mathbf{p}_1 = \mathbf{p}_3 = \mathbf{p}$ and

$$s = 4(\mathbf{p}^2 + m^2)$$
$$t = -2\mathbf{p}^2(1 - \cos \theta_s)$$
$$u = -2\mathbf{p}^2(1 + \cos \theta_s). \qquad (1.10)$$

The physical region for s-channel scattering is then given by $s \geq 4m^2$, $t \leq 0$ and $u \leq 0$. In this channel, s is an energy squared and each of t and u is a momentum transfer squared. Similarly the physical region for t-channel scattering is $t \geq 4m^2$, $u \leq 0$, $s \leq 0$; and for u-channel scattering it is $u \geq 4m^2$, $s \leq 0$, $t \leq 0$. The symmetry between s, t and u is readily demonstrated by plotting the physical regions in the s-t plane with the s and t axes inclined at $60°$, as shown in figure 1.2.

1.2 The cross section

For orthonormal states $\langle f|$ and $|i\rangle$, that satisfy $\langle f|f\rangle = \langle i|i\rangle$ and $\langle f|f'\rangle = \delta_{ff'}$, the S-matrix element $\langle f|S|i\rangle$ is defined such that

$$P_{fi} = |\langle f|S|i\rangle|^2 = \langle i|S^\dagger|f\rangle\langle f|S|i\rangle \qquad (1.11)$$

is the probability of $|f\rangle$ being the final state, given $|i\rangle$ as the initial state. If the set of orthonormal states $|f\rangle$ is complete,

$$\sum_f |f\rangle\langle f| = 1. \qquad (1.12)$$

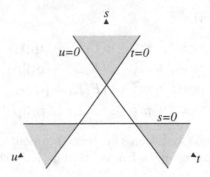

Figure 1.2.   Physical regions for equal-mass scattering such as $\pi\pi \to \pi\pi$

Starting from the initial state $|i\rangle$, the probability of ending up in some final state must be unity so

$$1 = \sum_f |\langle f|S|i\rangle|^2 = \sum_f \langle i|S^\dagger|f\rangle\langle f|S|i\rangle = \langle i|S^\dagger S|i\rangle. \qquad (1.13)$$

Since (1.13) must be true for any choice of the complete set of basis states $|i\rangle$ it follows that $S^\dagger S = 1$. Similarly the requirement that any final state $|f\rangle$ has originated from some initial state $|i\rangle$ yields $SS^\dagger = 1$. That is, S is unitary.

We now go over to the case of continuum states and specialise to a two-body initial state. The scattering matrix S is related to the transition matrix T by

$$\langle f|S|i\rangle = \langle P'_1 P'_2 \ldots P'_n|S|P_1 P_2\rangle = \delta_{fi} + i(2\pi)^4\delta^4(P^f - P^i)\langle f|T|i\rangle \qquad (1.14)$$

where P^i is the sum of the initial four-momenta and P^f the sum of the final four-momenta. The scattering amplitude is normalised such that the transition rate per unit time per unit volume from the initial state $|i\rangle = |P_1 P_2\rangle$ to the final state $|f\rangle = |P'_1 \cdots P'_n\rangle$ is

$$R_{fi} = (2\pi)^4\delta^4(P^f - P^i)\,|\langle f|T|i\rangle|^2. \qquad (1.15)$$

The total cross section for the reaction $12 \to n$ particles is

$$\sigma_{12\to n} = \frac{1}{4|\mathbf{p}_1|\sqrt{s}} \sum (2\pi)^4\delta^4(P^f - P^i)\,|\langle f_n|T|i\rangle|^2 \qquad (1.16)$$

where the sum is over the momenta of the particles in the n-particle state $\langle f_n|$. That is, with $\delta^+(p^2 - m^2) = \delta(p^2 - m^2)\,\theta(p^0)$,

$$\sigma_{12 \to n} = \frac{1}{4|\mathbf{p}_1|\sqrt{s}} \int \left(\prod_{i=1}^{n} \frac{d^4 P_i'}{(2\pi)^4} 2\pi \delta^+(P_i'^2 - m_i^2) \right)$$

$$\times (2\pi)^4 \delta^4 \left(\sum_{i=1}^{n} P_i' - P_1 - P_2 \right) |\langle P_1' \cdots P_n' |T| P_1 P_2 \rangle|^2$$

$$= \frac{1}{4|\mathbf{p}_1|\sqrt{s}} \int \left(\prod_{i=1}^{n} \frac{d^3 p_i'}{2E_i(2\pi)^3} \right) (2\pi)^4 \delta^4 \left(\sum_{i=1}^{n} P_i' - P_1 - P_2 \right)$$

$$\times |\langle P_1' \cdots P_n' |T| P_1 P_2 \rangle|^2. \tag{1.17}$$

Here, $\mathbf{p}_1$ is the initial momentum in the s-channel centre-of-mass frame. It is given by (1.7):

$$|\mathbf{p}_1|^2 s = (P_1.P_2)^2 - m_1^2 m_2^2 = \tfrac{1}{4}[s - (m_1 + m_2)^2][s - (m_1 - m_2)^2]. \tag{1.18}$$

We must use this in (1.17), which then gives the cross section in any frame: it is Lorentz invariant, and the momentum integrations may be performed in any frame.

We may calculate a differential cross section $d\sigma_{12 \to n}/d\omega$. Typically, ω will be a momentum transfer between an initial and a final particle, or the corresponding scattering angle, or the energy of one of the final particles. To calculate the differential cross section, we first express ω as a function $\omega(P_i, P_f')$ of the various momenta, and then include $\delta(\omega - \omega(P_i, P_f'))$ in the integrations in (1.17). For example, when the final state contains just two particles and t is the momentum transfer defined in (1.2),

$$\frac{d\sigma_{12 \to 34}}{dt} = \frac{1}{4|\mathbf{p}_1|\sqrt{s}} \int \frac{d^4 P_3}{(2\pi)^4} 2\pi \delta^+(P_3^2 - m_3^2) \frac{d^4 P_4}{(2\pi)^4} 2\pi \delta^+(P_4^2 - m_4^2)$$

$$\times (2\pi)^4 \delta^4(P_1 + P_2 - P_3 - P_4)|\langle P_3 P_4 |T| P_1 P_2 \rangle|^2 \delta(t - (P_1 - P_3)^2)$$

$$= \frac{1}{64\pi|\mathbf{p}_1|^2 s}|\langle P_3 P_4 |T| P_1 P_2 \rangle|^2 \, \delta(t - (P_1 - P_3)^2). \tag{1.19}$$

In the equal-mass case this gives

$$\frac{d\sigma}{dt} = \frac{1}{16\pi s(s - 4m^2)} |\langle P_3 P_4 |T| P_1 P_2 \rangle|^2. \tag{1.20}$$

The formulae in this section apply when the particles involved have no spin or, if they do have spin, when we average over initial spin states and sum over final spin states.

1.3 Unitarity and the optical theorem

Unitarity provides an important connection between the total cross section and the forward ($\theta_s = 0$) elastic scattering amplitude; this connection is known as the optical theorem. Because the operator S is unitary, so that $SS^\dagger = 1$, for any orthonormal states $\langle j|$ and $|i\rangle$

$$\delta_{ji} = \langle j|SS^\dagger)|i\rangle = \sum_f \langle j|S|f\rangle\langle f|S^\dagger)|i\rangle \qquad (1.21)$$

where we have used the completeness relation (1.12). With the definition (1.14) of the T-matrix, this is

$$\langle j|T|i\rangle - \langle j|T^\dagger|i\rangle = (2\pi)^4 i \sum_f \delta^4(P^f - P^i)\langle j|T^\dagger|f\rangle\langle f|T|i\rangle. \qquad (1.22)$$

For the particular case $j = i$,

$$2\,\mathrm{Im}\,\langle i|T|i\rangle = \sum_f (2\pi)^4 \delta^4(P^f - P^i)|\langle f|T|i\rangle|^2. \qquad (1.23)$$

The right-hand side is (1.15) summed over f: it is the total transition rate. This gives us the total cross section, which is (1.17) summed over n, the number of final-state particles:

$$\sigma_{12}^{\mathrm{Tot}} = \frac{1}{2|\mathbf{p}_1|\sqrt{s}}\,\mathrm{Im}\,\langle i|T|i\rangle. \qquad (1.24)$$

Here, $|\mathbf{p}_1|$ is again the magnitude of the initial centre-of-mass frame three-momentum, which is given by (1.18). $\langle i|T|i\rangle$ is the scattering amplitude for the reaction $1 + 2 \to 1 + 2$ with the direction of motion of the particles unchanged, that is it is the forward scattering amplitude, $\theta_s = 0$. For $m_3 = m_1$ and $m_4 = m_2$ the forward direction corresponds to $t = 0$. Then

$$\sigma_{12}^{\mathrm{Tot}} = \frac{1}{2|\mathbf{p}_1|\sqrt{s}}\,\mathrm{Im}\,A(s, t = 0) \qquad (1.25)$$

where $A(s, t)$ is the elastic scattering amplitude. Equation (1.24) or (1.25) is the optical theorem.

1.4 Crossing and analyticity

The basic principle of crossing is that the same function $A(s, t)$ analytically continued to the three physical regions of figure 1.2 gives the corresponding scattering amplitude there, with s, t, u related by (1.3). This is obviously

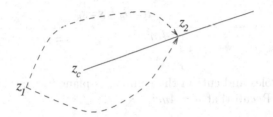

Figure 1.3. Paths of analytic continuation that pass round different sides of a branch point

true order by order for Feynman diagrams. For example Coulomb scattering $(e^-e^- \to e^-e^-)$ and Bhabha scattering $(e^+e^- \to e^+e^-)$ are described by the same Feynman diagrams.

It is necessary to make some assumption about the analytic structure of the scattering amplitude $A(s,t)$ in order to continue from one region to another. The assumption usually made is that any singularity has a dynamical origin. Poles are associated with bound states and thresholds give rise to cuts. For example in the s-plane a bound state of mass $m_B = \sqrt{s_B}$ will give rise to a pole at $s = s_B$ and there will be cuts with branch points corresponding to physical thresholds. These arise because of the unitarity condition (1.23). In this condition, $P^{f2} = s$ is the squared invariant mass of the state f, which shows that n-particle states contribute to the imaginary part of the amplitude if $\sqrt{s}$ is greater than the n-particle threshold energy. The threshold for producing a state in which the particles have masses $M_1, M_2, M_3, \ldots$ is at $s = (M_1 + M_2 + M_3 + \cdots)^2$. In a model with only one type of particle, of mass m, the thresholds are at $s = 4m^2, 9m^2, \ldots$. Each corresponds to a branch point of $A(s,t)$. When a function $f(z)$ of a complex variable z has a branch point at some point z_c, we attach a cut to the branch point, to remind us that continuing $f(z)$ from z_1 to z_2 along paths that pass to different sides of the branch point results in different values for the function: see figure 1.3. We say that $f(z)$ has a discontinuity across the cut. Since we may choose the point z_2 to lie in any direction relative to z_1, we must be prepared to draw the cut in any direction. It need not be a straight line. The only constraint is that one end of it is at $z = z_c$ and does not cross any other singularity. For $A(s,t)$, therefore, we need a cut attached to each branch point $s = 4m^2, 9m^2, 16m^2, \ldots$. By convention, we draw each cut along the real axis, so that the one attached to $s = 4m^2$ passes through all the other branch points and effectively all these branch points need only one cut, the right-hand one in figure 1.4.

A consequence of the assumption of analyticity is crossing symmetry. Con-

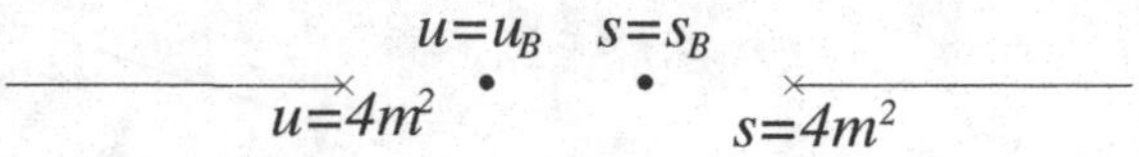

$$u=u_B \quad s=s_B$$

$$\frac{}{u=4m^2} \bullet \quad \bullet \quad \frac{}{s=4m^2}$$

Figure 1.4.   Poles and cuts in the complex s-plane for equal mass scattering for a given, fixed t. Recall that $u = 4m^2 - s - t$.

sider the scattering process

$$a + b \rightarrow c + d \tag{1.26}$$

and write its amplitude as $A_{a+b \rightarrow c+d}(s, t, u)$, reinstating the variable u for symmetry, but remembering that it is not independent being given in terms of s and t by (1.3). The physical region for the process (1.26) is $s > \max\{(m_a + m_b)^2, (m_c + m_d)^2\}$. In the equal-mass case, $t, u < 0$; in the unequal-mass case the constraint on t and u is more complicated, but most of the physical region lies in $t, u < 0$. The amplitude may be continued analytically to the region $t > \max\{(m_a + m_{\bar{c}})^2, (m_{\bar{b}} + m_d)^2\}$ and $s, u < 0$. This gives the amplitude for the t-channel process

$$a + \bar{c} \rightarrow \bar{b} + d \tag{1.27}$$

where $\bar{b}$ and $\bar{c}$ mean respectively the antiparticles of b and c. That is, we have

$$A_{a+\bar{c} \rightarrow \bar{b}+d}(t, s, u) = A_{a+b \rightarrow c+d}(s, t, u). \tag{1.28}$$

Similarly for the u-channel process

$$a + \bar{d} \rightarrow \bar{b} + d \tag{1.29}$$

we have

$$A_{a+\bar{d} \rightarrow \bar{b}+c}(u, t, s) = A_{a+b \rightarrow c+d}(s, t, u). \tag{1.30}$$

There are various mathematical results about the analytic properties of scattering amplitudes. Although these results are not complete, what is known is consistent with the assumption that the analytic structure in the complex s-plane for equal mass scattering is that shown in figure 1.4. The right-hand cut, from $s = 4m^2$ to ∞, arises from the physical thresholds in the s-channel. The pole at $s = s_B$ assumes that there is a bound state in the s-channel with mass $m_B = \sqrt{s_B}$. The left hand cut and pole arise respectively from the physical thresholds in the u-channel and an assumed u-channel bound state at $u = u_B$.   The position of the singularities in the s-plane arising from u-channel effects is given by the relation (1.3). Thus the presence of a threshold at $u = u_0$ for positive u means that the amplitude $A(s,t)$ must have a cut along the negative real axis with a branch point at $s = \bar{s}_0 = 4m^2 - t - u_0$, so that $\bar{s}_0 = -t$ when $u_0 = 4m^2$. Equally, a bound-state pole at $u = u_B$ will give rise to a pole at $s = 4m^2 - t - u_B$. In figure 1.4 we have drawn the u-channel bound-state pole and the u-channel cut to the left of the corresponding s-channel singularities. However, they move as t varies and for physical values of t, $t \le 0$, the u-channel pole is actually to the right of the s-channel pole, and when t is sufficiently large negative the two cuts actually overlap.

In perturbation theory, masses are assigned a small negative imaginary part, $m^2 \to m^2 - i\epsilon$, which is made to go to zero at the end of any calculation. The same $i\epsilon$ prescription is used outside the framework of perturbation theory; for example it makes Minkowski-space path integrals converge for large values of the fields. In figure 1.4, the $i\epsilon$ prescription pushes the branch point at $s = 4m^2$ downwards in the complex s-plane, and likewise the branch points corresponding to the higher thresholds, $s = 9m^2, s = 16m^2, \ldots$. As $\epsilon \to 0$, the branch points move back on to the real axis from below. That is, the physical s-channel amplitude is reached by analytic continuation down on to the real axis from the upper half of the complex s-plane. This is equivalent to saying that the physical amplitude is

$$\lim_{\epsilon \to 0} A(s + i\epsilon, t). \tag{1.31}$$

If we analytically continue it to real values of s between s_B and $4m^2$, there is no cut and the amplitude is real there[10]. The Schwarz reflection principle tells us that an analytic function $f(s)$ which is real for some range of real values of s satisfies

$$f(s^*) = [f(s)]^*.$$

So if we make a further continuation via the lower half of the complex plane, back to real values of s greater than $4m^2$, we obtain the complex conjugate of the physical amplitude:

$$A(s - i\epsilon, t) = [A(s + i\epsilon, t)]^*. \tag{1.32}$$

Therefore, for $s \ge 4m^2$ and $-s < t, u \le 0$,

$$2i \operatorname{Im} A(s + i\epsilon, t) = A(s + i\epsilon, t) - A(s - i\epsilon, t) \tag{1.33}$$

where it is understood in this equation that we have to take the limit $\epsilon \to 0$. (By convention the imaginary part of the amplitude is defined to be real, as is evident from the factor $2i$.) The right hand side of (1.33) is called the s-channel discontinuity, denoted by $D_s(s, t, u)$.

Similar arguments can be applied to the physical t-channel and u-channel processes $1 + \bar{3} \to \bar{2} + 4$ and $1 + \bar{4} \to 3 + \bar{2}$. Thus there must be cuts along the

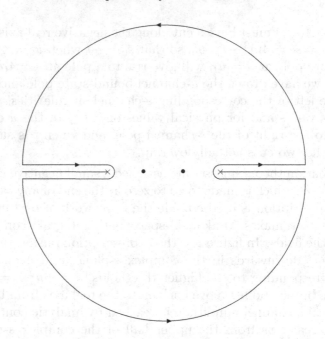

Figure 1.5.   Contour of integration in the complex s'-plane

real positive t and u axes, with branch points at the appropriate physical thresholds in these channels, and possibly poles as well. Equivalently to (1.33) we define the t-channel and u-channel discontinuities by

$$D_t(s,t,u) = A(s,t+i\epsilon) - A(s,t-i\epsilon) = 2i\,\mathrm{Im}\,A(s,t+i\epsilon)$$
$$t > 4m^2 \text{ and } u, s \leq 0$$
$$D_u(s,t,u) = A(s,u+i\epsilon) - A(s,u-i\epsilon) = 2i\,\mathrm{Im}\,A(s,u+i\epsilon)$$
$$u > 4m^2 \text{ and } s, t \leq 0 \qquad (1.34)$$

where again the limit $\epsilon \to 0$ is understood.

Knowing the analytic structure of an amplitude allows us to derive a "dispersion relation". We fix t and use the contour of integration shown in figure 1.5, which must be such that the point $s = s'$ is within it. Then $(s'-s)^{-1}A(s',t)$ is analytic within the contour except for a pole at $s' = s$, so that Cauchy's theorem tells us that the integral of this function is just the residue at the pole, which is $2\pi i A(s,t)$. Hence

$$A(s,t) = \frac{1}{2\pi i} \oint ds' \frac{A(s',t)}{s'-s} \qquad (1.35)$$

with u (u') given in terms of s (s') and t by (1.3). Assume for the moment

Figure 1.6. A contour of integration equivalent to that of figure 1.5

that $A(s,t)$ goes to zero like some negative power of s as $|s| \to \infty$, so that the contribution from the circle at infinity will vanish. So instead of the contour of figure 1.5 we may use the two-piece contour of figure 1.6. For the integration along each piece, we pick up the residue at the bound-state pole within the new contour, together with integrals in opposite directions along the upper and lower sides of the cut, which is just the integral of the discontinuity across the cut:

$$A(s,t) = \frac{g_s^2}{s - s_B} + \frac{g_u^2}{u - u_B} + \frac{1}{2\pi i} \int_{s_0}^{\infty} ds' \frac{D_s(s',t,u')}{s' - s}$$
$$+ \frac{1}{2\pi i} \int_{u_0}^{\infty} du' \frac{D_u(s',t,u')}{u' - u}. \qquad (1.36)$$

Here s_0, u_0 are the thresholds of the lowest states accessible to that channel. For example in nucleon-nucleon scattering $s_0 = 4m_N^2$ and $u_0 = 4m_\pi^2$.

We may write the dispersion relation (1.36) more compactly if we extend the definition of the discontinuities $D(s,t,u)$ to include any bound-state contributions. So we define

$$D_s(s,t,u) = -2\pi i g_s^2 \delta(s - s_B) \qquad s < 4m^2 \qquad (1.37)$$

with similar definitions for $D_t(s,t,u)$ and $D_u(s,t,u)$. Then (1.36) simplifies to

$$A(s,t) = \frac{1}{2\pi i} \int_0^{\infty} ds' \frac{D_s(s',t,u')}{s' - s} + \frac{1}{2\pi i} \int_0^{\infty} du' \frac{D_u(s',t,u')}{u' - u}. \qquad (1.38)$$

The denominator of the first integral vanishes for $s' = s$ and for physical values of s the s' integration passes through this value. We recall from (1.31) that we must give s a small positive imaginary part $i\epsilon$ to obtain the physical amplitude. This prevents the denominator from vanishing. We may write the resulting first denominator as

$$\frac{1}{s' - s - i\epsilon} = P\frac{1}{s' - s} + i\pi\delta(s' - s) \qquad (1.39)$$

where P denotes "principal value". The denominator of the second integral does not vanish and this term is real, so (1.38) is equivalent to

$$\text{Re } A(s,t) = \frac{1}{2\pi i}P \int_0^{\infty} ds' \frac{D_s(s',t,u')}{s' - s} + \frac{1}{2\pi i} \int_0^{\infty} du' \frac{D_u(s',t,u')}{u' - u}. \qquad (1.40)$$

1.7 The Froissart bound

In this section, we explain why the asymptotic ($s \to \infty$) behaviour of a scattering amplitude is limited by s-channel unitarity and the finite range of the forces.

To see this, start with the Froissart-Gribov formula (1.54) for s-channel partial waves. When $z > 1$, the behaviour of $Q_l(z'_t)$ for large l is given, up to a constant factor, by[9]

$$Q_l(z) \sim \frac{1}{l^{\frac{1}{2}}(z^2-1)^{\frac{1}{4}}} e^{-(l+\frac{1}{2})\log(z+\sqrt{z^2-1})}. \tag{1.55}$$

When $z < -1$, we may use the relation

$$Q_l(-z) = (-1)^l Q_l(z) \tag{1.56}$$

to deduce the large-l behaviour. Hence at fixed s, the large-l behaviour of $A_l(s)$ is controlled by the value z_0 of the singularity of $A(s, z_s(s,t))$ nearest to the origin in the z_s-plane. Usually this singularity is a t-channel or u-channel bound-state pole. In the neighbourhood of such a t-channel pole,

$$D_t(s,t,u) = -2\pi i g_t^2 \delta(t - t_B) \tag{1.57}$$

as in (1.37). We use (1.10), so that (1.57) is equivalent to

$$D_t(s,t(s,z'_s)) \sim 4\pi i g_t^2 |\mathbf{p}|^{-2} \delta(z'_s - z_0). \tag{1.58}$$

The expression for a u-channel bound state is exactly similar. So for large l

$$A_l(s) \sim |\mathbf{p}|^{-2} l^{-\frac{1}{2}} e^{-(l+\frac{1}{2})\zeta(z_0)}. \tag{1.59}$$

The value of z_s at the singularity is

$$z_0 \sim 1 + t_B/2\mathbf{p}^2 \quad \text{or} \quad -1 - u_B/2\mathbf{p}^2 \tag{1.60}$$

so that

$$\zeta(z_0) \sim \tfrac{1}{2}\sqrt{t_B}/|\mathbf{p}| \quad \text{or} \quad \tfrac{1}{2}\sqrt{u_B}/|\mathbf{p}|. \tag{1.61}$$

So for

$$l \geq 2|\mathbf{p}|/\sqrt{t_B} \quad \text{or} \quad 2|\mathbf{p}|/\sqrt{u_B} \tag{1.62}$$

$A_l(s)$ is exponentially small. This may be understood in physical terms: the range R of the force is given by $R^2 = t_B^{-1}$ or u_B^{-1}, and particles whose transverse separation or impact parameter b is greater than R are not scattered. Roughly speaking $l = b|\mathbf{p}|$.

The upper limit (1.62) on l applies at fixed s, but it may change if now s is allowed to be large. Then $\mathbf{p}^2 \sim \frac{1}{4}s$, so that at the bound-state pole $D_t(s, t(s, z_s')) \sim s^{-1}$, according to (1.58). It is possible, and indeed it is expected from the Regge theory described in the next chapter, that $D_t(s, t(s, z_s'))$ is larger for values of z_s' such that $t > 4m^2$ or $u < 4m^2$. In fact we expect it to be bounded by some power α of s, where α varies with t or u. This has the effect that for large l the dominant contribution to the Froissart-Gribov integral (1.54) will come from values of t and u such that s^α is as large as possible while still z_s' is close enough to ± 1 for the exponential in (1.55) not to provide too much damping. Hence instead of (1.59),

$$A_l(s) \sim l^{-\frac{1}{2}} \exp\left(M(l + \tfrac{1}{2})/\sqrt{s} + \alpha \log(s/s_0)\right) \qquad (1.63)$$

where s_0 is some fixed scale and the value of M depends on the relevant range of values of t or u. The physical reason for this change is that, as we shall see in the next chapter, at high energy the force is not the result of the exchange of a single particle, but rather the simultaneous exchange of whole families of particles. Hence $A_l(s)$ will be exponentially small for

$$l \geq \alpha M^{-1} \sqrt{s} \, \log(s/s_0) \qquad (1.64)$$

and the partial-wave series (1.45) may be truncated at this value. From (1.46), together with the unitarity constraint $0 \leq \eta_l \leq 1$,

$$\left| A_l(s) \right| = \left| \frac{\eta_l e^{2i\delta_l} - 1}{2i\rho_s} \right| \leq \frac{1}{\rho(s)} \qquad (1.65)$$

and $\rho(s) \to 1$ as $s \to \infty$. Also, $|P_l(z_s)| \leq 1$. So, for large s

$$|A(s, t(z_s = 1))| \leq \sum_{l=0}^{l_{\mathrm{MAX}}} (2l + 1)$$

$$l_{\mathrm{MAX}} = \alpha M^{-1} \sqrt{s} \, \log(s/s_0). \qquad (1.66)$$

After the arithmetic progression is summed, this gives

$$|A(s, t(z_s = 1))| \leq \text{constant} \times s \log^2(s/s_0). \qquad (1.67)$$

Applying the optical theorem (1.25) then gives, when s is large,

$$\sigma^{\mathrm{Tot}}(s) \leq \text{constant} \times \log^2(s/s_0) \qquad (1.68)$$

which is the Froissart bound[12].

Although the result (1.68) reproduces that of Froissart's original work[12] our derivation lacks its formal rigour. He assumed only that the dispersion relations require a finite number of subtractions and that the amplitude is polynomial bounded. From axiomatic field theory it proved possible to determine[13,14] the constant in (1.68):

$$\sigma^{\text{Tot}}(s) \leq \frac{\pi}{m_\pi^2} \log^2(s/s_0) \tag{1.69}$$

although the scale s_0 remains unspecified. However if one chooses a reasonable hadronic scale, $s_0 \sim 1 \text{ GeV}^2$ say, then the limit (1.69) is extremely high: 10 to 25 barns at Tevatron or LHC energies, that is in the range $\sqrt{s} = 1$ to 20 TeV.

A critical discussion of the formulation of asymptotic bounds in general and of their domain of validity can be found in [15].

1.8 The Pomeranchuk theorem

The Pomeranchuk theorem[16] asserts that, under certain quite strong assumptions, total cross sections for collisions of a particle and the corresponding antiparticle on the same target become asymptotically equal at high energy. For example, $\sigma^{\text{Tot}}(\pi^+ p)/\sigma^{\text{Tot}}(\pi^- p) \to 1$ or $\sigma^{\text{Tot}}(pp)/\sigma^{\text{Tot}}(\bar{p}p) \to 1$ as $s \to \infty$.

As we are comparing particle and antiparticle interactions we are concerned explicitly with s-channel $\leftrightarrow$ u-channel crossing. From the optical theorem (1.25), to calculate the total cross sections we need the amplitude at $t = 0$. It is convenient to use the variable $\nu = P_1.P_2$, in terms of which, when $t = 0$,

$$s = m_1^2 + m_2^2 + 2\nu$$
$$u = m_1^2 + m_2^2 - 2\nu \tag{1.70}$$

where m_1 and m_2 are the masses of the two particles. Thus the crossing simply takes $\nu \to -\nu$.

The forward scattering amplitude $A(\nu, t = 0)$ is analytic in the complex ν-plane cut along $(-\infty, -m_1 m_2)$ and $(m_1 m_2, \infty)$ with possible bound-state poles lying in the region $-m_1 m_2 < \nu < m_1 m_2$. For example in $\pi^\pm p$ scattering there will be poles at $\nu = \mp \frac{1}{2} m_\pi^2$ corresponding to the nucleon poles in the s- and u-channels.

For most physical scattering processes, amplitudes are neither symmetric nor antisymmetric under crossing. In general the process in the crossed channel is different from the one in the direct channel: the crossed channel for $\pi^+ \pi^+$ scattering is $\pi^- \pi^+$ scattering; the crossed channel for $\pi^+ p$ scattering is $\pi^- p$ or $\pi^+ \bar{p}$ scattering; and the crossed channel for pp scattering is $\bar{p}p$ scattering. But we may construct amplitudes which are symmetric or antisymmetric under crossing. Take $\pi^+\pi^+$ and $\pi^-\pi^+$ scattering as an example and define

$$A_+(\nu,0) = A(\pi^+\pi^+ \to \pi^+\pi^+) \quad A_-(\nu,0) = A(\pi^-\pi^+ \to \pi^-\pi^+). \quad (1.71)$$

Then the amplitudes which are symmetric and antisymmetric under crossing are given by

$$A^S(\nu) = \tfrac{1}{2}(A_+(\nu,0) + A_-(\nu,0))$$
$$A^A(\nu) = \tfrac{1}{2}(A_+(\nu,0) - A_-(\nu,0)). \quad (1.72)$$

We write fixed-t dispersion relations (1.38) for each of these. We introduce an integration variable ν', related linearly to s' and u' by equations similar to (1.70). For the symmetric amplitude the second integral in the dispersion relation for $A^S(\nu)$ is obtained from the first by changing the sign of ν; also, according to (1.34) the discontinuity $D_s(s',t=0,u')$ is just $2i\,\mathrm{Im}\,A^S(\nu'+i\epsilon)$. For $A^A(\nu)$ there are similar statements, except that we must in addition change the sign of the first integral to get the second one.

Because of the Froissart bound (1.69) the dispersion relations for the amplitudes $A_\pm$ or $A^{S,A}$ require at most two subtractions. We introduce a fixed value ν_1, as in (1.43), and then each dispersion relation contains two subtraction constants, the values of the amplitudes at $\nu = \nu_1$, together with their derivatives at the same point. But if we choose ν_1 to be 0, the antisymmetric amplitude A^A vanishes there, as does the derivative of the symmetric amplitude A^S. Hence the dispersion relation for each of these amplitudes has only one subtraction constant:

$$\mathrm{Re}\,A^S(\nu) - A^S(0) = \frac{\nu^2}{\pi}P\int_{m_\pi^2}^\infty d\nu' \frac{\mathrm{Im}\,A^S(\nu'+i\epsilon)}{\nu'^2(\nu'-\nu)} \quad + (\nu \to -\nu)$$

$$= \frac{2\nu^2}{\pi}P\int_{m_\pi^2}^\infty d\nu' \frac{\mathrm{Im}\,A^S(\nu'+i\epsilon)}{\nu'(\nu'^2-\nu^2)} \quad (1.73)$$

and

$$\mathrm{Re}\,A^A(\nu) - \nu\frac{d}{d\nu}A^A(0) = \frac{\nu^2}{\pi}P\int_{m_\pi^2}^\infty d\nu' \frac{\mathrm{Im}\,A^A(\nu'+i\epsilon)}{\nu'^2(\nu'-\nu)} \quad - (\nu \to -\nu)$$

$$= \frac{2\nu^3}{\pi}P\int_{m_\pi^2}^\infty d\nu' \frac{\mathrm{Im}\,A^A(\nu'+i\epsilon)}{\nu'^2(\nu'^2-\nu^2)}. \quad (1.74)$$

As in (1.40), we have written just the real part of each dispersion relation. We recall that the amplitudes are real at $\nu = 0$.

At sufficiently high energy the optical theorem (1.25) gives

$$\text{Im}\, A_{\pm}(\nu, 0) \sim 2\nu\, \sigma_{\pm}^{\text{Tot}}(\nu). \tag{1.75}$$

Now we know from the Froissart bound (1.69) that σ_{+}^{Tot} and σ_{-}^{Tot} are both bounded by a constant times $(\log \nu)^2$ as $\nu \to \infty$. Suppose that

$$\sigma_{+}^{\text{Tot}} - \sigma_{-}^{\text{Tot}} \sim C\,(\log \nu)^n \quad 0 < n \le 2 \tag{1.76}$$

so that, from (1.75),

$$\text{Im}\, A^{\text{A}}(\nu) \sim 2C\nu\,(\log \nu)^n, \quad 0 < n \le 2. \tag{1.77}$$

Then (1.74) gives, for large ν,

$$\text{Re}\, A^{\text{A}}(\nu) \sim -\frac{4C\nu}{\pi(n+1)}(\log \nu)^{n+1}. \tag{1.78}$$

From (1.77), the real part of the antisymmetric amplitude $A^{\text{A}}(\nu)$ exceeds the imaginary part by a factor $\log \nu$. This implies that the amplitudes become predominantly real at high energy. The original derivation of the Pomeranchuk theorem assumed that $\text{Re}\, A^{\text{A}}(\nu) \to 0$ at high energy, so that $C = 0$ and

$$\sigma_{+}^{\text{Tot}} - \sigma_{-}^{\text{Tot}} \to 0 \tag{1.79}$$

at high energy. More refined derivations with weaker assumptions have obtained[17,18] the weaker condition

$$\sigma_{+}(s)/\sigma_{-}(s) \to 1 \tag{1.80}$$

as $s \to \infty$, but it is still necessary to assume that a limit exists. It has not been possible to prove from field theory that this should be true[15].

2

Regge poles

In this chapter we introduce the concept of complex angular momentum and derive the basic Regge-pole formula as a description of high energy scattering processes. We give only what is sufficient for the purposes of this book. Further discussion can be found in appendix A and in [1].

2.1 Motivation

For the moment we restrict the discussion to the two-body process $1 + 2 \rightarrow 3 + 4$. The final state need not be the same as the initial state. We shall be interested in scattering at large s and small t. Most scattering processes exhibit a strong forward peak, and a study of the presence or absence of these peaks reveals a general rule. There is a correlation between the existence of a forward peak (small $t < 0$) in s-channel processes and the exchange of particles or resonances in the t-channel. This is connected directly to the quantum numbers of the exchange and reactions which do not have resonances allowed in the t-channel have appreciably smaller cross sections than those of similar ones which do.

Two examples of the latter point are provided by a comparison of the cross section $\sigma(K^-p \rightarrow \pi^-\Sigma^+)$ with $\sigma(K^-p \rightarrow \pi^+\Sigma^-)$, and $\sigma(\bar{p}p \rightarrow \bar{\Sigma}^-\Sigma^+)$ with $\sigma(\bar{p}p \rightarrow \bar{\Sigma}^+\Sigma^-)$. The data[19] for the K^-p reactions are shown in figure 2.1. It is obvious that $\sigma(K^-p \rightarrow \pi^-\Sigma^+) \gg \sigma(K^-p \rightarrow \pi^+\Sigma^-)$ and it seems that the latter cross section decreases more rapidly with increasing energy. A similar situation holds for the $\bar{p}p$ reactions with $\sigma(\bar{p}p \rightarrow \bar{\Sigma}^-\Sigma^+) \gg \sigma(\bar{p}p \rightarrow \bar{\Sigma}^+\Sigma^-)$[20]. Both $K^-p \rightarrow \pi^-\Sigma^+$ and $\bar{p}p \rightarrow \bar{\Sigma}^-\Sigma^+$ have a strong forward peak which contributes most of the cross section. In contrast the differential cross sections for $K^-p \rightarrow \pi^+\Sigma^-$ and $\bar{p}p \rightarrow \bar{\Sigma}^+\Sigma^-$ are isotropic within the

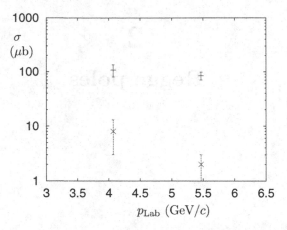

Figure 2.1.   Total cross sections for $K^-p \to \pi^-\Sigma^+$ (upper data) and $K^-p \to \pi^+\Sigma^-$ (lower data)[19]

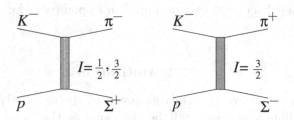

Figure 2.2.   Exchanges of different quantum numbers

errors.

The differences lie in the exchanged quantum numbers, as indicated in figure 2.2. Both the reactions $K^-p \to \pi^+\Sigma^-$ and $\bar{p}p \to \bar{\Sigma}^+\Sigma^-$ have double-charge exchange and isospin $\frac{3}{2}$ in the t-channel. No meson exists with these quantum numbers. However, for each of the reactions $K^-p \to \pi^-\Sigma^+$ and $\bar{p}p \to \bar{\Sigma}^-\Sigma^+$ there is zero charge exchange and the isospin in the t-channel can be $\frac{1}{2}$ or $\frac{3}{2}$. This means that $K\pi$ resonances, for example the $K^*(890)$ and the $K^*(1420)$, can be exchanged in the t-channel.

Particle and resonance exchange in the crossed channel thus appear as an important part of high-energy scattering. However, it is straightforward to show that the scattering amplitude cannot be dominated by the exchange of just a few t-channel resonances. Analogously to the s-channel partial wave series (1.45) we can write the t-channel partial-wave series for scattering of

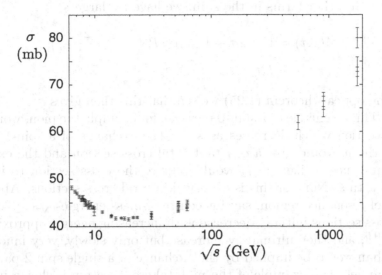

Figure 2.3. $\bar{p}p$ total cross section

spinless particles:

$$A(s,t) = 16\pi \sum_{l=0}^{\infty}(2l+1)A_l(t)P_l(z_t) \tag{2.1}$$

where

$$z_t = \cos\theta_t = 1 + \frac{2s}{t-4m^2} \tag{2.2}$$

in the equal-mass case. Although this series is a correct representation of scattering in the physical region of the t-channel, where $t > 4m^2$, $s < 0$ and $|z_t| \leq 1$, it is not readily applicable in this form to t-channel exchange for high-energy s-channel scattering. As $s \to \infty$, z_t becomes proportional to s and therefore large. For large z,

$$P_l(z) \sim z^l \tag{2.3}$$

so that $P_l(z_t) \sim s^l$ and the series diverges. In order to use it, we need to start in the region where it converges and cast it into a form that can be continued analytically to the region where we need it. In fact we shall write it as an integral over l.

However, for the moment let us suppose that only one resonance is exchanged so that the question of series convergence is avoided. As the resonance has a given spin, σ say, only one partial wave in (2.1) will contribute.

Dropping the other terms in the sum, we have for large s

$$A(s,t) = 16\pi(2\sigma + 1)A_\sigma(t)\,P_\sigma\left(1 + \frac{2s}{t - 4m^2}\right)$$
$$\sim f(t)s^\sigma. \tag{2.4}$$

Using the optical theorem (1.25) we see that this then gives $\sigma^{\text{Tot}} \sim s^{\sigma-1}$ at large s. Thus exchange of a spin-0 particle, for example the pion, would give a cross section which decreases as s^{-1}. The exchange of a spin-1 meson, such as the ρ, would give a constant total cross section and the exchange of a spin-2 meson like the f_2 would require the cross section to increase linearly with s. None of this is observed in total cross sections. Above the s-channel resonance region, say for centre-of-mass energies $\sqrt{s} \geq 2.5$ GeV, total cross sections initially decrease with increasing energy, approximately as $\sim s^{-0.5}$, and then ultimately increase but only slowly, very much more slowly than would be implied by the exchange of a single spin-2 particle in the t channel. The example of the $\bar{p}p$ total cross section is shown in figure 2.3. If we are to retain the picture of particle exchange then all the resonance contributions in the t channel must act collectively and combine in some way to give the observed energy dependence. Thus any meson, the f_2 for example, has to be considered as a member of a whole family of resonances of increasing spin and mass, and we must consider the exchange of all this family simultaneously and their contributions must be correlated with each other. The mathematical framework for adding the resonances together is based on a formalism initially developed by Regge[21–23] for nonrelativistic potential scattering. His formalism involved making the orbital angular momentum l, which is initially defined only for non-negative-integer values, into a continuous complex variable. He showed that the radial Schrödinger equation with a spherically-symmetric potential can be solved for complex l. That is, the partial wave amplitudes $A_l(t)$ can be considered as functions $A(l,t)$ of complex l, such that

$$A(l,t) = A_l(t) \qquad l = 0, 1, 2, \ldots. \tag{2.5}$$

Regge found that if the potential $V(r)$ is a superposition of Yukawa potentials then the singularities of $A(l,t)$ in the complex l-plane are poles whose locations vary with t:

$$l = \alpha(t). \tag{2.6}$$

These poles are known as Regge poles, or reggeons, and as t is varied they trace out paths defined by (2.6) in the complex l-plane. The functions $\alpha(t)$ are called Regge trajectories. In relativistic scattering theory they are associated with the exchanges of families of particles. Values of t such that

$\alpha(t)$ is a non-negative integer correspond to the squared mass of a bound state or resonance having that spin. The theory allows us to sum the whole family of exchanges corresponding to the particles associated with the Regge trajectory $\alpha(t)$.

2.2 The Sommerfeld-Watson transform

In order to develop Regge theory we first cast the t-channel partial-wave series (2.1) into an integral over l, called the Sommerfeld-Watson transform. In order to do this, we need to regard the t-channel orbital angular momentum l as a complex variable and introduce an amplitude $A(l, t)$ that coincides with the partial-wave amplitude $A_l(t)$ at physical values of l, as in (2.5). Note that the interpolation from non-negative-integer to complex l is not unique, as we can add any function $f(l, t)$ which is zero for physical values of l. For example (2.5) would also be satisfied by

$$A(l, t) \rightarrow A(l, t) + F(t)\sin(\pi l). \tag{2.7}$$

Making such a change affects the behaviour of $A(l, t)$ when $l \rightarrow \infty$ in most directions in the complex l-plane. As we describe in this section, and more fully in appendix A, we need to impose a requirement on the large-l behaviour of $A(l, t)$ which makes it unique.

In fact, in relativistic theory it is not possible to achieve this with a single amplitude $A(l, t)$. Instead, we need to introduce two amplitudes $A^\pm(l, t)$, such that

$$A^\pm(l, t) = \begin{cases} A_l(t) & l \text{ even} \\ A_l(t) & l \text{ odd}. \end{cases} \tag{2.8}$$

We write the t-channel partial-wave expansion (2.1) as

$$A(s, t) = A^+(s, t) + A^-(s, t) \tag{2.9}$$

with

$$A^\pm(s, t) = 8\pi \sum_{l=0}^{\infty} (2l + 1) A_l(t)(P_l(z_t) \pm P_l(-z_t)). \tag{2.10}$$

Because $P_l(-z) = (-1)^l P_l(z)$, $A^+(s, t)$ receives contributions only from even l, and $A^-(s, t)$ only from odd l. Therefore in $A^+(s, t)$ we can replace $A_l(t)$ by $A_l^+(t)$ and in $A^-(s, t)$ we can replace $A_l(t)$ by $A_l^-(t)$. The amplitudes $A^\pm$ are known as even- and odd-signatured amplitudes.

We require to continue the partial-wave expansions (2.10) from the region where the series converge, $t > 4m^2$, $s < 0$, to $t < 0$ and s large. To achieve this, we use Cauchy's theorem to rewrite the partial-wave expansions as integrals. We assume that $A^\pm(l,t)$ are analytic functions of l throughout the right-hand half of the l-plane with only isolated singularities. This is a fundamental assumption. It is found to be the case in potential theory and order-by-order in perturbation theory for relativistic quantum field theory. Understanding this in the framework of QCD will require a proper understanding of confinement. If the singularities are isolated we can easily continue past them. We first rewrite the partial-wave series as

$$A^\pm(s,t) = 8\pi i \int_C dl \, (2l+1) A^\pm(l,t) \frac{P_l(-z_t) \pm P_l(z_t)}{\sin(\pi l)} \qquad (2.11)$$

where the contour C surrounds the real axis from 0 to ∞ as in figure 2.4a. We have used the fact that $1/\sin(\pi l)$ has poles at $l = 0, 1, 2, \ldots$.

As we show in appendix A, the $A^+(l,t)$ are analytic for values of l such that Re l is sufficiently large. Therefore we may use Cauchy's theorem to deform the contour C into that of figure 2.4b. The difference between the integrals along the contours in figures 2.4a and 2.4b is an integral around the two closed contours shown in figure 2.5; within each of these two closed contours the integrand of (2.11) has no singularities, and therefore the integral around each of them vanishes. The dashed part of the integration contour in figure 2.4b is an infinite semicircle. We suppose that the behaviour of the amplitudes $A^\pm(l,t)$ along this infinite semicircle is such that it gives zero contribution to the integral. As we explain in appendix A, we achieve this by using the Froissart-Gribov formula (1.54) as a basis for defining the $A^\pm(l,t)$, and a theorem known as Carlson's theorem ensures that imposing this requirement on the $A^\pm(l,t)$ defines them uniquely.

In figure 2.4 we have imagined that $A^+(l,t)$ or $A^-(l,t)$ has a pole in the right-hand half of the complex l-plane, and also a branch point, with a cut attached to it. (We recall that a cut is attached to a branch point in order to remind us that, if we continue the amplitude around any closed curve that surrounds the branch point, we arrive at a different value for the amplitude. We may draw the cut along any line, provided only that it ends at the branch point.) We now deform the contour again so as to move it to the left past the pole of $A^\pm(l,t)$. The result, shown in figure 2.6, is that we must also integrate round the pole and so pick up its residue. In figure 2.7 we have shown how one picks up a contribution from a cut; however we shall postpone a discussion of cuts until section 2.4 and for the moment consider the case where the contour can be moved to the left to reach the line Re $l = -\frac{1}{2}$ with no interruptions other than poles. As it is moved across each pole, we pick up the residue at that pole. We then

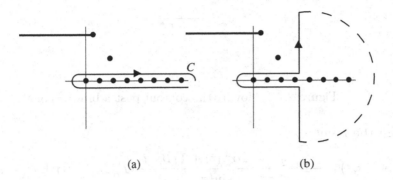

(a)   (b)

Figure 2.4. (a) An integration contour C in the complex l-plane surrounding the poles at integer values of l. A pole of $A^{\pm}(l,t)$ and a branch point with its attached cut are shown. (b) A deformed version of the contour C. The dashed curve is an infinite semicircle.

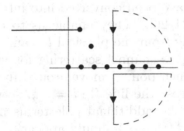

Figure 2.5. Two closed contours that represent the difference between the contours of figures 2.4a and 2.4b

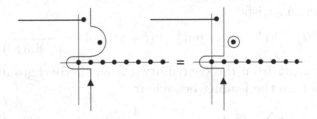

Figure 2.6. Result of moving the contour of figure 2.4b past the pole of $A^{\pm}(l,t)$

Figure 2.7.   Moving the contour past a branch point

obtain the result

$$A^{\pm}(s,t) = -16\pi^2 \sum_i \frac{(2\alpha_i^{\pm}(t) + 1)\,\beta_i^{\pm}(t)}{\sin(\pi\alpha_i^{\pm}(t))} (P_{\alpha_i^{\pm}(t)}(-z_t) \pm P_{\alpha_i^{\pm}(t)}(z_t))$$

$$+8\pi i \int_{-\frac{1}{2}-i\infty}^{-\frac{1}{2}+i\infty} dl \, \frac{(2l+1)A^{\pm}(l,t)\,(P_l(-z_t) \pm P_l(z_t))}{\sin(\pi l)} \qquad (2.12)$$

where the $\alpha_i^{\pm}(t)$ are the positions of the poles in the l-plane and the $\beta_i^{\pm}(t)$ are their residues. The integral along Re $l = -\frac{1}{2}$ is called the "background integral", for reasons we shall explain.

So the sums (2.10) have been converted into integrals, called Sommerfeld-Watson transforms[24–26]. They enable us to continue the partial-wave expansion analytically from the physical t-channel, for which $t > 4m^2$ and $s < 0$, to high-energy s-channel scattering for which $t < 0$. As we vary t, the poles (and branch points) move around in the complex l-plane, and some of them may cross the line Re $l = -\frac{1}{2}$, so we pick up their residues just as in figure 2.6. So additional pole terms may appear in (2.12): all those for which Re $\alpha_i^{\pm}(t) > -\frac{1}{2}$ should appear.

According to (2.2), when s is large positive with $t \le 0$, z_t is large negative. The asymptotic behaviour of the Legendre polynomials in (2.12) is

$$\sqrt{\pi}P_\alpha(z) \sim \begin{cases} \Gamma(\alpha + \frac{1}{2})/\Gamma(\alpha + 1)\,(2z)^\alpha & \text{Re } \alpha \ge \frac{1}{2} \\ \Gamma(-\alpha - \frac{1}{2})/\Gamma(-\alpha)\,(2z)^{-\alpha-1} & \text{Re } \alpha \le \frac{1}{2}. \end{cases} \qquad (2.13)$$

The Γ-function satisfies

$$l\Gamma(l) = \Gamma(l+1) \quad \text{and} \quad \Gamma(l+1)\Gamma(-l) = -\frac{\pi}{\sin(\pi l)}. \qquad (2.14)$$

Therefore, apart from the contribution from the background integral, the amplitudes have the leading behaviour

$$A^{\pm}(s,t) \sim -\pi \sum_i \beta_i^{\pm}(t) \frac{1}{\Gamma(\alpha_i^{\pm}(t) + 1)\sin(\pi\alpha_i^{\pm}(t))} (1 \pm e^{-i\pi\alpha_i^{\pm}(t)})\,s^{\alpha_i^{\pm}(t)}$$

$$= \sum_i \beta_i^{\pm}(t)\Gamma(-\alpha_i^{\pm}(t))\,(1 \pm e^{-i\pi\alpha_i^{\pm}(t)})\,s^{\alpha_i^{\pm}(t)} \qquad (2.15)$$

where we have redefined the $\beta_i^{\pm}(t)$ by absorbing factors:

$$32\,\pi\Gamma(\alpha_i^{\pm}(t)+\tfrac{3}{2})\,\beta_i^{\pm}(t) \to \beta_i^{\pm}(t)\,(2m^2-\tfrac{1}{2}t)^{\alpha_i^{\pm}(t)} \qquad (2.16)$$

and have used $P_l(z) = e^{-i\pi l}P_l(-z)$ for large negative z. The choice of $e^{-i\pi l}$ rather than $e^{+i\pi l}$ in (2.15) is due to the requirements of analytic continuation, as explained in appendix A. It is inelegant to raise to a power a quantity such as s that has dimensions of squared mass, so often we introduce a fixed scale s_0, with the dimensions of squared mass, and redefine $\beta_i^{\pm}(t)$ further to make (2.15) become

$$A^{\pm}(s,t) \sim \sum_i \beta_i^{\pm}(t)\Gamma(-\alpha_i^{\pm}(t))\,(1\pm e^{-i\pi\alpha_i^{\pm}(t)})\,(s/s_0)^{\alpha_i^{\pm}(t)}. \qquad (2.17)$$

The form (2.15) or (2.17) is the Regge representation for $A^{\pm}(s,t)$: it gives the behaviour of $A^{\pm}(s,t)$ for large s and small t. The background integrals must be added. Their integrands vanish as $1/\sqrt{s}$ for large s. We shall describe in appendix A how Mandelstam manipulated the background integrals so that when he pushed them further to the left in the l-plane he could make them arbitrarily small. Then all the poles, not just those for which $\mathrm{Re}\,\alpha_i^{\pm}(t) > -\tfrac{1}{2}$, should appear in the Regge representation (2.15), together with the contributions from branch points and their attached cuts. We consider the cut contributions in section 2.4.

The factors

$$\xi_\alpha^{\pm} = 1 \pm e^{-i\pi\alpha} \qquad (2.18)$$

are called signature factors. We argue in appendix A that for values of t in the s-channel physical region, the amplitudes $A^{\pm}(l,t)$ are real for real l; and that for these physical values of t both the positions $\alpha^{\pm}(t)$ of the poles of the amplitudes and their residues $\beta_i^{\pm}(t)$ are real, so the phase of the high-energy behaviour of a Regge-pole contribution to $A^{\pm}(s,t)$ is given by the signature factor $(1\pm e^{i\pi\alpha_i^{\pm}(t)})$.

2.3 Connection with particles

The $\Gamma(-\alpha_i^{\pm}(t))$ in (2.15) have poles for values of t such that an $\alpha_i^{\pm}(t)$ takes a non-negative integer value. The signature factors ξ_α^+ vanish when α is an odd integer, so that the even-signatured amplitude has poles at those values of t for which an $\alpha_i^+(t)$ is a non-negative even integer σ^+. Similarly, the odd-signatured amplitude has poles at the t-values for which an $\alpha_i^-(t)$ is a positive odd integer σ^-. We now identify these poles with the exchanges of particles of spin $\sigma^{\pm}$, whose squared mass is the corresponding value of t.

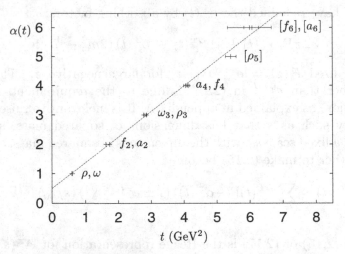

Figure 2.8. Four degenerate Regge trajectories: particle spins plotted against their squared masses t. The particles in square brackets are listed in the data tables[27], but there is some doubt about them. The straight line is $\alpha(t) = 0.5 + 0.9t$. See also figure 2.13.

Suppose that one of the signatured amplitudes, $A^+(l,t)$ say, has a pole in the complex l-plane, so that near the pole

$$A^+(l,t) \approx \frac{G(t)}{l - \alpha(t)}. \tag{2.19}$$

By construction, $A^+(l,t)$ coincides with the t-channel partial wave amplitude $A_l(t)$ when l is an even non-negative integer σ. So for values of t close to t_0, where t_0 is such that $\mathrm{Re}\,\alpha(t_0) = \sigma$,

$$A_\sigma(t) \approx \frac{G(t_0)}{\sigma - \alpha(t)}. \tag{2.20}$$

Make a Taylor expansion of $\mathrm{Re}\,\alpha(t)$ about $t = t_0$:

$$\mathrm{Re}\,\alpha(t) = \sigma + \alpha'(t - t_0) + \cdots \tag{2.21}$$

Near to $t = t_0$ it is sufficient to terminate the series after the linear term, so that

$$A_\sigma(t) \approx -\frac{G(t_0)/\alpha'}{t - t_0 + i\,\mathrm{Im}\,\alpha(t_0)/\alpha'}. \tag{2.22}$$

This has the Breit-Wigner form for a resonance of mass m_R such that $m_R^2 = t_0$, with width

$$\Gamma = \frac{\mathrm{Im}\,\alpha(m_R^2)}{\alpha' m_R}. \tag{2.23}$$

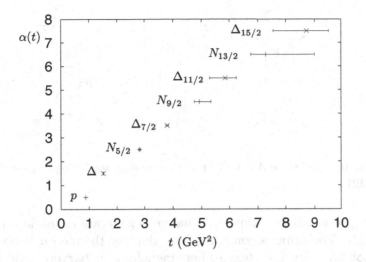

Figure 2.9. N and Δ trajectories

This interpretation requires that Im $\alpha(m_R^2) > 0$. If Im $\alpha(m_R^2) = 0$ then (2.22) has the form of a bound-state pole.

Similar considerations apply to the odd-signature amplitude. Thus we see that the poles in $A^\pm(l,t)$ at fixed integral l_0 can be identified with bound states and resonances in the t-channel with masses $m_R^\pm$ given by

$$\text{Re} \, \alpha_i^+(m_R^{+\,2}) = \sigma^+ \text{ (even non-negative integer)}$$
$$\text{Re} \, \alpha_i^-(m_R^{-\,2}) = \sigma^- \text{ (odd non-negative integer).} \quad (2.24)$$

Although the real parts of the Regge trajectories $\alpha(t)$ cannot be exactly linear, as above the physical threshold they have a non-zero imaginary part, it turns out that the linear approximation is remarkably good even away from integral values of $\alpha(t)$. The real part of the trajectories for some well-established meson states are shown in figure 2.8. A plot such as this is known as a Chew-Frautschi plot[28]. The plot shown has two remarkable features. The first is the linearity which allows a simple extrapolation from the t-channel physical region $t > 4m_\pi^2$ to the s-channel scattering region $t < 0$. The second is the near degeneracy of the four trajectories $\{f_2(1270), f_4(2050), \ldots\}$, $\{a_2(1320), a_4(2040), \ldots\}$, $\{\omega(780), \omega_3(1670), \ldots\}$, and $\{\rho(770), \rho_3(1690), \ldots\}$. The first pair are even signature, the second pair odd. Each pair contains an isospin zero and an isospin one trajectory, so effectively the "leading trajectory" contains four near-degenerate trajectories.

Linearity in the Chew-Frautschi plot is also apparent in the baryon tra-

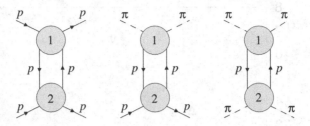

Figure 2.10.   Right-hand sides of the t-channel unitarity equations (2.26), (2.28) and (2.29)

jectories, of which two almost-exchange-degenerate examples are shown in figure 2.9. The slope is comparable to that for the meson trajectories. We shall not consider the Regge phenomenology of baryon trajectories. An excellent review of this topic can be found in [29].

It is familiar that the residue $G(t_0)$ of a particle pole in a partial-wave amplitude $A_\sigma(t)$ factorises. That is, when it occurs in an s-channel scattering amplitude $a + b \to c + d$,

$$G^{a+b \to c+d}(t_0) = g^{ab}g^{cd}. \qquad (2.25)$$

We now outline how the t-channel version of the unitarity relations (1.22) requires $G(t)$ in (2.19) to factorise similarly, even when t is not close to t_0. Consequently, the $\beta_i^\pm(t)$ in (2.15) also have a similar factorisation.

Strictly speaking, unitarity gives us the discontinuity of an amplitude across a whole collection of cuts, but it has been generalised[10] so as to give the discontinuities across single cuts. In the l-plane, t-channel two-body discontinuity formulae are rather simple and they impose significant constraints on the nature of the singularities in the l-plane. Ignore the (small) complications associated with the spin of the proton. Let $A^{pp}(l, t)$ be the elastic pp scattering amplitude. Denote by the suffix 1 the physical-sheet amplitudes, and by the suffix 2 their analytic continuations round the t-channel $\bar{p}p$ threshold branch point and back to the same value of t. Then[10]

$$A_1^{\pm pp}(l, t) - A_2^{\pm pp}(l, t) = i\rho(t)A_1^{\pm pp}(l, t)A_2^{\pm pp}(l, t) \qquad (2.26)$$

where

$$\rho(t) = \sqrt{\frac{t - 4m_p^2}{t}}. \qquad (2.27)$$

The right-hand side of (2.26) is shown diagrammatically as the first term in figure 2.10. Similarly, if we continue the πp and $\pi\pi$ elastic scattering amplitudes around the same $\bar{p}p$ threshold branch point, they change by

$$A_1^{\pm\pi p}(l,t) - A_2^{\pm\pi p}(l,t) = i\rho(t)A_1^{\pm\pi p}(l,t)A_2^{\pm pp}(l,t) \qquad (2.28)$$

and

$$A_1^{\pm\pi\pi}(l,t) - A_2^{\pm\pi\pi}(l,t) = i\rho(t)A_1^{\pm\pi p}(l,t)A_2^{\pm\pi p}(l,t). \qquad (2.29)$$

We solve (2.28) for $A_2^{\pm\pi p}(l,t)$ and insert the result into (2.29):

$$A_1^{\pm\pi\pi}(l,t) - A_2^{\pm\pi\pi}(l,t) = \frac{i\rho(t)[A_1^{\pm\pi p}(l,t)]^2}{1 + i\rho(t)A_1^{\pm pp}(l,t)}. \qquad (2.30)$$

Suppose now that $A_1^{+pp}(l,t)$, say, has a pole as in (2.19):

$$A_1^{+pp}(l,t) \approx \frac{G^{pp}(t)}{l - \alpha(t)}. \qquad (2.31)$$

Then (2.26) tells us that if we continue in t round the $\bar{p}p$ threshold branch point the position of the pole must change; that is, the Regge trajectory $\alpha(t)$ must have the $\bar{p}p$ threshold branch point, so that the position of the l-plane pole of $A_2^{+pp}(l,t)$ is different from that of $A_1^{+pp}(l,t)$. If this were not so, the right-hand side of (2.26) would have a double pole at $l = \alpha(t)$, while the left-hand side would only have a single pole, which would be inconsistent.

In particular, for hadron-hadron scattering there need not be a fixed pole at some point $l = l_0$, independent of t. (This statement is a slight simplification: see [10].) In section 2.8 we explain that unitarity does not impose such a restriction on amplitudes some or all of whose external legs are currents rather than hadrons.

Suppose now that $A_1^{\pm\pi p}(l,t)$ and $A_1^{\pm\pi\pi}(l,t)$ also have the pole at $l = \alpha(t)$, with non-zero residues $G^{\pi p}(t)$ and $G^{\pi\pi}(t)$, which will be the case if the particles on the Regge trajectory couple to the pion as well as to the proton. Again the pole will have a different position in $A_2^{\pm\pi\pi}(l,t)$, so that when we match the residues of the poles at $l = \alpha(t)$ on either side of (2.30) we obtain

$$G^{\pi\pi}(t) = \frac{[G^{\pi p}(t)]^2}{G^{pp}(t)}. \qquad (2.32)$$

This is an example of the factorisation property of Regge-pole residues.

We note that it is mathematically possible[30], though there is no experimental evidence for it, that the t-channel bound states and resonances manifest themselves not as simple poles $(t - t_B)^{-1}$ or $(t - t_R)^{-1}$, but as higher-order poles $(t - t_B)^{-n}$ or $(t - t_R)^{-n}$. Likewise, there may be higher-order poles in the l-plane, $(1 - \alpha(t))^{-n}$. Since such a behaviour of $A^+(l,t)$

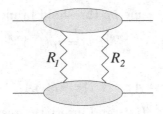

Figure 2.11.   Exchange of two reggeons R_1 and R_2

corresponds to differentiating (2.19) $n-1$ times with respect to $\alpha(t)$, the contribution to $A^{\pm}(s,t)$ is found by similarly differentiating (2.15). The result is a combination of terms in which $s^{\alpha(t)}$ is multiplied by powers of $\log s$,

$$(\log s), (\log s)^2, \ldots, (\log s)^{n-1}. \tag{2.33}$$

2.4 Regge cuts

We saw in section 2.2 that if a branch point, with its attached cut, is encountered during the deformation of the contour in the complex l-plane then it also contributes to the asymptotic behaviour. As figure 2.7 suggests, its contribution is an integral involving the discontinuity of $A(l,t)$ across the l-plane cut:

$$A^c(s,t) = \int^{\alpha_c(t)} dl \, (2l+1) \frac{\text{disc}\,(A(l,t))}{\sin(\pi l)} s^l. \tag{2.34}$$

Its leading behaviour for large s is of the form

$$A^c(s,t) \sim s^{\alpha_c(t)}(\log(s))^{-\gamma(t)} \tag{2.35}$$

where $\gamma(t)$ depends on the behaviour of $\text{disc}\,(A(l,t))$ near $l = \alpha_c(t)$.

As we do not know the discontinuity, in general (2.34) is of little use as a basis for a parametrisation. Also (2.34) gives us no insight into what a Regge cut corresponds to physically. In spite of a great effort in the early years of Regge theory, the theory of Regge cuts is still not well developed. What understanding we have is based mainly on perturbation theory, beginning with the important work of Mandelstam[31], but even in perturbation theory the situation is far from simple[32].

The simplest case is when the cut corresponds to the exchange of two reggeons R_1 and R_2, which is represented diagrammatically in figure 2.11.

We do know[33] that the exchange of two reggeons, each having a linear trajectory, yields a cut with a linear $\alpha_c(t)$:

$$\alpha_c(t) = \alpha_c(0) + \alpha'_c t \qquad (2.36)$$

with

$$\alpha_c(0) = \alpha_1(0) + \alpha_2(0) - 1$$
$$\alpha'_c = \frac{\alpha'_1 \alpha'_2}{\alpha'_1 + \alpha'_2} . \qquad (2.37)$$

This means that, unless one of the intercepts $\alpha_1(0)$ or $\alpha_2(0)$ is greater than or equal to one, the cuts have lower intercepts than the poles. Even then, for sufficiently large $|t|$ the cut contribution to the large-s behaviour of the amplitude will dominate over those from the poles, because the slope α'_c is smaller than either of α'_1 and α'_2. So cuts become important in large-t elastic scattering, as we shall find in chapter 3. The cut contributions do not factorise and the logarithmic factor in (2.35) is highly model-dependent.

If we want to calculate $\alpha_c(t)$ for trajectories that are not necessarily linear, we introduce two-dimensional Euclidean vectors $\mathbf{q}$ and $\mathbf{q}'$. Fix $\mathbf{q}$ such that $\mathbf{q}^2 = -t$. Then as $\mathbf{q}'$ varies over all possible values we have

$$\alpha_c(t) = \max\{\alpha_1(-\mathbf{q}'^2) + \alpha_2(-(\mathbf{q} - \mathbf{q}')^2) - 1\}. \qquad (2.38)$$

In addition to two-reggeon cuts, there are multi-reggeon cuts. If all the reggeons have linear trajectories, the n-reggeon cut has intercept

$$\alpha_{12\ldots n}(0) = \alpha_1(0) + \alpha_2(0) + \cdots + \alpha_n(0) - n + 1 \qquad (2.39)$$

and its slope is calculated as in (2.37); for example

$$\alpha'_{123} = \frac{\alpha'_{12}\alpha'_3}{\alpha'_{12} + \alpha'_3} = \frac{\alpha'_1 \alpha'_2 \alpha'_3}{\alpha'_2 \alpha'_3 + \alpha'_3 \alpha'_1 + \alpha'_1 \alpha'_2}. \qquad (2.40)$$

Although there is no well-formulated theory that allows one to calculate cut contributions, there are models. The most popular model is the eikonal model. This introduces the impact parameter $\mathbf{b}$, such that

$$A(s, -\mathbf{q}^2) = 4 \int d^2 b \, e^{-i\mathbf{q}\cdot\mathbf{b}} \tilde{A}(s, \mathbf{b}^2). \qquad (2.41)$$

This Fourier transform may be inverted:

$$\tilde{A}(s, \mathbf{b}^2) = \frac{1}{16\pi^2} \int d^2 q \, e^{i\mathbf{q}\cdot\mathbf{b}} A(s, -\mathbf{q}^2). \qquad (2.42)$$

The angular integration is easy to perform:

$$\tilde{A}(s, \mathbf{b}^2) = \frac{1}{8\pi} \int q \, dq \, J_0(bq) A(s, -\mathbf{q}^2). \tag{2.43}$$

For large l,

$$P_l(\cos\theta_s) \approx J_0(l\theta_s) \tag{2.44}$$

and, for large s and small t, $z_s = \cos\theta_s$ may be approximated by $1 - 2q^2/s$. So $\theta_s \approx 2q/\sqrt{s}$ and $dz_s \approx 4q \, dq/s$. Hence, unless b is so small that $b\sqrt{s}$ is not large, the integral (2.43) for $A(s, \mathbf{b}^2)$ is approximately s times the partial-wave amplitude (1.51) with l the integer $l_0(s, b)$ closest to $\frac{1}{2}b\sqrt{s}$. We use the complex-phase-shift representation (1.48) and the fact that $\rho(s) \to 1$ for large s:

$$A_l(s) \sim -\tfrac{1}{2}i(e^{2i\delta_l(s)} - 1). \tag{2.45}$$

Hence

$$\tilde{A}(s, \mathbf{b}^2) \approx \tfrac{1}{2}is(1 - e^{-\chi(s,b)}) \tag{2.46}$$

where

$$i\chi(s, b) = 2\delta_{l_0(s,b)}(s). \tag{2.47}$$

Then, according to (1.49),

$$\mathrm{Re}\,\chi(s, b) \geq 0. \tag{2.48}$$

So we obtain the eikonal form for $A(s, t)$:

$$A(s, -\mathbf{q}^2) = 2is \int d^2 b \, e^{-i\mathbf{q}\cdot\mathbf{b}} (1 - e^{-\chi(s,b)}). \tag{2.49}$$

If we now expand in powers of χ,

$$A(s, -\mathbf{q}^2) = 2is \int d^2 b \, e^{-i\mathbf{q}\cdot\mathbf{b}} \left(\chi - \frac{\chi^2}{2!} + \frac{\chi^3}{3!} \cdots - \frac{(-\chi)^n}{n!} \cdots \right). \tag{2.50}$$

If we identify the first term with single reggeon exchange, that is, from (2.15),

$$\beta(-\mathbf{q}^2)\Gamma(-\alpha(-\mathbf{q}^2))(1 \pm e^{-i\pi\alpha(-\mathbf{q}^2)})s^{\alpha(-\mathbf{q}^2)} = is \int d^2 b \, e^{-i\mathbf{q}\cdot\mathbf{b}} \chi(s, b) \tag{2.51}$$

then the second term corresponds to the two-reggeon-exchange cut, the third term to three-reggeon exchange, and so on. If on the left-hand side of (2.51) we include contributions from two single-reggeon exchanges, R_1 and R_2, then the χ^2 term in (2.50) will generate three two-reggeon-exchange cuts, (R_1R_1), (R_2R_2) and $(R_1R_2) = (R_2R_1)$.

We may easily show that the eikonal model then reproduces the general expressions for $\alpha_c(t)$. It also allows us to calculate the cut contributions explicitly in terms of the single-reggeon-exchange amplitude. However, we must warn that (2.51) is an assumption which is a pure guess and has little or no theoretical foundation.

The eikonal model does enable one to understand how the Froissart bound may be achieved, even if there is a Regge pole for which $\alpha(0) > 1$. Suppose that $\alpha(t)$ is even-signature and that for small t it may be approximated by

$$\alpha(t) = 1 + \epsilon_0 + \alpha' t \tag{2.52}$$

where we have written $\alpha(0) = 1 + \epsilon_0$. The inverse of the two-dimensional Fourier transform (2.51) then is

$$\chi(s,b) = \frac{1}{2\pi i} s^{\epsilon_0} \int d^2 q \, e^{i\mathbf{q}\cdot\mathbf{b}} \beta(-\mathbf{q}^2) \Gamma(-\alpha(-\mathbf{q}^2))(1 + e^{-i\pi\alpha(-\mathbf{q}^2)}) e^{-\mathbf{q}^2 \alpha' \log s}. \tag{2.53}$$

Because of the last exponential factor, the dominant contribution to the integral arises from values of $\mathbf{q}^2$ of order $1/\log s$, that is $\mathbf{q}^2$ close to 0. So we have

$$\chi(s,b) \sim \frac{1}{2\pi i} s^{\epsilon_0} \beta(0) \Gamma(-1 - \epsilon_0)(1 + e^{-i\pi(1+\epsilon_0)}) \int d^2 q \, e^{i\mathbf{q}\cdot\mathbf{b}} e^{-\mathbf{q}^2 \alpha' \log s}$$

$$= C e^{-\frac{1}{2}i\pi\alpha(0)} \exp\left(-\frac{b^2}{4\alpha' \log s} + \epsilon_0 \log s - \log \log s\right) \tag{2.54}$$

where the constant $C = \cos(\frac{1}{2}\pi\epsilon_0)\beta(0)\Gamma(-1 - \epsilon_0)/\alpha'$. The optical theorem (1.25) gives at high energy

$$\sigma^{\text{Tot}} = s^{-1} \text{Im}\, A(s, t = 0) = \text{Re} \int d^2 b \,(1 - e^{i\chi(s,b)}). \tag{2.55}$$

Change the integration variable to $b' = b/\log s$. When $s \to \infty$, Re $\chi \to +\infty$ unless $b' > \sqrt{4\alpha'\epsilon_0}$, in which case $\chi \to 0$. Hence in the limit

$$\sigma^{\text{Tot}} = (\log s)^2 \int d^2 b' \, \theta(4\alpha'\epsilon_0 - b'^2) = 4\pi\alpha'\epsilon_0(\log s)^2. \tag{2.56}$$

Thus the Froissart bound (1.69) is satisfied provided that $4\alpha' m_\pi^2 \epsilon_0 < 1$.

2.5 Signature and parity of cuts

If Regge cuts make a significant contribution to a reaction it is clearly important to know which amplitudes receive contributions from particular cuts, that is we need to know the quantum numbers associated with a two-reggeon system. Obviously internal quantum numbers, such as isospin and G-parity,will combine exactly as if the reggeons were elementary particles. For example, the exchange of the f_2 Regge pole ($I = 0, G = +1$) and the a_2 Regge pole ($I = 1, G = -1$) will give a cut with $I = 1$, $G = -1$. However one would expect a Regge cut to appear in amplitudes of both parities because of the angular momentum associated with the two-reggeon system, and there is also the important question of signature.

The most general discussion of the signature of Regge cuts is that of [34], in which it was shown, for external particles both with spin and without spin, that the signature of the cut is

$$\tau = \tau_1 \tau_2 \eta \tag{2.57}$$

where τ_1 and τ_2 are the signatures of the two exchanged reggeons and $\eta = -1$ if both reggeons are fermions and $\eta = +1$ otherwise.

A Regge trajectory is said to have natural parity, or naturality $+1$, if the spin and parity of the mesons on it are given by $J^P = J^{(-1)^J}$, for example $0^+, 1^-, 2^+, 3^-, \ldots$, and to have unnatural parity, or naturality -1 if the spin and parity of the mesons on it are given by $J^P = J^{(-1)^{J+1}}$, for example $0^-, 1^+, 2^-, 3^+, \ldots$. The parity and naturality of Regge cuts are discussed in [35]. It was shown that if the exchanged reggeons have naturalities n_1 and n_2 then amplitudes of naturality $-n_1 n_2$ are suppressed relative to amplitudes of naturality $n_1 n_2$ and that this suppression grows with increasing energy. As a consequence, for cuts where the two reggeons are any of ρ, ω, a_2 or f_2, all of which are natural parity, the natural-parity cut will dominate over the unnatural parity. The only exception to this is in reactions for which there is a constraint relation between amplitudes at $t = 0$ which makes the natural and unnatural parity amplitudes equal there. Then the suppression of one is accompanied, at small t, by the suppression of the other, although away from $t = 0$ the $+n_1 n_2$ contribution can recover from this suppression. This is phenomenologically important for reactions involving pion exchange, since in these reactions all of the strong single-reggeon exchanges are kinematically suppressed in the forward direction and the cuts are visible. The specific example of charged-pion photoproduction is considered in section 5.6.

2.6 Reggeon calculus

The bubbles in figure 2.11 are the amplitudes that couple reggeons to particles. They are just one example of amplitudes that couple reggeons

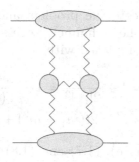

Figure 2.12.   A diagram of the reggeon calculus

to particles or to each other. Another example is shown in figure 2.12, which involves triple-reggeon couplings. A formalism was developed by Gribov[36,37] for calculating such diagrams. It is known as the reggeon calculus. In it, each reggeon line has a two-dimensional momentum associated with it, and an "energy" $(l-1)$ fed in from the external lines. A reggeon line corresponds to a propagator

$$\frac{1}{l_i - \alpha_i(\kappa_i^2)} \tag{2.58}$$

where κ_i is the two-dimensional momentum on the line and α_i is the corresponding trajectory. The triple-reggeon couplings $r_{l_1 l_2 l_3}(\kappa_1^2, \kappa_2^2, \kappa_3^2)$ vanish unless the product of the reggeon signatures at the vertex is $+1$. As in old-fashioned time-ordered perturbation theory, the ordering of the vertices in a diagram is significant, so that one should imagine each reggeon as being either absorbed or emitted. For each pair of emitted reggeons of signatures ξ_1 and ξ_2 there is a certain signature factor. One integrates over the transverse momentum and the "energy" round each loop, and so obtains the contribution to the Regge amplitudes $A^{\pm}(l, t)$.

Although the reggeon calculus is useful for analysing the analytic structure of the Regge amplitudes, and the nature of their singularities, it has not as yet produced useful quantitative output. In particular, it has not told us how large the two-reggeon cut contributions are.

2.7 Daughter trajectories

A problem arises when we try to repeat the argument of section 2.2 for unequal masses in the t-channel initial or final states, for example for the s-channel process $\pi^- \pi^+ \to K^- K^+$. It does not arise when we merely have unequal masses in the s-channel, for example $\pi^- K^+ \to \pi^- K^+$.

Here we shall only indicate the problem and its resolution, and refer to [1] for full details. For the s-channel process $\pi^- \pi^+ \to K^- K^+$ the simple expression (2.2) for z_t is replaced with

$$z_t = 1 + \frac{s}{2|\mathbf{p}_t|^2} = 1 + \frac{st}{T(t)}$$
$$2T(t) = t^2 - 2t(m_1^2 + m_2^2) + (m_1^2 - m_2^2)^2 \qquad (2.59)$$

where we have applied (1.6), (1.7) and (1.8) to the t-channel. Consider a Regge trajectory $\alpha_0(t)$. The origin of the problem is that we redefined $\beta_0(t)$ in (2.16) in order to achieve the simple asymptotic Regge expression (2.15). Now we must instead make the change

$$32\pi \Gamma(\alpha_0(t) + \tfrac{3}{2}) \, \beta_0(t) \to \beta_0(t) \, (-T(t)/t)^{\alpha_0(t)}. \qquad (2.60)$$

This again gives the leading term (2.15). However, when we raise z_t, given by (2.59), to the power $\alpha_0(t)$, as the asymptotic form (2.13) for the Legendre polynomial $P_{\alpha_0(t)}(z_t)$ requires, we also obtain a next-to-leading term;

$$s^{\alpha_0(t)} + \alpha_0(t) s^{\alpha_0(t)-1} \, T(t)/t. \qquad (2.61)$$

The pole at $t = 0$ in the nonleading term is absent in the equal-mass case. It is unphysical and must be cancelled. This can be achieved by introducing a second reggeon $\alpha_1(t)$ provided that the corresponding $\beta_1(t)$ has a simple pole at $t = 0$ and that $\alpha_1(t) = \alpha_0(t) - 1$. However, we cannot stop there. We also require a third trajectory $\alpha_2(t) = \alpha_0(t) - 2$ with $\beta_2(t)$ having a double pole at $t = 0$ so that its leading term cancels the third term of $\alpha_1(t)$ and the second term of $\alpha_2(t)$. This process must be continued indefinitely, requiring an infinite set of trajectories with ever more singular $\beta_i(t)$. The trajectory $\alpha_1(t)$ is known as the first daughter of $\alpha_0(t)$; $\alpha_2(t)$ is the second daughter and so on.

The concept of daughter trajectories was developed at a time when little was known about meson spectroscopy. There was no experimental evidence for them and no theoretical proof of their existence. Alternative viewpoints were developed including the possibility that the required cancellations could come from the background integral in (2.12) along the line $\mathrm{Re}\, l = -\tfrac{1}{2}$. However, we explain in appendix A that stopping the deformation of the contour at $\mathrm{Re}\, l = -\tfrac{1}{2}$ is not necessary: the contour can be moved as far to the left as we wish. Further, meson spectroscopy is now much more developed. Present data for non-strange mesons are summarised in the Chew-Frautschi plots of figure 2.13. In each plot the successive lines are parallel and one unit down in angular momentum from the one above.

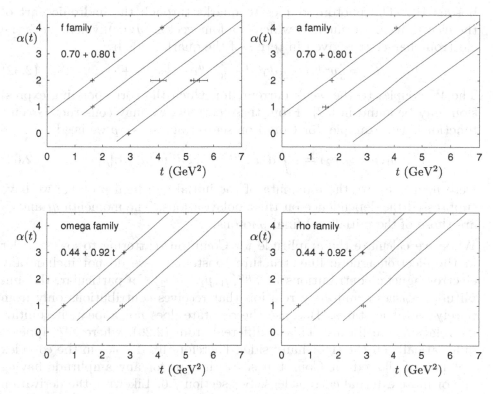

Figure 2.13. Four families of daughter Regge trajectories. Only confirmed states[27] are shown.

The evidence for daughter trajectories looks very convincing. However, because they are low-lying, the contributions from daughter trajectories to cross sections decrease much more rapidly than do the contributions from the leading trajectories and so normally are unimportant except at very low energies.

In figure 2.13, unlike in figure 2.8, we have included only particles which are well-established, according to the data tables[27]. The plots show that the degeneracy suggested by figure 2.8 is only approximate. We return to this point in the next chapter.

2.8 Fixed poles

If we trace back through the mathematics earlier in this chapter, we can show that we may apply Regge theory not only to purely-hadronic reactions, but also to those involving real or virtual photons or weak vector bosons W. The photon couples to quarks through the hadronic part of the electromagnetic current, which is a four-vector $eJ^\mu(x)$, where $-e$ is the electron charge. It is given in terms of the quark fields by

$$J^\mu = \tfrac{2}{3}\bar{u}\gamma^\mu u - \tfrac{1}{3}\bar{d}\gamma^\mu d + \tfrac{2}{3}\bar{c}\gamma^\mu c - \tfrac{1}{3}\bar{s}\gamma^\mu s + \cdots. \qquad (2.62)$$

The W couples to the weak current, for which the corresponding expression may be found in [27]. From these currents we may construct Green's functions. For example, for Compton scattering $\gamma p \to \gamma p$ we need

$$T^{\mu\nu}(q_1, p_1; q_2, p_2) = i \int d^4x \, e^{iq_2 \cdot x} \langle p_2 | \mathrm{T} J^\mu(x) J^\nu(0) | p_1 \rangle. \qquad (2.63)$$

Here p_1 and p_2 are the momenta of the initial and final protons; we have suppressed the dependence on their polarisations. The momenta q_1 and q_2 are those of the initial and final photons.

When we calculate the amplitude for Compton scattering to lowest order in the electromagnetic fine-structure constant α, we do not include any electromagnetic contributions in $T^{\mu\nu}(q_1, p_1; q_2, p_2)$. In particular, the amplitude satisfies a unitarity relation that receives contributions only from purely-hadronic states. Because the γp state does not appear, its unitarity relations are linear. This is different from (2.26), where A^{pp} appears quadratically on the right-hand side. Therefore fixed poles in the complex l-plane are allowed for Compton scattering, or for any amplitude having one or more external current legs. See section 7.6. Likewise, the derivation of the Froissart bound in section 1.7 depends on the presence of a quadratic term in the unitarity relation. Therefore, the Froissart bound holds only for purely-hadronic processes. In particular, it imposes no constraint on either real or virtual Compton scattering. We return to this in section 7.4, where we draw attention to the lack of a Froissart-bound constraint on the small-x behaviour of the proton structure function $F_2(x, Q^2)$.

During the late 1960s there was a great deal of research on the algebra of currents, in particular of the weak current. By expressing it in terms of quark fields, as (2.62) does the electromagnetic current, and using the canonical equal-time anticommutation relations for these fields, one may deduce equal-time commutation relations for the currents and various useful consequences follow[38]. In particular, one finds that various amplitudes involving currents must have fixed poles in the complex l-plane. An example[39–41] is the amplitude for the scattering of the charged electroweak vector current on a hadron, say a pion for simplicity. The amplitude $T_\pm^{\mu\nu}$, where the suffix $\pm$ indicates the charge of the current being scattered, has a fairly-complicated tensor structure, a part of which is of the form

$$A_\pm(s, t, q_1^2, q_2^2)\, t^{\mu\nu}(p_1, p_2)$$

$$t^{\mu\nu}(p_1, p_2) = \tfrac{1}{4}(p_1 + p_2)_\mu\,(p_1 + p_2)_\nu. \tag{2.64}$$

This part of the amplitude contributes to the scattering of longitudinally-polarised currents. This is most simply seen for the case $p_1 = p_2 = p$ and $q_1 = q_2 = q$, when $t^{\mu\nu} = p^\mu p^\nu$. Work in the rest frame of p, so that $p = (m, 0, 0, 0)$ and $q = (\nu/m, 0, 0, \sqrt{\nu^2/m^2 + Q^2})$, where $\nu = p.q$ and $Q^2 = -q^2$. Then transverse and longitudinal polarisation vectors are

$$\epsilon_T = (0, 1, 0, 0) \quad \text{or} \quad (0, 0, 1, 0)$$

$$\epsilon_L = Q^{-1}(\sqrt{\nu^2/m^2 + Q^2}, 0, 0, \nu/m) \tag{2.65}$$

and $\epsilon_T^\mu\, t_{\mu\nu} = 0 = t_{\mu\nu}\,\epsilon_T^\nu$, so that the amplitude $A_\pm$ does not contribute to the scattering of transversely-polarised currents. But

$$\epsilon_L^\mu\, t_{\mu\nu}\,\epsilon_L^\nu = \nu^2/(m^2 Q^2) + 1 \tag{2.66}$$

so that $A_\pm$ does contribute to longitudinal scattering.

The squared centre-of-mass energy is $s = 2\nu + m^2 - Q^2$ so that $s \sim 2\nu$ for $s \gg Q^2, m^2$. Regge theory leads one to conclude that, as in (2.15), both $\epsilon_{L\mu}\, T_+^{\mu\nu}\,\epsilon_{L\nu}$ and $\epsilon_{L\mu}\, T_-^{\mu\nu}\,\epsilon_{L\nu}$ should behave for large s as a sum of powers $s^{\alpha(t)}$. Because (2.66) behaves as $\tfrac{1}{4}s^2$, $A_\pm(s, t, q_1^2, q_2^2)$ behave as a sum of powers $s^{\alpha(t)-2}$ rather than $s^{\alpha(t)}$, with coefficients that depend on q_1^2 and q_2^2 as well as on t. The pomeron contribution is independent of whether the current is positively or negatively charged, and so cancels in the difference

$$A(s, t, q_1^2, q_2^2) = A_+(s, t, q_1^2, q_2^2) - A_-(s, t, q_1^2, q_2^2). \tag{2.67}$$

Thus the dominant behaviour of $A(s, t, q_1^2, q_2^2)$ apparently comes from a_2 exchange and so might be expected to be approximately $s^{-\frac{3}{2}}$ at $t = 0$. However this is not correct; it turns out[42] that, as a consequence of analyticity and the current commutation relations, the large-s behaviour of the amplitude A is

$$A(s, t, q_1^2, q_2^2) \sim \frac{2F_\pi(t)}{s} \tag{2.68}$$

where $F_\pi(t)$ is the pion form factor. That is, A has a singularity fixed at $l = 1$ in the complex angular-momentum plane. The singularity is a simple pole and its residue depends only on t, not on q_1^2 and q_2^2. In fact, A_+ and A_- both have this fixed pole, with equal and opposite residues. It is in addition to the moving poles that are present in hadron-hadron scattering amplitudes.

In [40] it was shown that in a field-theoretical model fixed poles with the form factor as residue do indeed occur. It is the class of diagrams shown

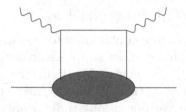

Figure 2.14.   Class of Feynman diagrams that give a fixed pole at $l = 1$

in figure 2.14 which gives rise to fixed singularities. The essential feature
of this figure is that there is a single line joining the two current vertices,
and the couplings of the currents at these vertices are pointlike. For targets
with spin the model even leads[43,44] to nonanalytic terms in the complex
angular-momentum plane, Kronecker deltas δ_{l1}.

2.9 Spin

The formalism we have described so far applies to the scattering of spinless
particles. It also applies to reactions involving particles that carry spin,
when their spin states are not measured. However, even then there may be
additional features that arise because of the spins. Spin effects may impose
particular behaviour on the amplitudes, particularly in the forward direction. In later chapters, we consider data for which we must take spin effects
into account; here we describe some essential features of the formalism.

The most convenient way to handle spin is to use the helicity formalism[45].
Each particle is labelled by its helicity λ, which is the spin projection along
its direction of motion. That is, λ is an eigenvalue of

$$\frac{\mathbf{S}.\mathbf{p}}{|\mathbf{p}|} \tag{2.69}$$

where $\mathbf{S}$ is the spin operator of the particle. The helicity of a particle
depends on the Lorentz frame chosen to define it. To begin with, we choose
the s-channel centre-of-mass frame. We consider matrix elements taken
between helicity states, that is helicity amplitudes, and for the process
$1 + 2 \rightarrow 3 + 4$ write

$$\langle P_3, \lambda_3; P_4, \lambda_4 | T | P_1, \lambda_1; P_2, \lambda_2 \rangle = T_{\lambda_3 \lambda_4; \lambda_1 \lambda_2}(s, t) \tag{2.70}$$

with

$$\frac{d\sigma_{\lambda_3 \lambda_4; \lambda_1 \lambda_2}}{d\Omega} = \frac{|\mathbf{p}_3|}{64\pi^2 |\mathbf{p}_1| s} |T_{\lambda_3 \lambda_4; \lambda_1 \lambda_2}|^2 \ . \tag{2.71}$$

If the spins of the particles are s_1, s_2, s_3, s_4 then there appear to be $(2s_1 + 1)(2s_2 + 1)(2s_3 + 1)(2s_4 + 1)$ amplitudes. However because of parity and time-reversal invariance the number of independent amplitudes will be less than this. For example for πp scattering there are two, for $\gamma N \to \pi N$ there are four, and for pp or $p\bar{p}$ elastic scattering there are five.

If no helicity measurements are made then the differential cross section is obtained by averaging over the initial helicities and summing over the final helicities:

$$\frac{d\sigma}{d\Omega} = \frac{1}{(2s_1 + 1)(2s_2 + 1)} \sum_{\lambda_1 \lambda_2 \lambda_3 \lambda_4} \frac{d\sigma_{\lambda_3 \lambda_4; \lambda_1 \lambda_2}}{d\Omega} . \qquad (2.72)$$

Apart from the simplicity of this formula, with no cross terms, the usefulness of the helicity formalism stems from the fact that the partial-wave series is a straightforward generalisation of the series for the scattering of two spin-zero particles. It is

$$T_{\lambda_3 \lambda_4; \lambda_1 \lambda_2}(s, t) = 16\pi \sum_{J \geq \mu}^{\infty} (2J + 1) T_{\lambda_3 \lambda_4; \lambda_1 \lambda_2}^{J}(s) \, d_{\lambda \lambda'}^{J}(\theta_s) \qquad (2.73)$$

where J is the total angular momentum in the t channel,

$$\lambda = \lambda_1 - \lambda_2 \quad \lambda' = \lambda_3 - \lambda_4 \quad \mu = \max\left(|\lambda|, |\lambda'|\right) \qquad (2.74)$$

and $d_{\lambda \lambda'}^{J}(\theta_s)$ is an element of the rotation matrix[46]. Using $d_{00}^{J}(\theta_s) = P_J(\cos \theta_s)$ we immediately recover the partial wave series for spinless particles, with $J = l$.

The relevance of λ and λ' becomes transparent if we recall that in the centre-of-mass system particles 1 and 2, and also 3 and 4, are moving in opposite directions. Thus λ and λ' are, respectively, the projections of the total angular momentum in the direction of motion before and after the collision, since the orbital angular momentum has no component in this direction. Hence for physical J we must have $J \geq \mu$ as in (2.73). In the forward direction $\theta_s = 0$, λ and λ' are projections of the total angular momentum in the same direction and the conservation of angular momentum demands that the amplitude vanishes unless $\lambda = \lambda'$. For $\theta_s \approx 0$, the behaviour of $d_{\lambda \lambda'}^{J}(\theta_s)$ makes $T_{\lambda_3 \lambda_4; \lambda_1 \lambda_2}(s, t)$ vanish at least as fast as

$$T_{\lambda_c \lambda_d; \lambda_a \lambda_b} \sim (\sin \tfrac{1}{2} \theta_s)^n \qquad (2.75)$$

where $n = |\lambda - \lambda'|$ is known as the net helicity flip.

Finally we give the relationships between the partial-wave helicity amplitudes imposed by parity and time-reversal invariance:

$$T_{-\lambda_3, -\lambda_4; -\lambda_1, -\lambda_2}^{J}(s) = \eta(-1)^{s_3 + s_4 - s_1 - s_2} T_{\lambda_3 \lambda_4; \lambda_1 \lambda_2}^{J}(s) \qquad (2.76)$$

where η is the product of the intrinsic parities of the particles, and

$$T^J_{\lambda_1\lambda_2;\lambda_3\lambda_4}(s) = T^J_{\lambda_3\lambda_4;\lambda_1\lambda_2}(s) \qquad (2.77)$$

Similar relations hold between the helicity amplitudes as functions of s and t.

We must now carry out the Reggeisation procedure described in section 2.2 for each independent amplitude. Because helicity amplitudes have such a simple partial wave expansion, they are the most convenient to use. Since it is the t-channel partial-wave expansion that we use we need the t-channel helicity amplitudes $T_{\tilde\lambda_2\tilde\lambda_4;\tilde\lambda_1\tilde\lambda_3}(s,t)$. These are related to the s-channel helicity amplitudes we have used so far by a complicated matrix. Since Regge poles have definite parity, it is appropriate to use t-channel amplitudes of definite parity, so that only poles of that parity contribute to them. This is done by taking sums and differences of partial-wave helicity amplitudes:

$$T^{J\pm}_{\tilde\lambda_2\tilde\lambda_4;\tilde\lambda_1\tilde\lambda_3}(t) = T^J_{\tilde\lambda_2\tilde\lambda_4;\tilde\lambda_1\tilde\lambda_3}(t) \pm \eta_1\eta_3(-1)^{-s_1-s_3}T^J_{\tilde\lambda_2\tilde\lambda_4;-\tilde\lambda_1,-\tilde\lambda_3}(t) \qquad (2.78)$$

where the superscript $(\pm)$ specifies the parity and η_1 and η_3 are intrinsic parities of the particles concerned. Note that this is in addition to constructing amplitudes of definite signature. To take account of kinematic constraints for s-channel scattering, for example (2.75), it is easiest to go back to the s-channel helicity amplitudes, obtained from the t-channel helicity amplitudes using the appropriate helicity crossing matrix. It is found that a Regge pole makes s-channel helicity amplitudes behave for large s as $s^{\alpha(t)}$, as in the case of spinless particles, though it might not contribute to some of the amplitudes.

We have seen in (2.75) that the conservation of angular momentum makes s-channel helicity amplitudes vanish at $\theta_s = 0$ at least as fast as a certain power of $\sin\theta_s$. However, parity conservation may give a more stringent requirement on the behaviour of the contribution from a given Regge pole. A contribution from any t-channel reggeon has definite parity in the t-channel. Therefore it satisfies

$$T_{\lambda_3\lambda_4;\lambda_1\lambda_2} = \pm T_{-\lambda_3\lambda_4;-\lambda_1\lambda_2}. \qquad (2.79)$$

The behaviour (2.75) is therefore not achievable and must be modified to

$$T_{\lambda_3\lambda_4;\lambda_1\lambda_2} \sim (\sin\tfrac{1}{2}\theta_s)^{n+n'} \qquad (2.80)$$

where

$$\begin{aligned} n + n' &= |\lambda_1 - \lambda_3| + |\lambda_2 - \lambda_4| \\ &= \max\left[|\lambda_1 - \lambda_3 - \lambda_2 + \lambda_4|, |\lambda_1 + \lambda_2 - \lambda_3 - \lambda_4|\right]. \end{aligned} \qquad (2.81)$$

3

Introduction to soft hadronic processes

In this chapter we apply the Regge-pole concepts of chapter 2 initially to total cross sections. This requires the introduction of a new Regge exchange, the pomeron. It is not associated with the meson trajectories and has the quantum numbers of the vacuum, that is isospin 0 and $C = +1$. Other phenomenological properties of the pomeron are derived from high energy pp and $\bar{p}p$ total and differential-cross-section data. The discussion is extended to diffraction dissociation and to central production at high energy.

We shall use the word reggeon to describe an exchange of a family of ordinary mesons, composed of a quark-antiquark pair. There is a belief that the pomeron rather is associated with the exchange of a family of glueballs. It is possible that there exists also an odderon, a $C = -1$ partner of the pomeron, which too is associated with glueball exchange.

3.1 Total cross sections

The optical theorem (1.25) relates the total cross section σ^{ab} for the scattering of a pair of hadrons a and b to the amplitude $T^{ab}(s,t)$ for elastic $a\,b$ scattering. When the centre-of-mass energy $\sqrt{s}$ is large, the theorem reads

$$\sigma^{ab}(s) \sim s^{-1} \operatorname{Im} T^{ab}(s, t = 0) \tag{3.1}$$

and so according to (2.15) a Regge trajectory $\alpha(t)$ contributes a term to $\sigma^{ab}(s)$ that behaves as

$$c\, s^{\alpha(0)-1}. \tag{3.2}$$

If the particles whose exchange is described by the trajectory have even C-parity, the contribution to the total cross section $\sigma^{\bar{a}b}$, for which the initial

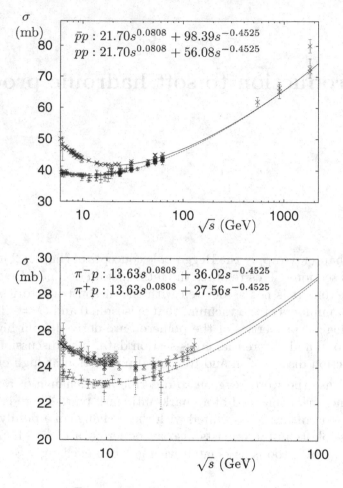

Figure 3.1. Total cross sections

particle a is replaced with its antiparticle, is just the same. If they have odd C-parity, it changes sign.

To a good approximation the two $C = -1$ trajectories ρ, ω and the two $C = +1$ trajectories f_2, a_2 are all degenerate (see figure 2.8) with $\alpha(0) \approx \frac{1}{2}$, so they all contribute terms that behave approximately like $1/\sqrt{s}$. It is observed that all total cross sections rise with s at high energy, which is why a soft-pomeron-exchange term is needed, in addition to ρ, ω, f_2, a_2. For many years fits to data were strongly influenced by the Froissart-Lukaszuk-Martin bound[47]:

$$\sigma^{ab}(s) \leq \frac{\pi}{m_\pi^2} \log^2(s/s_0) \qquad (3.3)$$

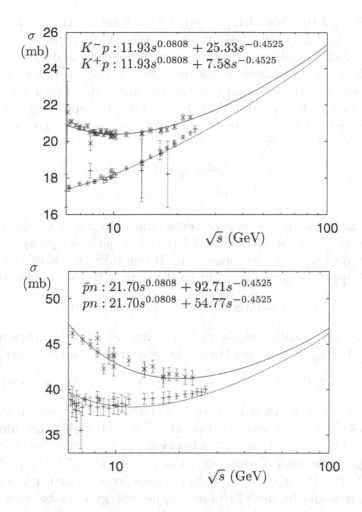

Figure 3.2. Total cross sections *continued*

for some fixed and unknown value of s_0. This bound applies in the limit $s \to \infty$, but it is often assumed that in practice currently-accessible energies are high enough. However, at present-day energies the bound is more than 100 times greater than measured values, taking s_0 in the range 1 to 10 GeV^2, so it is not really a significant constraint. Nevertheless, for many years it was customary[48] to parametrise the data with expressions of the form

$$\sigma^{ab} = A + BP^n + C(\log P)^2 + D \log P \qquad (3.4)$$

where P is the momentum of the particle a in the rest frame of particle b (so that at high centre-of-mass energy $\sqrt{s}$ it is proportional to s). We explained in section 2.3 that logarithmic terms arise from multiple poles in the complex angular-momentum plane: see (2.33). While a parametrisation such as (3.4) can nowadays be made both successfully and systematically[49], it has long been realised[50] that one may fit the data at least as successfully, more simply, and even more systematically[51], with a simple pole at a value of l a little greater than 1. This can be seen from figures 3.1 and 3.2 in which all available hadronic total cross sections are compared with expressions of the form

$$\sigma^{ab} = X^{ab}s^\epsilon + Y^{ab}_+ s^{\eta_+} + Y^{ab}_- s^{\eta_-}$$
$$\sigma^{\bar{a}b} = X^{ab}s^\epsilon + Y^{ab}_+ s^{\eta_+} - Y^{ab}_- s^{\eta_-}. \tag{3.5}$$

In each case, the first term represents pomeron exchange, which is $C = +1$. The power η_+ refers to the $C = +1$ meson trajectories f_2, a_2 and the power η_- to the $C = -1$ trajectories ρ, ω. If one takes $\eta_+ = \eta_-$, as degeneracy would imply, one gets a good fit to the data with the choices[51]

$$\epsilon \approx 0.081 \qquad \eta_+ = \eta_- = -0.45. \tag{3.6}$$

Relaxing the condition $\eta_+ = \eta_-$ introduces additional parameters into the fits and therefore improves them; the preferred values then are[49]

$$\epsilon \approx 0.096 \qquad \eta_+ = -0.35 \qquad \eta_- = -0.56. \tag{3.7}$$

As is evident in figure 3.1, there is a conflict[52,53] between the measurements of the $\bar{p}p$ total cross section at $\sqrt{s} = 1800$ GeV. The value $\epsilon \approx 0.096$ is derived from a fit that passes between the two 1800 GeV data points; if the upper point should turn out to be correct, then[53] the preferred value is $\epsilon \approx 0.112$. The πp and Kp data perhaps favour a higher value than 0.08, though it would be useful to have higher-energy data because the parameter values do not reach stability if data at too-low energies are included in the fits[49]. There are higher-energy data[54] for pion scattering on a nuclear target that seem to confirm the need for a larger value of ϵ, but the extraction of nucleon-target cross sections from nuclear-target data is subject to considerable uncertainty.

For the pp elastic amplitude, where the two initial particles are the same, the factorisation property (2.25) implies that $\beta(t) > 0$. Since $\Gamma(\alpha(0)) < 0$ when $\alpha(0)$ is near to $\frac{1}{2}$, the signature factor (2.18) makes the contributions from the positive-signature f_2 and a_2 positive for σ^{pp}, and negative from the negative-signature ω and ρ. For $\sigma^{\bar{p}p}$ both the positive and the negative-signature contributions are positive. So we understand why $\sigma^{\bar{p}p} > \sigma^{pp}$. According to the data in figures 3.1 and 3.2, $\sigma^{\bar{p}n} \approx \sigma^{\bar{p}p}$ and $\sigma^{pn} \approx \sigma^{pp}$. This implies that the contributions from the $I = 1$ exchanges ρ and a_2 are very much less than those from the $I = 0$ exchanges f_2 and ω. Within very large errors, the couplings to the proton at $t = 0$ satisfy

$$g^{f_2} \approx 2g^\omega \qquad g^\rho \approx g^{a_2} \qquad g^\omega \approx 4g^\rho. \qquad (3.8)$$

(Remember that the squares of these couplings are needed for the total cross sections.) Thus, given that the degeneracy of the trajectories is not exact, as is evident from figure 2.13, $\eta_+ \approx 1 - \alpha_{f_2}(0)$ and $\eta_- \approx 1 - \alpha_\omega(0)$. From figure 2.13, we see that $\alpha(0)$ is rather greater than $\frac{1}{2}$ for the f_2 trajectory and rather less than $\frac{1}{2}$ for the ω. This is in agreement with the values of η_+ and η_- in (3.7). We note that the fact that g^ρ and g^{a_2} are small compared with g^{f_2} and g^ω may be understood in the framework of $SU(3)$ symmetry[55].

Of course fits such as those shown in figures 3.1 and 3.2, with a positive power of s, will violate the Froissart bound (3.3) when the energy becomes extremely large. This implies that the power ϵ cannot be exactly constant: it must decrease as s increases, though probably very slowly indeed. There is a ready explanation for how this occurs: if single-pomeron exchange corresponds to a simple pole at $l = \epsilon_{I\!P}$, resulting in power behaviour $s^{\epsilon_{I\!P}}$, as the energy increases its effect becomes more and more moderated by the exchanges of two or more pomerons. So the power ϵ in (3.6) or (3.7) is just an effective power: it is a little less than $\epsilon_{I\!P}$ and decreases as s increases.

We showed at the end of section 2.4 how this effect can be seen explicitly in the eikonal expansion (2.50). If the first term is identified with pomeron exchange as in (2.51) then $\chi(s, b)$ is almost real, and positive because of (2.48). The second term then has the opposite sign to the first. Although at present-day energies it is small at $t = 0$, it tends to give a cancellation which becomes increasingly strong as s increases because according to (2.54) $\chi(s, b)$ rises like a positive power $s^{\epsilon_{I\!P}}$ for small b.

However, the view that at present-day energies cut effects in the total cross section are small is controversial: there are those who argue[56,57] that the multiple exchanges have a rather strong effect, so that $\epsilon_{I\!P}$ is considerably greater than ϵ. Against this, and favouring the view that the multiple exchanges are weak at present-day energies, is the fact that the pomeron contribution satisfies the additive-quark rule.

This rule relates the magnitude of the contribution from pomeron exchange to the total cross section σ^{ab} to the number of valence quarks in the hadrons a and b:

$$X^{ab} \propto n_a n_b. \qquad (3.9)$$

For example, it may be seen in figure 3.1 that

$$X^{\pi p} : X^{pp} \approx 2 : 3. \qquad (3.10)$$

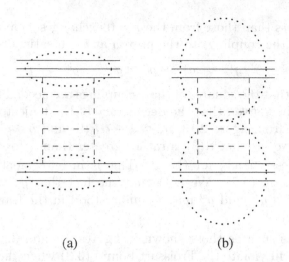

(a)   (b)

Figure 3.3. (a) Exchange of a ladder between two relatively-small hadrons, and
(b) the same process viewed as the direct interaction of two larger hadrons

The rule extends to hadrons containing strange valence quarks if one accepts
that the coupling of the pomeron to strange quarks is some 70% as strong
as its coupling to the light quarks. The origin of the additive-quark rule
is not fully understood: the most obvious interpretation is that it directly
gives information on how the pomeron couples to the quarks in a hadron[7]
and suggests that it is a single exchange that dominates.

An alternative view, which is discussed more fully in chapter 8, is that the
strength of the pomeron's coupling to a hadron is determined by the radius
of the hadron. This happens to reproduce the same relative strengths for
couplings to the hadrons composed of u, d quarks as is given by the additive
quark rule[58]. Such geometric approaches, in various forms, have been
formulated by many authors. In order to reproduce the rise of the total
cross sections with s, one may suppose[5,59] that hadron radii effectively
increase with s. Geometric approaches are most conveniently formulated
by introducing the impact parameter b as in (2.41). The data on total
cross sections and on elastic scattering then reveal that, as s increases,
hadrons become larger, and blacker in their centres. Of course, this is
just a statement of the data; in order to turn it into an explanation, we
must understand in dynamical terms why hadrons effectively change in this
fashion as the energy increases.

Notice that, to some extent, whether one describes the increase with s of
the cross section as being a radius effect, or as being a consequence of
an exchange process, is a matter of language. As may be seen in figure

3.3, the same process may be viewed either as the exchange of a ladder between two hadrons which are relatively small and simple in structure, or as a direct interaction between two hadrons which become larger and more complicated in structure as s increases.

3.2 Elastic scattering

If it is correct, the additive-quark rule implies that the pomeron couples to the separate valence quarks in a hadron, rather than to the hadron as a whole. This particular assumption forms the basis of an internally consistent and extremely successful model. It is consistent with experiment to assume[60] that the pomeron has a γ^μ coupling to quarks, so that the pomeron-exchange contribution to quark-quark scattering $q_1 q_2 \to q_3 q_4$ is, instead of (2.17),

$$\bar{\beta}(t)\,(\bar{u}_3\gamma^\mu u_1)(\bar{u}_4\gamma_\mu u_2)\,e^{-\frac{1}{2}i\pi\alpha_{I\!P}}(s/s_0)^{\alpha_{I\!P}-1}. \tag{3.11}$$

Here, the factor $e^{-\frac{1}{2}i\pi\alpha_{I\!P}}$ has the same phase as the signature factor (2.18) and therefore the function $\bar{\beta}(t)$ is real. For want of any information about it, we suppose that it is a constant

$$\bar{\beta}(t) = \beta_{I\!P}^2 \tag{3.12}$$

and we will see that this agrees well with experiment. The quantity s_0 is then expected to be of the order of a GeV^2, because this is the typical scale associated with high-energy soft hadronic reactions. To check that $\alpha_{I\!P} - 1$ is the correct power of s, we calculate from the optical theorem (3.1) the total cross section for unpolarised quark-quark scattering corresponding to (3.11):

$$\sigma^{qq} = s^{-1}\beta_{I\!P}^2\,\left(\tfrac{1}{2}\mathrm{tr}\,(\gamma.p_1 + m_q)\gamma^\mu\right)\left(\tfrac{1}{2}\mathrm{tr}\,(\gamma.p_2 + m_q)\gamma_\mu\right)$$
$$\times \cos(\tfrac{1}{2}\pi(\alpha_{I\!P}(0) - 1))\,(s/s_0)^{\alpha_{I\!P}(0)-1} \tag{3.13}$$

so that

$$\sigma^{qq} \sim 2\beta_{I\!P}^2\cos(\tfrac{1}{2}\pi(\alpha_{I\!P}(0) - 1))\,(s/s_0)^{\alpha_{I\!P}(0)}. \tag{3.14}$$

The γ^μ coupling introduced in (3.11) is similar to that for photon exchange. However, there is the important difference that photon exchange is $C = -1$, and therefore the coupling changes sign when a quark is replaced with an antiquark, while it does not for pomeron exchange because it is $C = +1$.

Just as for photon exchange, when the coupling is to a quark that is bound within a hadron, the wave function of the hadron is taken into account by introducing two form factors, the Dirac elastic form factor $F_1(t)$ and the Pauli form factor $F_2(t)$. We do not know these form factors for $C = +1$ exchange, so we must assume that, at least to a reasonable approximation, they are the same as for $C = -1$ exchange. Pomeron exchange is isosinglet, so we need the sum of the elastic electromagnetic form factors of the proton and the neutron. In the case of the form factor $F_1(t)$, experiment finds that for the neutron it is very small, so we use just the proton form factor. For $F_2(t)$, the value at $t = 0$ is the sum of the anomalous magnetic moments of the proton and the neutron. It happens that these are almost equal and opposite, so the form factor F_2 that we need is small at $t = 0$. It remains small away from $t = 0$; thus to a good approximation there is no F_2 form factor for pomeron exchange. If it were present, it would correspond to s-channel helicity flip, and there is indeed good evidence[61] that the helicity-flip amplitude for pomeron exchange is small at small t.

So the contribution from single-pomeron exchange to the differential cross section for unpolarised pp or $\bar{p}p$ scattering is

$$\frac{d\sigma}{dt} = \frac{(3\beta_{I\!P} F_1(t))^4}{4\pi} \left(\frac{s}{s_0}\right)^{2\alpha_{I\!P}(t)-2}. \tag{3.15}$$

The factor 3 counts the number of valence quarks in each proton. A fit to the data for $F_1(t)$ in the region $|t| < 1$ GeV^2 is provided by the dipole form

$$F_1(t) = \frac{4m_p^2 - 2.79t}{4m_p^2 - t} \frac{1}{(1 - t/0.71)^2}. \tag{3.16}$$

This form is in excellent agreement with electron-scattering data for values of $|t|$ up to about 1 GeV^2. For calculational purposes it is sometimes more convenient to use

$$(F_1(t))^2 = Ae^{at} + Be^{bt} + Ce^{ct} \tag{3.17}$$

with

$$A = 0.27 \quad a = 8.38 \quad B = 0.56 \quad b = 3.78 \quad C = 0.18 \quad c = 1.36 \tag{3.18}$$

which is in close numerical agreement with the dipole form (3.16) for $|t| < 1$ GeV^2. It seems to be consistent with experiment to assume that the pomeron trajectory $\alpha_{I\!P}(t)$ is linear in t, like the ρ, ω, f_2, a_2 trajectory but with a different slope:

$$\alpha_{I\!P}(t) = 1 + \epsilon_{I\!P} + \alpha'_{I\!P} t. \tag{3.19}$$

The value of $\alpha'_{I\!P}$ may be determined by fitting to the shape of the small-t data at some fixed energy. Very accurate data are available from the

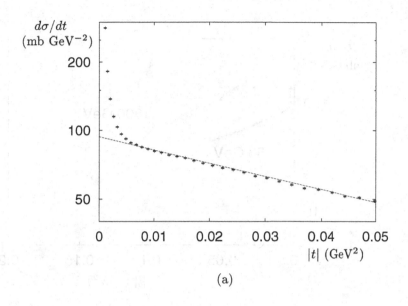

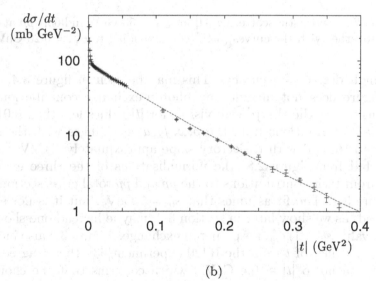

Figure 3.4. *pp* elastic scattering data at $\sqrt{s} = 53$ GeV from the CERN ISR[62,63] with (a) the fit that determines the value of $\alpha'_{I\!P}$ and (b) the fit extended to larger values of $|t|$

R211 experiment[62] at the CERN Intersecting Storage Rings at energy $\sqrt{s} = 52.8$ GeV for $|t| < 0.06$ GeV^2, which determine[66]

$$\alpha'_{I\!P} = 0.25 \text{ GeV}^{-2} \tag{3.20}$$

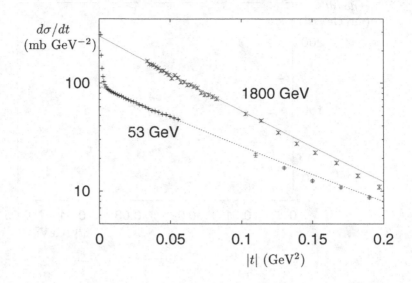

Figure 3.5.   *pp* elastic scattering[64] at $\sqrt{s}$ = 53 GeV and[65] $\bar{p}p$ at $\sqrt{s}$ = 1800 GeV, together with the curves (3.15) corresponding to $\alpha'_{I\!P}$ = 0.25 GeV^{-2}

to a high degree of accuracy. This may be seen in figure 3.4. The fit in this figure does not include any photon-exchange contribution, which is responsible for the sharp peak visible for $|t|$-values less than 0.01 GeV^2. It includes contributions from the ρ, ω, f_2, a_2 exchanges with the same form factor (3.16) and with trajectory slope approximately 1 GeV^{-2} as may be extracted from figure 2.8; the normalisations of the three exchanges are fixed from their contributions to the *pp* and $\bar{p}p$ total cross sections, as given in figure 3.1. The fit assumes that $s_0 = 1$ GeV^2, but it is not sensitive to this and, as we show later in section 3.2, maybe instead one should use the larger value $s_0 = 1/\alpha'_{I\!P}$ for pomeron exchange. Figure 3.4 also includes data at larger values of t from the R420 experiment[63]; the fit agrees well with these data up to $|t| \approx 0.4$ GeV^2, which confirms that the choice of form factor is correct. As the fit includes $F_1(t)$ raised to the fourth power, it is very sensitive to the shape of this function. This also provides evidence that the pomeron, like the photon, does indeed couple to single quarks within a hadron and that double-pomeron exchange is relatively small at small t at this energy. Notice, however, that the form (3.15) for the differential cross section, corresponding to an amplitude proportional to the square of the $C = -1$ form factor $F_1(t)$ times the $C = +1$ signature factor, cannot be exact. If we were to continue this amplitude analytically in t, we would find a pole at each $C = -1$ t-channel resonance, when really the amplitude

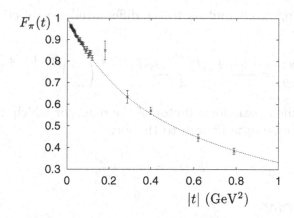

Figure 3.6. Data[68,69] for the pion elastic form factor, with the fit (3.22)

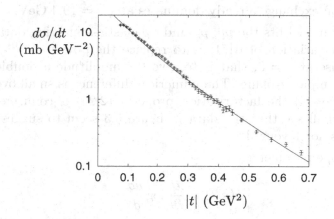

Figure 3.7. πp elastic scattering data[70] at $\sqrt{s} = 19.4$ GeV with the curve (3.21)

should have only the $C = +1$ t-channel poles. Furthermore, the analytic continuation has double poles while there should only be single poles.

With $\alpha'_{I\!P}$ determined from the data at one energy, (3.15) predicts how the forward peak shrinks as the energy increases. This is compared with data in figure 3.5. The fit in figure 3.4 combines single and double-pomeron exchange into a single term. If one separates them, which means that one constructs a simple model of the double exchange, and supposes that the double exchange contributes only a few per cent to the amplitude at $t = 0$, the fit does not become significantly different and the value extracted for α' does not change[67].

An obvious modification of (3.15) gives what we might expect to be the pomeron-exchange contribution to the differential cross section for πp elastic scattering:

$$\frac{d\sigma}{dt} = \frac{(2\beta_{I\!P}F_\pi(t))^2(3\beta_{I\!P}F_1(t))^2}{4\pi}\left(\frac{s}{s_0}\right)^{2\alpha_{I\!P}(t)-2}. \tag{3.21}$$

Here, $F_\pi(t)$ is the elastic form factor of the pion, for which data are shown in figure 3.6. These data fit well to the form

$$F_\pi(t) = \frac{1}{1-t/m_0^2} \tag{3.22}$$

with $m_0^2 = 0.5$ GeV^2.

If we use this and take $s_0 = 1/\alpha'$, we obtain the curve for $d\sigma/dt$ shown in figure 3.7, which agrees remarkably well with the data and shows that single-pomeron exchange already dominates at $\sqrt{s} = 19.4$ GeV.

So the simple model fits the pp, $\bar{p}p$ and πp elastic-scattering data surprisingly well. A variation of it[71] is to replace the power $s^{\alpha_{I\!P}(0)-1}$ with a logarithmic rise at $t = 0$, that is to give the amplitude a double pole at $l = 1$ in the complex l-plane. The numerical difference is small over a wide range of energy. If the factorisation property (2.25) is confirmed, as for example we shall see that the data in figure 5.3 seem to suggest, then a double pole is not favoured.

Define the slope parameter

$$b(s,t) = \frac{\partial}{\partial t}\log\frac{d\sigma}{dt}. \tag{3.23}$$

Then locally the t dependence of $d\sigma/dt$ is $\exp(-b(s,t)|t|)$. The simple model has

$$\frac{\sigma^{\text{Elastic}}}{\sigma^{\text{Tot}}} \sim \text{const}\,\frac{s^{\alpha_{I\!P}(0)-1}}{b(s)} \tag{3.24}$$

where $b(s)$ is an average of $b(s,t)$ over the the range of t-values that contribute significantly to σ^{Elastic}. Because $d\sigma/dt$ contains the factor $(F_1(t))^4$ and, according to (3.17), $F_1(t)$ is not just a single exponential, the exponential slope certainly varies with t. It also varies with s, but only linearly in $\log s$. Experiment[53] finds that for $\bar{p}p$ collisions $\sigma^{\text{Elastic}}/\sigma^{\text{Tot}}$ rises from 0.21 at $\sqrt{s} = 546$ GeV to 0.25 at 1800 GeV. Of course the rise cannot continue indefinitely, as the elastic cross section would otherwise become greater than the total cross section, and there are arguments[72] that in fact the ratio cannot rise above $\frac{1}{2}$. It is the double exchange that causes the rise to slow down and, at very high energies, also the multiple exchanges.

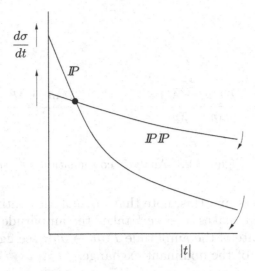

Figure 3.8. Single and double-pomeron exchange contributions to $d\sigma/dt$; the arrows indicate how they change as the energy increases

We do not have sufficient theoretical knowledge of the double exchange to be able to calculate it – which is why there is disagreement about how large it is – but its general properties are known. They are indicated in figure 3.8. The contribution to the amplitude from double exchange has energy dependence $s^{\alpha_c(t)}$ divided by some unknown function of $\log s$. For a linear pomeron trajectory (3.19), we have from (2.36) and (2.37)

$$\alpha_c(t) = 1 + 2\epsilon_{I\!P} + \tfrac{1}{2}\alpha'_{I\!P}t. \tag{3.25}$$

The double exchange is flatter in t than single exchange and so its relative importance becomes greater as one goes away from $t = 0$. It increases more rapidly with energy at $t = 0$ than the single exchange, and it becomes steeper more slowly, so that the t-value at which it becomes important decreases as the energy increases. Note that the single and double-exchange terms must be added in the amplitude, so that interference between them is important. The details of this are complicated, as they do not have the same phase.

The phase of an elastic-scattering amplitude at each value of t is related to its energy variation at that t. This is a rather general property, derived from the analyticity and crossing properties of the amplitude. If the energy variation is parametrised as an effective power $s^{\alpha(t)}$, the phase $e^{i\phi(\alpha(t))}$ depends on the C-parity of the dominant exchange:

$$\phi(\alpha(t)) = \begin{cases} -\tfrac{1}{2}\pi\alpha(t) & C = +1 \\ -\tfrac{1}{2}\pi(\alpha(t) - 1) & C = -1. \end{cases} \tag{3.26}$$

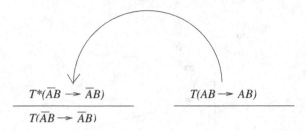

$$\frac{T^*(\overline{AB} \to \overline{AB})}{T(\overline{AB} \to \overline{AB})} \qquad\qquad T(AB \to AB)$$

Figure 3.9.   Analytic continuation $s \to se^{i\pi}$

In order to understand this, note that an analytic continuation in the complex s-plane that makes $s \to se^{i\pi}$ takes the amplitude $T(ab \to ab)$ to the complex conjugate of the amplitude $T(\bar{a}b \to \bar{a}b)$: see figure 3.9. Depending on the C-parity of the dominant exchange,

$$T(\bar{a}b \to \bar{a}b) = \pm T(ab \to ab). \qquad (3.27)$$

It is very difficult to measure the phase of an amplitude, though at extremely small t the phase of an elastic-scattering amplitude may, in principle, be deduced from interference effects between photon exchange and the strong interaction – see the sharp peak at extremely small t in the pp elastic-scattering data shown in figure 3.4. In practice, there are significant theoretical problems with this; for example, we do not know how to allow for the effect of electromagnetic and strong interactions occurring simultaneously. However, the published results[73,74] for the ratio of the real to imaginary part of the pp and $\bar{p}p$ forward amplitudes as functions of s are in agreement[49] with what is expected from (3.26) and what we know about the relative weights of the $C = \pm 1$ contributions from the fits to the total cross sections shown in figure 3.1.

The question of the phase is important for understanding the remarkable dip structure seen in the data at larger t: see figure 3.10. It leads one to conclude[2,75,76] that, while to the left of the dip $C = +1$ exchange dominates, to the right it is rather $C = -1$. With the inclusion of both C-parities, it is possible to achieve an excellent description of the dip structure at all energies[77]. Both to the right and to the left of the dip, the energy dependence at each t is weak, so that the effective power $\alpha(t)$ is close to 1. According to (3.26), the amplitude changes from almost imaginary on the left to almost real on the right. In the dip region it changes rapidly between these two phases; this is the region where, at each fixed t, there is rapid variation with s because the dip has steep sides and, as may be seen from figure 3.10, it moves to the left as s increases.

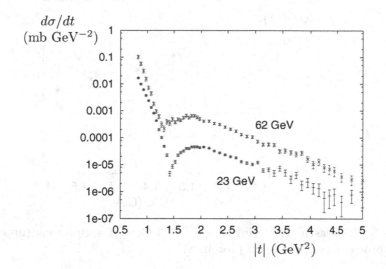

Figure 3.10. *pp* elastic scattering data at large *t* (CHHAV collaboration[64]). The 62 GeV data are multiplied by 10.

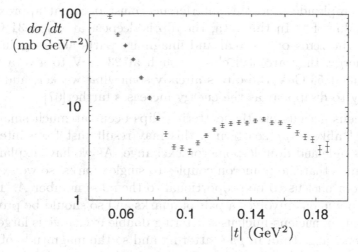

Figure 3.11. Elastic $\alpha\alpha$ scattering[78] at $\sqrt{s} = 126$ GeV

If one takes proper account of the phase, it is not easy to model the dip structure. If the amplitude is assumed to be pure imaginary for all *t*, as in simple geometric models[79], it is easy to generate dips, but these models do not correctly take account of the basic theory of the phase. Of course, they may be refined so as to take correct account of the phase[80,71]. When,

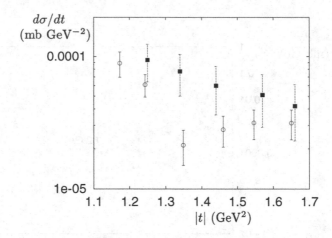

Figure 3.12.   Elastic scattering at $\sqrt{s} = 53$ GeV of antiprotons (upper points) and protons (lower points) on protons[81]

as is required by the theory, the amplitude is not pure imaginary, there needs to be destructive interference in both the real and imaginary parts of the amplitude, and this interference must occur at approximately the same value of t. In the data, the dip is deepest at $\sqrt{s} = 31$ GeV; at this energy the zeros of the real and imaginary parts almost coincide. Below this energy, they are still close enough at 23 GeV to give a healthy dip; above it at 53 GeV the dip is already somewhat weaker, and is predicted steadily to disappear as the energy increases further[67].

In nucleus-nucleus elastic scattering dips occur at much smaller t (figure 3.11). Unlike in pp scattering, this may result just from interference between single and double-pomeron exchange. As we have explained, experiment finds that the pomeron couples to single quarks, so we expect its coupling to a nucleus to be proportional to the mass number A. The coupling of two pomerons involves a pair of quarks and so should be proportional to A^2. Thus in nucleus-nucleus scattering double exchange is larger relative to single exchange than in pp scattering and so the magnitude of the two becomes equal at a smaller value of t. At small t the numerical values of $\alpha_{I\!P}(t)$ and $\alpha_c(t)$, the single and double-exchange trajectories (3.19) and (3.25), are not very different and so the phases of their contributions to the amplitude are similar. So if the imaginary parts of their contributions cancel, the real parts almost cancel too and there is a dip. But the dips in figure 3.10 are at such a large t that the phases of the single and double exchanges are very different. So if their imaginary parts cancel and help to generate a dip, something else is needed to cancel their combined real part and give dips

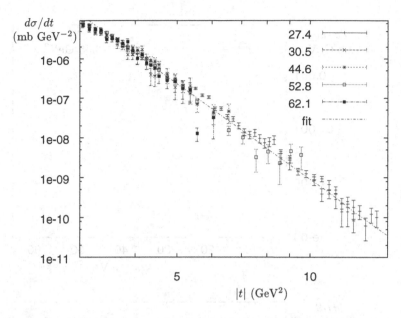

Figure 3.13. *pp* elastic scattering data[64,82] at the largest available t, at various energies indicated in the figure as $\sqrt{s}$ in GeV. The line is $0.09\,t^{-8}$.

as well-defined as are seen in the data. It was suggested[2,67,77] that this additional contribution is triple-gluon exchange. This is $C = -1$ and so, unlike $I\!P$ and $I\!P I\!P$ exchange, it changes sign if we replace one of the initial protons with an antiproton. So if it helps to give a dip in *pp* scattering, it does the opposite in $\bar{p}p$ scattering. There is evidence[81] that $\bar{p}p$ scattering does indeed have a much less pronounced dip than does *pp* scattering: see figure 3.12. A $C = -1$ contribution that remains significantly large at high energy is called an odderon contribution: see section 3.9.

Figure 3.13 shows that, when $|t|$ is larger than about 3 GeV^2, and the energy is high enough, the differential cross section for *pp* scattering becomes rather independent of energy. The line drawn in the figure corresponds to

$$d\sigma/dt = 0.09\,t^{-8} \tag{3.28}$$

(in mb GeV^{-2} units). A t^{-8} behaviour is what is calculated[83] from perturbative triple-gluon exchange at large $|t|$ to lowest order in the QCD coupling. There are many unanswered questions about this[84]: why does this simple behaviour set in already at such a small t, why is it not significantly altered by higher-order perturbative corrections, and are the data really energy-independent? It will be interesting to check this at LHC energies. A possibility is that triple-gluon exchange will be replaced at higher ener-

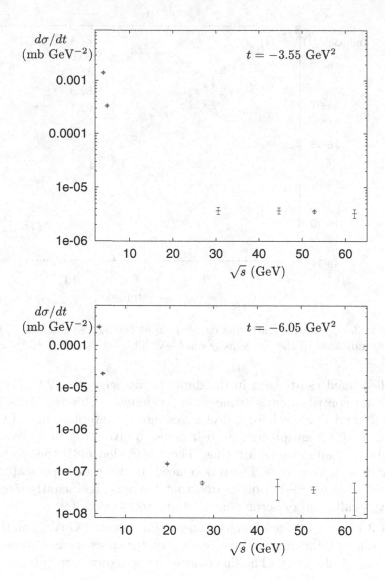

Figure 3.14. Energy dependence of *pp* elastic scattering data at two values of *t*

gies with triple-BFKL exchange, if the BFKL pomeron exists (see section 7.3) so that the large-*t* differential cross section actually rises with increasing energy. This contrasts with the prediction of [85], which explains the data of figure 3.13 not with an energy-independent form, but rather with a $C = -1$ odderon Regge pole. Odderons are discussed in section 3.9.

Note that the behaviour (3.28) sets in only for $|t|$ larger than about 3 GeV^2

and much less than s. Also, $\sqrt{s}$ must be larger than about 28 GeV. For $\sqrt{s}$ below about 8 GeV, a rather different behaviour has been found[86], approximately

$$d\sigma/dt \approx t^{-7}s^{-3}. \tag{3.29}$$

The transition from the form (3.29) to the form (3.28) occurs rather rapidly, as $\sqrt{s}$ is increased[83]. Figure 3.14 shows the transition at two values of t, from the steep energy dependence of (3.29) to the energy-independent behaviour illustrated in figure 3.13. The behaviours (3.28) and (3.29) may each be written in the form $s^{-n}f(\theta)$, where θ is the centre-of-mass scattering angle and with $n = 8$ for (3.28) and 10 for (3.29). The value $n = 10$ is realised[87] in a model known as the constituent-interchange model, so it seems that this mechanism dominates at low energies and then gives way to three-gluon exchange at higher energies.

3.3 Spin dependence of high energy proton-proton scattering

The phenomenological γ_μ coupling of the pomeron to quarks postulated in (3.11) has the implication that pomeron exchange conserves quark helicity. This does not necessarily imply that pomeron exchange conserves nucleon helicity as the quarks in a nucleon have intrinsic transverse momentum, but it does imply that helicity-flip amplitudes should be relatively small. The polarisation data available at present in high-energy proton-proton scattering are compatible with s-channel helicity conservation[61].

In proton-proton scattering there are five independent s-channel helicity amplitudes[88]:

$$\phi_1(s,t) = \langle + + |T| + + \rangle$$
$$\phi_2(s,t) = \langle + + |T| - - \rangle$$
$$\phi_3(s,t) = \langle + - |T| + - \rangle$$
$$\phi_4(s,t) = \langle + - |T| - + \rangle$$
$$\phi_5(s,t) = \langle + + |T| + - \rangle. \tag{3.30}$$

The differential cross section is given in terms of these amplitudes by

$$\frac{d\sigma}{dt} = \frac{1}{16\pi s(s - 4m^2)}\left(|\phi_1|^2 + |\phi_2|^2 + |\phi_3|^2 + |\phi_4|^2 + 4|\phi_5|^2\right) \tag{3.31}$$

and the total cross section by

$$\sigma^{\text{Tot}} = \frac{1}{\sqrt{s(s - 4m^2)}}\Big(\text{Im}\,\phi_1(s, t = 0) + \text{Im}\,\phi_3(s, t = 0)\Big) \tag{3.32}$$

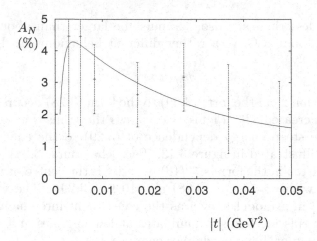

Figure 3.15.   Data[91] for the asymmetry A_N in polarised pp scattering; the curve is from the calculation of [92]

as ϕ_1 and ϕ_3 are the only helicity non-flip amplitudes. Note that here the normalisation of the $\phi_i(s,t)$ differs from that of [88] by a factor of 4π to be consistent with the amplitude normalisation we defined in chapter 1.

Only ϕ_5 corresponds to unique t-channel quantum numbers[89,90], and it is necessary to take linear combinations $\phi_1 \pm \phi_3$ and $\phi_2 \pm \phi_4$ to obtain specific t-channel exchanges. The pomeron, ρ, ω, f_2 and a_2 exchanges can all contribute to $\phi_1 + \phi_3$, ϕ_5 and $\phi_2 - \phi_4$, but not to $\phi_1 - \phi_3$ or $\phi_2 + \phi_4$. Only nonleading exchanges can contribute to $\phi_1 - \phi_3$ and $\phi_2 + \phi_4$: the a_1 to the former and the π, b_1 to the latter. Thus at high energy we expect only $\phi_1 + \phi_3$, $\phi_2 - \phi_4$ and ϕ_5 to be relevant. Further, we would expect the approximate equalities $\phi_1 \approx \phi_3$ and $\phi_2 \approx -\phi_4$. These conclusions are strictly true only in the case of pole dominance and might be modified in the presence of cuts. However, we have seen in section 2.4 that the signature of the cut is the same as that of the pole[34] and that[35] the "wrong" parity cut, that is a cut such that $P_{\text{cut}} = -P_{\text{pole}}$, is suppressed relative to the "right" parity cut, that is one with $P_{\text{cut}} = P_{\text{pole}}$, so our comments on the high-energy amplitudes are of more general validity.

There is obviously a multiplicity of polarisation measurements that can be made, even if they are restricted to initial-state polarisation[61]. However the only data currently available at high energy are for the transverse single-spin asymmetry A_N, which is given by

$$A_N \frac{d\sigma}{dt} = -\frac{1}{4\pi s(s-4m^2)} \operatorname{Im}\left(\phi_5^*((\phi_1+\phi_3)+(\phi_2-\phi_4))\right). \qquad (3.33)$$

As ϕ_2 and ϕ_4 are both helicity double-flip amplitudes $\phi_2 - \phi_4$ is negligible at small t compared with $\phi_1 + \phi_3$. Thus at small t A_N effectively measures the interference between the dominant non-flip helicity amplitude $\phi_1 + \phi_3$ and the single-flip helicity amplitude ϕ_5.

If t is sufficiently small then Coulomb scattering has also to be considered. Photon exchange contributes to $\phi_1 + \phi_3$, $\phi_2 - \phi_4$ and ϕ_5. So in the absence of any hadronic spin-flip amplitude, at small t we may expect A_N to be given by the interference between the dominant non-flip hadronic amplitude $\phi_1^{\text{h}} + \phi_3^{\text{h}}$ and the single-flip Coulomb amplitude ϕ_5^{em}. Data in the relevant t-range exist[91] only at $p_{\text{Lab}} = 200$ GeV/c, $|t| \leq 0.05$ GeV^2. These are shown in figure 3.15 together with the prediction[92] of a purely Coulomb-nuclear interference asymmetry. The prediction agrees well with the data, but within the experimental uncertainty the possibility of a non-zero spin-flip hadronic amplitude cannot be discounted[92,61]. If the phase of ϕ_5^{h} is assumed to be the same as that of $\phi_1^{\text{h}} + \phi_3^{\text{h}}$ then the magnitude of ϕ_5^{h} is found to be consistent with zero[61]. Specifically, with r_5 defined by

$$r_5 = \frac{m\phi_5}{\sqrt{-t}\,\text{Im}\,(\phi_1 + \phi_3)} \tag{3.34}$$

it is found that $|r_5| = 0.0 \pm 0.16$. If the relative phase between ϕ_5^{h} and $\phi_1^{\text{h}} + \phi_3^{\text{h}}$ is left free in a fit, then $r_5 = 0.2 \pm 0.3$ and the phase is found to be 0.15 ± 0.27 radians. That is, the data are consistent with s-channel helicity conservation, although they do not demand it at the present level of precision.

At larger values of $|t|$ data are available on A_N over a wider energy range, 10 GeV $\leq p_{\text{Lab}} \leq 300$ GeV, and show that A_N decreases rapidly with increasing energy. This general behaviour is what is expected in terms of Regge exchange, but the data are again insufficiently precise to provide an unambiguous conclusion at the highest energy. They are consistent with the expected interference between ϕ_5^{em} and $\phi_1^{\text{h}} + \phi_3^{\text{h}}$, although a contribution from the helicity-flip amplitude cannot be excluded.

3.4 Soft diffraction dissociation

We now consider inelastic events, in which additional particles are produced. In the centre-of-mass frame at high energy $\sqrt{s}$, the magnitude of the longitudinal momentum of each initial particle is almost $\frac{1}{2}\sqrt{s}$. The longitudinal momentum P_L of a final-state particle is often written as

$$P_L = \tfrac{1}{2}x_F\sqrt{s}. \tag{3.35}$$

The variable x_F, first introduced by Feynman, has maximum and minimum values close to ± 1. An alternative variable that is often used to describe the longitudinal momentum of a final-state particle is called its rapidity, defined as

$$Y = \tfrac{1}{2} \log \frac{E + P_L}{E - P_L} \tag{3.36}$$

where E is the energy of the particle. In the centre-of-mass frame, Y has maximum and minimum values close to $\pm \frac{1}{2} \log(s/m^2)$, where m is the mass of the particle. The rapidity Y is a useful variable because the rapidity difference between any pair of particles remains unchanged if we go from the centre-of-mass frame to any other frame reached from it by a longitudinal boost. If

$$E \to E \cosh \alpha + P_L \sinh \alpha \qquad P_L \to E \sinh \alpha + P_L \cosh \alpha \tag{3.37}$$

then simply $Y \to Y + 2\alpha$. When $E \gg m$, the rapidity Y is almost equal to the pseudorapidity

$$\eta = \log \cot \tfrac{1}{2}\theta \tag{3.38}$$

where θ is the angle at which the particle emerges in the centre-of-mass frame.

In a small number of inelastic events, one of the initial particles a loses only a very small fraction ξ of its momentum. That is, it has $x_F = 1 - \xi$ very close to its maximum value 1. Because it emerges very fast, energy-momentum conservation excludes the possibility that it is accompanied by other very fast particles. Therefore, there is a large rapidity gap between the particle a and the next fastest particle. The detection of a very-fast final-state particle requires an in-beam-pipe detector. If this is not available, experimentalists often instead look for large-rapidity-gap events. However, in such events there need not be just a single undetected fast particle a in the beam pipe; instead there may be several.

Mueller[93,94] proposed a generalisation of the optical theorem so as to provide a way to calculate single-particle inclusive differential cross sections

$$a(P_1) + b(P_2) \to c(P) + X \tag{3.39}$$

in terms of certain discontinuities of multiparticle amplitudes. Mueller's generalised optical theorem may be combined with Regge theory so as to give a well-defined way to calculate such inclusive differential cross sections $P^0 \, d^3\sigma/d^3\mathbf{P}$. It may be extended also to two-particle inclusive cross sections

$$a(P_1) + b(P_2) \to c(P) + d(P') + X \tag{3.40}$$

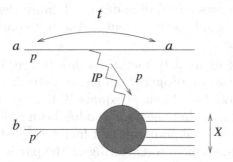

Figure 3.16.   Pomeron exchange in an inelastic diffractive event

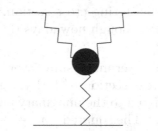

Figure 3.17.   The squared amplitude of figure (3.16) summed over X, in the large-M^2 limit

or to any number of detected final-state particles. There is no equivalent well-defined theory of events for which instead the experimental signature is that there is a rapidity gap; apart from anything else, even if one identifies a mechanism that might produce a gap, there is always an unknown probability that it will be filled in by final-state interactions among the various final-state particles.

Suppose that the particle c in (3.39) is the initial particle a and that the energy is so high that its mass is negligible compared with $\sqrt{s}$. Suppose also that the momentum transfer

$$t = (P_1 - P)^2 \qquad (3.41)$$

is no larger than about 1 GeV^2. Then to a first approximation the four-vector P is in the same direction as P_1 and we may write $P = x_F P_1$. When $\xi = (1 - x_F)$ is extremely small, less than 0.001 or so, the dominant mechanism is the pomeron-exchange mechanism shown in figure 3.16. The particle a "radiates" a pomeron, which then strikes the other particle b and breaks it up into a system X of hadrons. Because the pomeron is involved, this is known as a diffractive event. Another common notation for the variable ξ is $x_{I\!P}$.

The amplitude of figure 3.16 factorises in a Regge sense that is an extension of the factorisation property of elastic-scattering amplitudes which we described in section 2.3. When we square it, to calculate a cross section, we obtain the probability that the pomeron has been radiated from the particle a, which we may call the pomeron flux factor $D^{I\!P/a}(t, \xi)$, multiplied by the cross section for the pomeron scattering on the particle b. This does *not* imply that the pomeron is a particle, but it may be thought of as almost being like a particle in this context because of the Regge factorisation. However, because it is not a particle, the pomeron flux is not uniquely defined; one is free to change its definition and make a compensating change in the definition of the pomeron-scattering cross section. Different conventions may be found in the literature, though nowadays the usual convention is that of [95].

An inclusive diffractive experiment sums over all possible systems X and so measures the total cross section $\sigma^{I\!Pb}$. By a generalisation of the optical theorem (3.1), this is related to the imaginary part of the forward amplitude for $I\!Pb$ elastic scattering. The squared centre-of-mass energy corresponding to this amplitude is

$$M^2 = (\xi P_1 + P_2)^2 \sim \xi s. \tag{3.42}$$

The Regge factorisation corresponds to

$$\frac{d^2\sigma}{dt\,d\xi} = D^{I\!P/a}(t, \xi)\, \sigma^{I\!Pb}(M^2, t). \tag{3.43}$$

The definition of the factors adopted in [95] is such that if one continues in t to a value such that the pomeron trajectory $\alpha_{I\!P}(t)$ takes the value 1 the normalisation of the cross section $\sigma^{I\!Pb}$ is the usual one for the scattering of a spin-1 particle. In the notation of (3.15) this results in

$$D^{I\!P/a}(t, \xi) = \frac{9\beta_{I\!P}^2}{4\pi^2}\, (F_1(t))^2\, \xi^{1-2\alpha_{I\!P}(t)}. \tag{3.44}$$

For large M^2 the pomeron-particle total cross section should have a behaviour analogous to (3.5):

$$\sigma^{I\!Pb}(M^2, t) = X^{I\!Pb}(t)\, (M^2)^\epsilon + Y_+^{I\!Pb}(t)\, (M^2)^{\eta_+} \tag{3.45}$$

where again the first term represents pomeron exchange and the second represents f_2, a_2 exchange. The pomeron-exchange term contributes to

$d^2\sigma/dt\,d\xi$ a term of the structure shown in figure 3.17, in which three pomerons are coupled together. The lower one carries zero momentum transfer and, if the lower hadron is a proton or antiproton, its coupling has strength $3\beta_{I\!\!P}$. The upper two pomerons carry momentum transfer t and, if the upper hadron is a proton or antiproton, they couple with strength $3\beta_{I\!\!P}F_1(t)$. The vertex in the centre is called the triple-pomeron coupling. It is different from the triple couplings in figure 2.12, though it is related to them[96].

Unless M^2 is very large, we cannot ignore the last term in (3.45). That is, we must add another term in which the lower pomeron $I\!\!P$ of figure 3.17 is replaced with nonleading exchanges. Also, unless ξ is very small, we must include terms in which either or both of the upper pomerons are replaced with nonleading exchanges. Thus we need a whole series of terms:

$$
\begin{array}{cccccc}
I\!\!P\,I\!\!P & I\!\!P\,I\!\!P & f_2\,I\!\!P & I\!\!P\,f_2 & f_2\,I\!\!P & \omega\,I\!\!P \\
I\!\!P & f_2 & I\!\!P & I\!\!P & f_2 & \omega
\end{array}
\quad \cdots
$$
(3.46)

There are also contributions from a_2 and ρ exchange, but these couple more weakly to the proton, according to (3.8). When $|t|$ is of the order of m_π^2 or less, one must also take account of pion exchange. Fits to data [97,95,98] confirm the great importance of these nonleading terms. A term $\binom{12}{3}$ contributes to $d^2\sigma/dt\,d\xi$

$$
f_1^a(t)f_2^a(t)f_3^b(0)G_3^{12}(t)\,e^{i(\phi(\alpha_1(t))-\phi(\alpha_2(t)))}\xi^{1-\alpha_1(t)-\alpha_2(t)}\left(\frac{M^2}{s_0}\right)^{\alpha_3(0)-1}.
$$
(3.47)

Here, $f_1^a(t)$ and $f_2^a(t)$ are the couplings of the reggeons 1 and 2 to the hadron a, while $f_3^b(t)$ is the coupling of the reggeon 3 to the hadron b. $G_3^{12}(t)$ is the triple-reggeon vertex and $\phi(\alpha(t))$ is the phase (3.26) arising from the signature factor $1 \pm e^{-i\pi\alpha(t)}$ associated with the trajectory $\alpha(t)$. The complex exponential is replaced with $2\cos(\phi(\alpha_1(t)) - \phi(\alpha_2(t)))$ when we add the term $\binom{21}{3}$.

Existing data do not fit well with the simple theory. An example is shown in figure 3.18, which shows data for $d^2\sigma/dt\,d\xi$ in pp collisions at $\sqrt{s} = 35$ GeV and in $\bar{p}p$ collisions at $\sqrt{s} = 630$ GeV. The curves are from an updated version of an old simple model[95]; the lower curve is for $\sqrt{s} = 23$ GeV and the upper for 630 GeV. Independently of the model, at fixed t and ξ we expect $d^2\sigma/dt\,d\xi$ to rise with energy, a feature which is not seen in the data. The rise is expected because, while the pomeron flux (3.44) does not depend on s, at each ξ the argument $M^2 = \xi s$ of $\sigma^{I\!\!P b}$ increases with s and, according to (3.45), at large enough M^2 we expect $\sigma^{I\!\!P b(M^2)}$ to increase

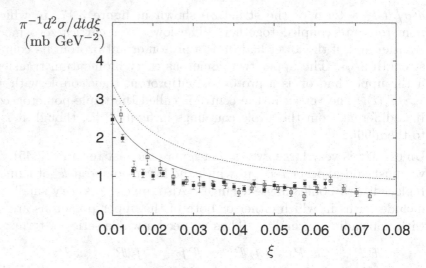

Figure 3.18.   Diffraction dissociation data at $\sqrt{s} = 23$ GeV (open points[101]) and 630 GeV (black points[102]) at $t = -0.75$ GeV^2. The lower curve is from a simple model and is for 23 GeV; the upper curve is the prediction for 630 GeV.

with M^2. Problems such as this have led authors[99,100] to suggest that the pomeron flux (3.44) does depend on s, though there is no theoretical justification for this.

Experimentalists often quote results for the "total diffractive cross section"

$$\sigma^{\text{Diff}}(s) = \int dt \int_{\xi_{\min}}^{\xi_{\max}} d\xi \; \frac{d^2\sigma}{dt \, d\xi}. \tag{3.48}$$

In practice, the integration will only extend over a finite range of t, but if the range is large enough the result will not be sensitive to it. The upper limit on the ξ integration is usually chosen to be 0.05 or 0.1, which is thought to be the largest value for which the Regge theory is still applicable. This means that, at low energies, there are important contributions from terms for which either $\alpha_1(t)$ or $\alpha_2(t)$ (or both) is a nonleading trajectory. But the lower limit $\xi_{\min}$ in practice causes most problems. Usually it is chosen to correspond to a fixed lower limit M_0^2 on M^2:

$$\xi_{\min} = M_0^2/s \tag{3.49}$$

but it is rather unsafe to apply Regge theory when M is less than 5 GeV or so: that is, (3.45) and (3.47) cannot be used.

Notice that (3.48) rises rather rapidly with s at high s if the ξ dependence of the integrand is given by (3.44). Indeed, it eventually becomes larger than

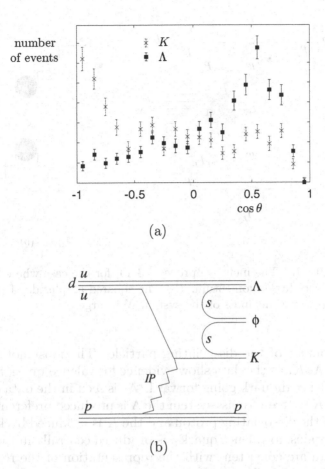

(a)

(b)

Figure 3.19. $pp \to p + \Lambda\phi K$: (a) uncorrected data[104] for the angular distribution in the $\Lambda\phi K$ centre-of-mass frame; (b) quark-flow interpretation of the data

the total cross section. This is because, according to (3.49), as s increases the integration extends down to smaller and smaller values of ξ. When ξ is very small, it is necessary to include terms in which the exchange of the single pomeron $\alpha_{I\!P}(t)$ is replaced with a pair of pomerons (or more). This slows down the rise with s and is often said to be a unitarity correction[103].

As well as inclusive diffractive processes, one may study exclusive ones in which, instead of summing over all possible hadronic systems X in figure 3.16, one picks out a particular system. These have revealed some interesting effects[104], which support the hypothesis that the pomeron interacts with only one quark in a hadron; they appear to reflect directly the

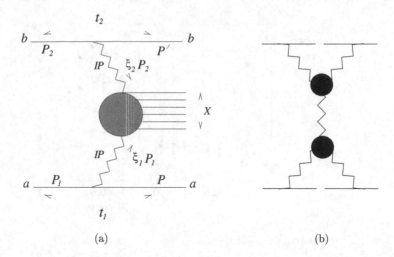

Figure 3.20.   (a) The inclusive process (3.40) for the case where both initial particles lose very little momentum. (b) The squared amplitude of (a) summed over X when the invariant mass of the system X is large.

parton content of the dissociating particle. The most notable reaction is $pp \rightarrow p + \Lambda\phi K$, as the data show evidence for valence-quark back-scattering with a valence diquark going forward. As is seen in the data of figure 3.19a, in the $\Lambda\phi K$ centre-of-mass system the Λ is produced preferentially in the direction of the dissociating proton and the K is produced backwards. The ϕ, which contains no valence quarks, is produced centrally in the $\Lambda\phi K$ system. These data are consistent with the representation of the reaction given in figure 3.19b. A similar picture is seen in $pp \rightarrow p + \bar{\Lambda}\Lambda p$. In this case the $\bar{\Lambda}$, which contains no valence quarks, is produced centrally in the $p\bar{\Lambda}\Lambda$ centre-of-mass system and the p and Λ are produced both forward and backwards. Similar evidence has been found[105] in the reaction $pp \rightarrow p + \Lambda K$,

So far we have discussed reactions where one of the initial particles loses only a very small fraction of its initial momentum. Suppose now that this is true of both initial particles, that is consider the inclusive process (3.40) where the final-state particle c is of the same type as a, with $P = (1 - \xi_1)P_1$ and $\xi_1 \ll 1$; similarly d is of the same type as b, with $P' = (1 - \xi_2)P_2$ and $\xi_2 \ll 1$. This is shown in figure 3.20a. If ξ_1 and ξ_2 are small enough, then both reggeons in the figure will be pomerons. If we square the amplitude and sum over possible systems X, we may calculate

$$\frac{d^4\sigma}{dt_1 d\xi_1 dt_2 d\xi_2} = D^{I\!P/a}(t_1, \xi_1)\, D^{I\!P/b}(t_2, \xi_2)\, \sigma^{I\!P I\!P}(M^2, t_1, t_2). \qquad (3.50)$$

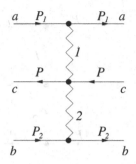

Figure 3.21. Mueller-Regge mechanism for inclusive central production

For large M^2, the total cross section $\sigma^{I\!P I\!P}(M^2, t_1, t_2)$, which may be thought of as that for pomeron-pomeron scattering, will have Regge behaviour, like $\sigma^{I\!P b}(M^2, t)$ in (3.45). If M^2 is so large that only pomeron exchange need be included, then the right-hand side of (3.50) corresponds to figure 3.20b, which contains two triple-pomeron vertices. The diagram factorises: the triple-pomeron vertex is the same as in figure 3.17b. Thus the inclusive cross section for two extremely-fast final-state particles may be written in terms of that for one extremely-fast one:

$$\sigma^{\text{Tot}}(s)\frac{d^4\sigma}{dt_1 d\xi_1 dt_2 d\xi_2} = \frac{d^2\sigma}{dt_1 d\xi_1}\frac{d^2\sigma}{dt_2 d\xi_2}. \tag{3.51}$$

3.5 Central production

Mueller's generalised optical theorem does not only apply to the case when the particle $c(P)$ in (3.39) is very fast. Another case of interest is that of central production, when instead the particle is comparatively slow. It need not be a particle of the same type as either of the initial particles. The inclusive differential cross section then corresponds to terms of the type shown in figure 3.21. This diagram involves the exchange of two trajectories α_1 and α_2, both evaluated at zero momentum transfer, as in the normal optical theorem. As with the normal optical theorem, it is necessary to take an appropriate discontinuity of the diagram: it is to be cut right down the middle. The reggeons have some structure such that cutting through them reveals the particles making up the undetected system X of hadrons in (3.39).

Define the momentum transfers

$$t_1 = (P_1 - P)^2 \qquad t_2 = (P_2 - P)^2. \tag{3.52}$$

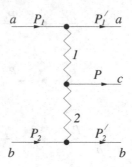

Figure 3.22. Exclusive central production

The production mechanism corresponding to figure 3.21 applies only if both of these are large. When the longitudinal momentum component $P_L = 0$ in the centre-of-mass frame each is equal to $-m_T\sqrt{s}$, where $m_T^2 = m^2 + P_T^2$ with P_T the transverse momentum component. Most of the particles produced are pions, for which m_T is a few hundred MeV, so that $m_T\sqrt{s}$ does not become large until s is very large. When P_L is moved away from 0, the product of t_1 and t_2 remains fixed at $m_T^2 s$ and one of t_1 and t_2 increases, but the other decreases rapidly. So very high initial energy is needed before the theory becomes applicable over any appreciable range of rapidity. Even higher energy is needed if one wants the dominant contribution to arise from the case where both the trajectories α_1 and α_2 are the pomeron trajectory. When this is achieved, figure 3.21 contributes when a and b are protons,

$$\frac{9\beta_{I\!P}^2}{4\pi^2 s} V_{I\!P I\!P}(m_T^2) \left(\frac{-t_1}{s_0}\right)^{\alpha_{I\!P}(0)-1} \left(\frac{-t_2}{s_0}\right)^{\alpha_{I\!P}(0)-1} = \frac{9\beta_{I\!P}^2}{4\pi^2 s} \tilde{V}_{I\!P I\!P}(m_T^2) \left(\frac{s}{s_0}\right)^{\alpha_{I\!P}(0)-1}$$

(3.53)

to $P^0 d^3\sigma/d^3 P$. Here, V is the vertex that couples the particle c to the two pomerons, and $\tilde{V}_{I\!P I\!P}(m_T^2) = V_{I\!P I\!P}(m_T^2)(m_T^2/s_0)^{\alpha_{I\!P}(0)-1}$. The fixed scale s_0 is introduced as in (2.17), in order to define V to be dimensionless. There are nonleading contributions, in which either or both pomeron trajectories are replaced with nonleading ones, with appropriate couplings. Note that experiment finds that the vertex V decreases exponentially with increasing m_T; when m_T is larger than of the order of a GeV, the Regge contribution (3.53) becomes insignificant compared with other mechanisms related to perturbative QCD.

Instead of the inclusive process (3.39), we may study the exclusive process

$$a(P_1) + b(P_2) \rightarrow a(P_1') + c(P) + b(P_2').$$

(3.54)

The final-state particle P might be an ordinary hadron, or something more

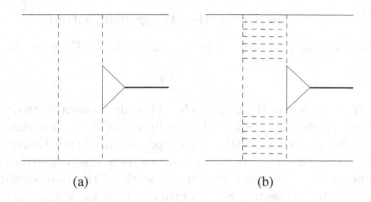

(a)   (b)

Figure 3.23. An explicit simple nonperturbative gluon-exchange model for the diffractive Higgs-production mechanism of figure 3.22: (a) first approximation, (b) corrections that turn the gluons into pomerons

exotic, for example the Higgs[106–108]. Define the momentum transfers

$$t_1 = (P_1 - P_1')^2 \qquad t_2 = (P_2 - P_2')^2 \qquad (3.55)$$

and the final-state subenergy variables

$$s_1 = (P_1' + P)^2 \qquad s_2 = (P_2' + P)^2. \qquad (3.56)$$

When s_1 and s_2 are both large, but t_1 and t_2 are not large, the diagram of figure (3.22) dominates the amplitude and contributes

$$\beta_1(t_1)\Gamma(-\alpha_1(t_1))\xi_{\alpha_1(t_1)}\beta_2(t_1)\Gamma(-\alpha_2(t_1))\xi_{\alpha_2(t_2)}f_{12}(\eta)$$
$$\times (s_1/s_0)^{\alpha_1(t_1)}(s_2/s_0)^{\alpha_2(t_2)} \qquad (3.57)$$

to it. The reggeon-hadron coupling functions $\beta(t)$ are the same as in (2.17) and the signature factors ξ_α are defined in (2.18). The function $f_{12}(\eta)$ couples the two reggeons to the particle P. The variable η is defined as

$$\eta = s_1 s_2/s. \qquad (3.58)$$

One may verify that, when s_1 and s_2 are both large with t_1 and t_2 not large, η also cannot be large. It is related to the Toller angle[109,110], the angle in the centre-of-mass frame between the production planes of the final-state particles. See [33] for a review of what is known about the dependence of $f_{12}(\eta)$ on η.

3.6 Diffractive Higgs production

If we identify the particle c in figure 3.22 with the Higgs we obtain[106]

$$pp \to pHp \qquad\qquad (3.59)$$

where H denotes the Higgs particle. This mechanism is known as diffractive Higgs production. An explicit simple model[107] for this is shown in figure 3.23a, where the dotted lines represent nonperturbative gluon propagators. This figure is to be thought of as a first approximation to the exchange of the two pomerons, in the spirit of the Landshoff-Nachtmann model described in section 8.1. In order to turn the simple gluon exchange into pomeron exchange, so as to give the diagram of figure 3.22, we must presumably add to it ladder diagrams such as shown in figure 3.23b, and more complicated diagrams. The coupling of the Higgs to a pair of gluons is through a t-quark loop; its magnitude may be related[107] to the similar vertex where the gluons are replaced with photons.

After the original calculation[107] of figure 3.23 has been corrected and improved[111], it gives for a Higgs mass of 120 GeV a cross section of a little less than 1 pb at LHC energy, $\sqrt{s} = 14$ TeV. The Higgs is likely to be accompanied by gluon emission; requiring that this is absent introduces a Sudakov-suppression factor which is[108] about an order of magnitude. This cross section is large enough for it to be suggested[112] that the exclusive reaction (3.59), with two very fast protons, might be the best way to discover the Higgs particle. This is because the momenta of the two final protons (or proton and antiproton) may be measured very accurately, so that the missing mass, presumed to be that of the Higgs, may be determined very accurately, and in consequence the signal-to-background ratio is much enhanced. The reason for this is as follows. The best way to detect the Higgs is through its decay to a heavy quark-antiquark pair. The background is direct production of the pair, not through a Higgs. It turns out[113] that direct $q\bar{q}$ production is suppressed in the mechanism of figure 3.23, though one does have to worry about the production of gluon jets. However, because of the good mass resolution, the background is integrated over only a small mass range.

Although the simple diffractive mechanism of figure 3.22 gives a cross section large enough to be measurable, there is the issue of possible screening corrections. These correspond to exchanging an additional pomeron, as is shown in figure 3.24. This figure represents a sum over initial and final-state interactions, plus interference between them[114]. There are differing views about how significant such corrections are: some authors believe that they suppress the production by a factor that is about an order of magni-

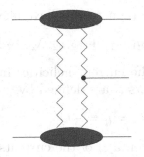

Figure 3.24. Screening correction to figure 3.22

tude[115,108]. This would make the cross section too small to be interesting. However, this view is the one taken by those who believe that cut contributions to the total cross section are large. As we have discussed, this view is controversial and we do not share it. Our belief is that, likewise, the screening correction to Higgs production is rather insignificant.

3.7 Helicity structure of the pomeron

We have seen in section 3.2 that phenomenologically the pomeron behaves like a $C = +1$ photon, coupling to quarks with a vector γ^μ coupling. However unlike the electromagnetic vector current J_μ^{em} the pomeron vector current $J_\mu^{I\!\!P}$ need not be conserved. Indeed there are theoretical reasons[116] to believe that current nonconservation is important: the phenomenological pomeron discussed in this chapter behaves as a nonconserved vector current. It was proposed in [116] that the diffractive dissociation reaction $ep \rightarrow epX$ can be used to study this question. By defining a suitable azimuthal angle ϕ between the lepton and hadron planes it was shown that interference terms among the photon helicity states for the reaction $\gamma^* p \rightarrow pX$ can be used to test models of diffraction. Effectively the reaction is considered as a γ^*-pomeron interaction, the proton merely serving as the source of the pomeron. This work was generalised[117,118] to include $\gamma^*\gamma^*$, γ^*-pomeron and pomeron-pomeron interactions. The production of mesons in the central region of high-energy proton-proton collisions, that is the reaction $pp \rightarrow ppM$, provides unambiguous tests[117,118] of whether or not the pomeron couples like a conserved vector current.

Consider the case of reaction (3.54), where c is a J^{P+} meson of mass M produced centrally in the high-energy scattering of two protons. We examine the hypothesis that the pomerons behave like vector currents:

$$V_1^\dagger(q_1, \lambda_1) + V_2(q_2, \lambda_2) \to M(J, J_z). \tag{3.60}$$

Here $\lambda_i = \pm 1, 0$ are the current helicities in the meson rest frame. The four-momentum transfers t_i are defined by

$$t_i = q_i^2 = -Q_i^2, \qquad i = 1, 2 \tag{3.61}$$

and the fractional momenta ξ_i of the currents by

$$\xi_1 = \frac{p_2 \cdot q_1}{p_2 \cdot p_1} \qquad \xi_2 = \frac{p_1 \cdot q_2}{p_2 \cdot p_1}. \tag{3.62}$$

The two other relevant kinematic variables are the fractional longitudinal momentum x_F of the meson, $x_F = p_\parallel / p^{\max}$, and the azimuthal angle ϕ, the angle between the two proton scattering planes in the centre-of-mass system. Strictly the correct azimuthal angle to consider is the angle $\hat\phi$ between the two proton scattering planes in the meson rest frame. However in the kinematic regime of experimental interest $\hat\phi \approx \phi$, and the effect of any extra kinematic factors does not significantly affect the conclusions[117, 118]. Finally, as we are interested in the dominant region of phase space $|t_i| \ll M^2$, $\xi_i \approx \frac{1}{2}(\sqrt{x_F^2 + 4M^2/s} \pm x_F) \ll 1$ and $M^2 = \xi_1\xi_2 s$. Further we can use $Q_i \approx |\mathbf{q}_{i\perp}| = q_{i\perp}$, where $\mathbf{q}_{i\perp}$ is the transverse momentum of current i.

The ϕ dependence provides a rather direct probe of the dynamics[117,118]. The reaction can be described in terms of the magnitudes $A_{\lambda_1\lambda_2}$ of the helicity amplitudes for the reaction (3.60), which are functions of t_1, t_2 and M. There are five independent amplitudes, A_{++}, A_{+-}, A_{+0}, A_{0+} and A_{00}. The cross section may be written as[117,118]

$$\frac{d^4\sigma}{dt_1 dt_2 d\phi\, dx_F} = \mathcal{A}(x_F, t_1, t_2)\, t_1 t_2\, (\sigma_2 + \sigma_1 + \sigma_0) \tag{3.63}$$

where σ_n denotes the differential cross section for the production of a meson with $|J_z| = n$. The function $\mathcal{A}(x_F, t_1, t_2)$ is concentrated at $x_F = 0$; it does not vanish for t_i approaching $t_i^{\min}$, $i = 1, 2$, and falls steeply with increasing $|t_i|$, being approximately proportional to $\exp(b(t_1 + t_2))$. The σ_n are given in terms of the helicity amplitudes by[118]

$$\sigma_2 = \tfrac{1}{2} A_{+-}^2$$
$$\sigma_1 = A_{+0}^2 + A_{0+}^2 - 2\eta S_2 A_{+0} A_{0+} \cos\phi$$
$$\sigma_0 = \begin{cases} (A_{00} - S_1 A_{++} \cos\phi)^2 & (\eta = +1) \\ A_{++}^2 \sin^2\phi & (\eta = -1). \end{cases} \tag{3.64}$$

Here the subscripts $\pm, 0$ refer to the pomeron helicities; η is the product of the naturality of the two currents and the meson, $\eta = \eta_1 \eta_2 \eta_M = \eta_M$ for two-pomeron exchange; and the S_i are sign factors[117,118]. Naturality for currents is defined as is the naturality of mesons, which we introduced in section 2.5. The electromagnetic current has naturality $\eta = 1$ and it is assumed that the pomeron currents also have naturality $\eta_1 = \eta_2 = 1$.

In the standard model of the phenomenological pomeron discussed in section 3.2 the pomeron carries helicities $\lambda = \pm 1$ and 0. The helicity state of a pomeron of momentum $\mathbf{q}$ is described by a density matrix[116]

$$\rho^{\mu\nu} = \sum_{\lambda\lambda'} \epsilon_\lambda^\mu \, \rho_{\lambda\lambda'} \, \epsilon_{\lambda'}^{*\nu} \tag{3.65}$$

where the ϵ_λ are the polarisation vectors

$$\epsilon_0 = \frac{1}{\sqrt{-t}}\left(|\mathbf{q}|, q^0 \frac{\mathbf{q}}{|\mathbf{q}|}\right) \qquad \epsilon_\pm = \left(0, \boldsymbol{\epsilon}_\pm\right) \tag{3.66}$$

and

$$\epsilon_\pm . \mathbf{q} = 0 \qquad \epsilon_+^* . \epsilon_+ = \epsilon_-^* . \epsilon_- = 1 \qquad \epsilon_+^* . \epsilon_- = 0 \,. \tag{3.67}$$

The matrix elements $\rho_{\lambda\lambda'}$ are all of the same order of magnitude[116], but note the factor $1/\sqrt{-t}$ in ϵ^0 which is crucial for the following discussion.

Consider first the production of a pseudoscalar meson $J^P = 0^-$, $\eta = -1$. In this case rotational and parity invariance imply that

$$A_{00} = A_{+-} = A_{+0} = A_{0+} = 0$$
$$\sigma_2 = \sigma_1 = 0 \,. \tag{3.68}$$

From (3.64), the cross section is then

$$\frac{d^4\sigma(0^-)}{dt_1 dt_2 d\phi \, dx_F} = \mathcal{A}(x_F, t_1, t_2) \, t_1 t_2 \, A_{++}^2 \sin^2 \phi \tag{3.69}$$

both for conserved and non-conserved currents. The experimental cross sections for η and η' production exhibit the $\sin^2 \phi$ and $t_1 t_2 \exp(b(t_1 + t_2))$ behaviour[119] expected, confirming the vector nature of the pomeron current. The ϕ distributions for η and η', normalised to unity, are shown in figures 3.25a and 3.25b. To distinguish between conserved and non-conserved currents it is necessary to consider other J^P states which do provide a test.

Consider now the production of a meson with $J^P = 1^+$. For simplicity consider the case $t_1 = t_2 = t$, so that rotational invariance and Bose symmetry of the two pomeron currents require

$$A_{++} = 0 \qquad A_{+0} = A_{0+} \qquad \sigma_0 = \sigma_2 = 0. \tag{3.70}$$

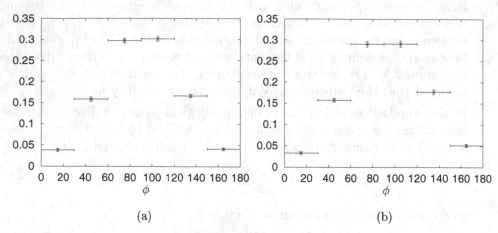

(a)   (b)

Figure 3.25. ϕ distributions for central η and η' production[120] in pp collisions at $\sqrt{s} = 29.1$ GeV

The sign of S_2 in σ_1 can be shown to be negative[121,122]. As η_M is also negative, this then gives

$$t_1 t_2 \sigma_1 = t_1 t_2 \, 4 A_{+0}^2 \sin^2 \tfrac{1}{2}\phi. \tag{3.71}$$

This holds both for conserved and non-conserved pomeron currents. However, the t dependence of A_{+0} is very different in these two cases. We have

$$A_{+0} = \epsilon_+^{*\mu} \, A_{\mu\nu} \, \epsilon_0^\nu \tag{3.72}$$

where $A_{\mu\nu}$ is a covariant amplitude satisfying

$$A_{\mu\nu} \, q_2^\nu \neq 0 \tag{3.73}$$

if the pomeron current is not conserved and

$$A_{\mu\nu} \, q_2^\nu = 0 \tag{3.74}$$

if the pomeron current is conserved.

If the pomeron current is not conserved, then with ϵ_0 from (3.66) we find from (3.73) that

$$A_{+0} \propto \frac{1}{\sqrt{-t}} \tag{3.75}$$

as $t \to 0$. On the other hand, if the pomeron current is conserved, then we find from (3.74) that

$$A_{+0} \propto \sqrt{-t} \tag{3.76}$$

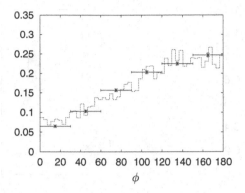 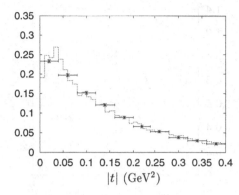

Figure 3.26. ϕ and t distributions for central production of $f_1(1285)$ in pp collisions at $\sqrt{s} = 29.1$ GeV. The data from the WA102 collaboration[120] are compared with the theoretical prediction[123] (histogram) for the case of the pomeron coupling like a non-conserved current.

as $t \to 0$, leading to a strong suppression of the cross section for small t.

Although data are not only for the symmetric case $t_1 = t_2$, as both t_1 and t_2 are small in the experimental configuration it is a reasonable approximation and the qualitative features are unchanged. The ϕ and t distributions[123] for the $f_1(1285)$ are shown in figure 3.26, together with the results of a realistic model, taking into account the experimental configuration. The ϕ distribution clearly shows the expected dominance of the $\sin^2 \frac{1}{2}\phi$ behaviour, and that the contribution to mesons with helicity zero, which is proportional to $\sin^2 \phi$, is very small. Thus the ϕ distribution is compatible both with current conservation and with non-conservation. However the t distribution shows no evidence of a strong suppression at small t, which clearly favours the pomeron coupling like a non-conserved vector current.

3.8 Glueball production

As pomeron exchange is believed to be the exchange of a system of gluons (see chapters 6 and 8), the production of mesons in the central region of proton-proton scattering has long been regarded as a potential source of glueballs[121,122]. However, gluons also couple to quarks, so one can also produce quark-antiquark states. Well-established quark-antiquark states are known to be produced, which has led to searches for a mechanism to distinguish between glueballs and quark-antiquark states. It has been shown that the pattern of resonances produced in the central region of proton-proton collisions depends on the vector difference $\mathbf{k}_\perp = \mathbf{q}_{1\perp} - \mathbf{q}_{2\perp}$, where

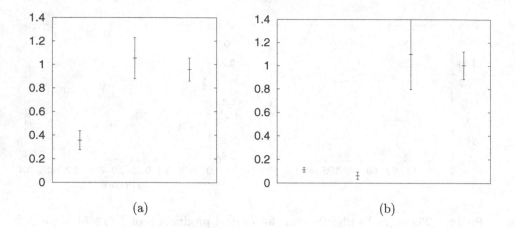

Figure 3.27.   Ratios of number of events with $k_T \leq 0.2$ GeV to the number with $k_T \geq 0.5$ GeV for central production[120] of, from left to right, (a) $f_0(1370), f_0(1500), f_0(1710)$ and (b) $f_2(1270), f_2'(1520), f_2(1910), f_2(1950)$ in pp collisions at $\sqrt{s} = 29.1$ GeV

$\mathbf{q}_1$ and $\mathbf{q}_2$ are the momenta of the pomerons. When $k_T = |\mathbf{k}_\perp|$ is small all well-established quark-antiquark states are observed[119] to be suppressed. Those which are not suppressed include potential glueball states, or more accurately, states which are believed to have a large glueball component.

A good example of the latter is provided by the scalar states $f_0(1370)$, $f_0(1500)$ and $f_0(1710)$. These states are seen by the WA102 collaboration[124] in central production in proton-proton collisions, a complete set of decay branching ratios of these states to all pseudoscalar meson pairs has been obtained. To investigate the glueball and quark-antiquark content of the $f_0(1370)$, $f_0(1500)$ and $f_0(1710)$ these data have been analysed[125] in the $|G\rangle = |gg\rangle$, $|S\rangle = |s\bar{s}\rangle$, $|N\rangle = |u\bar{u} + d\bar{d}\rangle/\sqrt{2}$ basis. Each of the three scalar states is found to have a strong glueball component, and the solution is compatible with the relative production strengths of the $f_0(1370)$, $f_0(1500)$ and $f_0(1710)$ in $\bar{p}p$ and J/ψ radiative decay as well as in pp central production. The physical states are found to be

$$|f_0(1370)\rangle = 0.60|G\rangle - 0.13|S\rangle - 0.79|N\rangle$$
$$|f_0(1500)\rangle = -0.69|G\rangle + 0.37|S\rangle - 0.62|N\rangle$$
$$|f_0(1710)\rangle = 0.39|G\rangle + 0.91|S\rangle + 0.14|N\rangle. \tag{3.77}$$

The glueball mass is found to be about 1.44 GeV, which is at the lower end of the range found[126,127] in the quenched approximation to lattice QCD. Figure 3.27a shows the ratio for these states of the number of events with

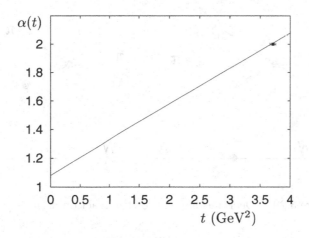

Figure 3.28. A 2^{++} glueball candidate[128], with the line $\alpha(t) = 1.08 + 0.25t$

$k_T \le 0.2$ GeV to the number with $k_T \ge 0.5$ GeV in proton-proton central production[120]. The $f_0(1500)$ and $f_0(1710)$ satisfy the "no suppression" rule at small k_T rather well, but that is not the case for the $f_0(1370)$. Figure 3.27b shows the same ratio for the isoscalar tensor states $f_2(1270)$, $f_2'(1520)$, $f_2(1910)$ and $f_2(1950)$ in pp central production. The first two are well-established quarkonia states and are clearly suppressed at small k_T. It is still not known[27] whether the $f_2(1910)$ and $f_2(1950)$ are distinct states, but if they are then their non-suppression at small k_T suggests that they are mixtures of two bare states of which at least one is a glueball. The near degeneracy of these states implies that the glueball mass is close to the same value. The Chew-Frautschi plot for the pomeron trajectory is shown in figure 3.28, together with the glueball at spin 2. The coincidence is remarkable and suggests that the pomeron lies on a glueball trajectory.

3.9 The Gribov-Morrison rule

Morrison[129] observed in 1968 that in certain quasi-elastic reactions of the type

$$pp \to pN^*_{1/2} \qquad (3.78)$$

where $N^*_{1/2}$ is a baryon resonance of isospin $\frac{1}{2}$, the energy dependence of the cross sections is characteristic of diffraction. The particular $N^*_{1/2}$ resonances observed were the $N^*(1440)$, $J^P = \frac{1}{2}^+$; the $N^*(1520)$, $J^P = \frac{3}{2}^-$; the $N^*(1680)$, $J^P = \frac{5}{2}^+$; and the $N^*(2190)$, $J^P = \frac{7}{2}^-$. Morrison proposed empirically that for these diffractive-like events the change in spin ΔJ is re-

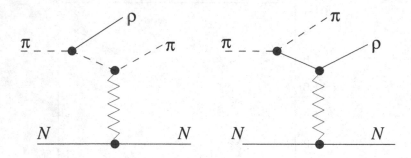

Figure 3.29.   Deck mechanism for the reaction $\pi^- p \to (\pi^- \pi^+ \pi^-)p$

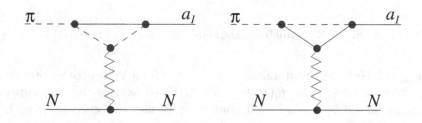

Figure 3.30.   Deck mechanism followed by rescattering

lated to the change in parity between the incident particle and the outgoing resonance by

$$P_{\text{out}} = (-1)^{\Delta J} P_{\text{in}} \tag{3.79}$$

where P_{in} is the parity of the incident particle and P_{out} the parity of the outgoing resonance. The quasi-elastic production of baryon resonances with isospin $\frac{1}{2}$ for which (3.79) is not satisfied, such as the $N^*(1535)$, $J^P = \frac{1}{2}^-$ and the $N^*(1675)$, $J^P = \frac{5}{2}^-$, is not diffractive-like, the cross sections decreasing with increasing energy. The same result holds for the quasi-elastic production of baryon resonances with isospin $\frac{3}{2}$. Although the rule (3.79) was introduced empirically by Morrison, it was suggested independently by Gribov[130]. The rule (3.79) is known as the Gribov-Morrison rule for the production of a resonance by pomeron exchange. Some specific examples are discussed in [131].

A physical mechanism for resonance production had been proposed[132] in 1961. It was suggested that the reaction mechanism could be the dissociation of a proton into a baryon together with a virtual pion which scatters on the other proton, and then recombines with the baryon to produce the N^* resonance. This suggestion was generalised by Deck[133] for the quasi-elastic production of any resonance. Taking the reaction

$$\pi^- p \to (\pi^- \pi^+ \pi^-) p \tag{3.80}$$

as an example, in the Deck model the process proceeds by the incident pion transforming into a ρ meson and a virtual pion which is then diffractively scattered into a real state on the target nucleon. Subsequently it was shown[134] that the scattering of the heavier virtual constituent is just as important as that of the lighter. The pion and the ρ meson may then rescatter to form a different final state. These processes are illustrated in figures 3.29 and 3.30.

The Deck model was applied[135] to the reaction (3.80) and a partial-wave analysis of the $\rho\pi$ system performed. It was shown that, in addition to the $J^P = 0^-$ partial wave, the dominant waves are $J^P = 1^+$, 2^- and 3^+, in agreement with the Gribov-Morrison rule. The 1^+ partial wave is the largest of the three and 3^+ the smallest. Partial waves with $J^P = 0^+$, 1^-, $2^+, \cdots$, which do not satisfy the rule, are extremely small. The Deck mechanism produces broad low-mass enhancements which can simulate resonance production, an effect which bedevilled attempts[136] to extract the parameters of the $a_1(1260)$ meson which is produced in the reaction (3.80) and decays to $\rho\pi$. Interference among the Deck mechanism, rescattering and direct meson production is important[137]. A particularly notable example is provided by ρ photoproduction which is discussed in section 5.3.

3.10 The odderon

The amplitudes for scattering of hadrons ab and $\bar{a}b$ can be decomposed into $C = +1$ and $C = -1$ exchange parts

$$T_\pm(s, t) = T^{ab}(s, t) \pm T^{\bar{a}b}(s, t). \tag{3.81}$$

We have seen in sections 3.1 and 3.2 that $T^{ab}, T^{\bar{a}b}$ are well described at large s and small t by a dominant $C = +1$ term, the pomeron, and a nonleading $C = -1$ term T_- due to the exchange of the ρ, ω trajectories such that, at most,

$$\left| \frac{T_-(s, t)}{T_+(s, t)} \right|_{t=0} = O\left(\frac{1}{\sqrt{s}} \right). \tag{3.82}$$

The odderon is generally defined to be a $C = -1$ contribution that for large s either does not vanish relative to the pomeron contribution, or vanishes relative to it at most like a small power of s or a power of $\log s$.

The odderon concept was introduced in terms of Regge language[138,139]. The pomeron was assumed to saturate the Froissart bound (1.68):

$$\sigma^{ab}(s) + \sigma^{\bar{a}b}(s) \sim (\log s)^2 \tag{3.83}$$

and so, according to (2.33), it was associated with a triple pole at $l = 1$. In order to conform to the Pomeranchuk theorem (1.80), the odderon contribution was assumed to rise less rapidly with energy; specifically, it was associated with a double pole at $l = 1$, giving for large s

$$\sigma^{ab}(s) - \sigma^{\bar{a}b}(s) \sim \log s. \tag{3.84}$$

Later also simple poles with trajectory intercepts $\alpha_O(t = 0) \approx 1$ were considered and extensive fits to pp and $\bar{p}p$ data including odderon contributions were made; see for instance [140,141]. Indeed, we have already seen in section 3.2 that pp and $\bar{p}p$ scattering differ in the dip region and thus both T_+ and T_- must be non-zero there. Also, we have argued that pp scattering at large t is dominated by three-gluon exchange with $C = -1$. It is then natural to ask about odderon exchange at small t. There, the odderon amplitude would be almost real and have the opposite signs for pp and $\bar{p}p$ scattering. Thus a quantity sensitive to odderon exchange is the ratio of real to imaginary parts of the amplitude at $t = 0$:

$$\rho(s) = \left. \frac{\mathrm{Re}\ T(s,t)}{\mathrm{Im}\ T(s,t)} \right|_{t=0}. \tag{3.85}$$

There is a measurement[74] for this ratio in $\bar{p}p$ scattering at $\sqrt{s} = 540$ GeV:

$$\rho_{\bar{p}p}(s) = 0.135 \pm 0.015 \tag{3.86}$$

but pp data at a similar energy are unavailable and so if we wish to know $\rho_{pp}(s)$ at this energy it is necessary to use dispersion relations to calculate it. The results are usually interpreted[142] as giving tight bounds for an odderon contribution:

$$|\rho_{pp}(s) - \rho_{\bar{p}p}(s)| < 0.05 . \tag{3.87}$$

A measurement[73] at $\sqrt{s} = 1800$ GeV does not tighten this constraint.

In the Regge framework, the phase of the amplitude is given by the signature factor (2.18), where the + sign is applicable if odderon exchange is indeed negligible. If, as usual, we write $\alpha(0) = 1 + \epsilon$, the result (3.86) corresponds to

$$\epsilon = 0.086 \pm 0.01 \tag{3.88}$$

If we compare this with the values given in (3.6) or (3.7), we see that the rise of the total cross section is explained well by $C = +1$ exchange, with very little room for odderon exchange. The conclusions found in [141] are similar, that no odderon is evident in the amplitude at $t = 0$, but inclusion of it for $t \neq 0$ and especially beyond the dip region improves the fits considerably.

The search for the odderon need not be restricted to elastic scattering. Other reactions which have been suggested include central J/ψ and ϕ production in high-energy pp and $p\bar{p}$ collisions[143]; spin dependence of high-energy pp scattering[142]; photoproduction[144] of $f_2(1270)$ and $a_2(1320)$; photoproduction of exclusive neutral pseudoscalar mesons[145–147]; photoproduction and electroproduction of heavy $C = +1$ quarkonia[148,149]; and the asymmetry in the fractional energy of charm versus anticharm jets, which is sensitive to odderon-pomeron interference[150].

Our present understanding of QCD provides no obvious reason to have a high suppression of $C = -1$ relative to $C = +1$ exchange at small t in high-energy scattering. The simplest diagrams giving $C = +1$ exchange have two-gluon exchange in the t-channel. But in general diagrams with three-gluon exchange in the t-channel contribute already both to the $C = +1$ and the $C = -1$ amplitudes, and for small t the exchange of an additional gluon is not suppressed by a small coupling constant. Thus, it is quite mysterious why odderon exchange should be highly suppressed at small t. We will return to this odderon puzzle in sections 7.12 and 8.9.

3.11 Scattering on nuclei

Hadron-nucleus and nucleus-nucleus scattering deserve to be treated in a separate book. Here we can only make a few remarks and indicate some relevant references.

First consider hadron-nucleus scattering, treating the nucleus simply as a bound state of nucleons. This approximation ignores the partonic degrees of freedom implied by QCD, in which a nucleus should be considered as a bound state of quarks and gluons which shows clustering in nucleon-like subunits. However this approximation allows the calculation of hadron-nucleus scattering following the prescription proposed by Glauber[151]. In essence, it is assumed that the phase shifts produced in the scattering of the hadron on each nucleon can be added up. This is strictly applicable only in the limit that the nucleons in the nucleus are far apart and only loosely bound together. Although reality is rather far from this limit, Glauber theory has been widely applied, as it is simple and easy to use, and gives a good qualitative description of many effects[152].

An advance on such simple models, which have no solid theoretical basis, is provided by effective theories. These are not derived from first principles, but are self-consistent and calculable. One such[155] is based on the reggeon calculus described in section 2.6 and another is the eikonalised parton model[153,154] which is formally quite similar. In the first approach one typically[155] uses a phenomenological hadron-nucleon scattering amplitude, parametrised in terms of pomeron exchange, and constructs the hadron-nucleus and nucleus-nucleus amplitudes from multiple-pomeron exchanges using the reggeon calculus described in section 2.6 and the AGK cutting rules[156] discussed in section 6.2. Technically this is rather complex. For practical applications Monte Carlo generators based on these principles have been developed[157]. In the second approach, Monte Carlo generators are based on the Lund string model[158] to which some unitarization procedure is applied. The problems and shortcomings of existing calculations of nucleus-nucleus reactions are discussed in [159].

Thus at present one takes the pomeron properties from hadron-hadron scattering as an input for more-or-less sophisticated calculations of hadron-nucleus and nucleus-nucleus scattering. For the study of the properties of the pomeron itself, which is the subject of this book, these reactions have so far not been very useful.

4

Duality

In this chapter we introduce the concepts of finite-energy sum rules and duality, and their realisation in the Veneziano model. They are then applied to pion-nucleon scattering.

4.1 Finite-energy sum rules

In (1.74) we have written a subtracted dispersion relation for the crossing-odd amplitude $A^A(\nu, t = 0)$ defined in (1.72). Here, ν is the variable defined in (1.70), that is $\nu = \frac{1}{4}(s - u)$. We may write similar dispersion relations for non-zero t. For values of t such that $A^A(\nu, t) \to 0$ as $\nu \to \infty$, we do not need subtractions and the dispersion relation reads

$$\text{Re}\, A^A(\nu, t) = \frac{2\nu}{\pi} P \int_0^\infty d\nu' \frac{\text{Im}\, A^A(\nu' + i\epsilon, t)}{\nu'^2 - \nu^2}. \tag{4.1}$$

We have chosen the lower limit of the integration to be 0 for the same reason as for the dispersion relation (1.38): apart from a δ-function contribution from any bound state, $\text{Im}\, A^A(\nu' + i\epsilon, t)$ is zero for values of ν' below the lowest-ν' branch point of $A^A(\nu' + i\epsilon, t)$. If $A^A(\nu, t)$ falls faster than ν^{-1} as $\nu \to \infty$, that is

$$\lim_{\nu \to \infty} \nu A^A(\nu, t) = 0 \tag{4.2}$$

then we can take the limit $\nu \to \infty$ in (4.1) to obtain the superconvergence relation

$$\int_0^\infty d\nu'\, \text{Im}\, A^A(\nu', t) = 0. \tag{4.3}$$

We can now use this relation in conjunction with the fact that Regge theory provides information on the behaviour of $A^A(\nu, t)$ as $\nu \to \infty$. Suppose that the large-ν behaviour of $A^{\mathrm{A}}(\nu, t)$ is dominated by a sum of Regge poles, as in (2.15):

$$A^{\mathrm{A}}(\nu, t) \sim -\frac{\sum_i \pi \beta_i(t)}{\Gamma(\alpha_i(t) + 1) \sin(\pi \alpha_i(t))} \left(1 - e^{-i\pi\alpha_i(t)}\right) (2\nu)^{\alpha_i(t)}. \qquad (4.4)$$

As $A^{\mathrm{A}}(\nu, t)$ is odd under crossing the sum is over poles of odd signature only, so that the right-hand side of (4.4) is also crossing-odd. Suppose that we include in the sum all those Regge poles for which $\alpha_i(t) \geq -1$, and neglect the contributions from any cuts. Then

$$\bar{A}^{\mathrm{A}}(\nu, t) = A^{\mathrm{A}}(\nu, t) + \sum_i \frac{\pi \beta_i(t)}{\Gamma(\alpha_i(t) + 1) \sin(\pi \alpha_i(t))} \left(1 - e^{-i\pi\alpha_i(t)}\right) (2\nu)^{\alpha_i(t)}$$

$$(4.5)$$

decreases faster than ν^{-1} as $\nu \to \infty$ and satisfies the superconvergence relation

$$\int_0^\infty d\nu \ \mathrm{Im} \, \bar{A}^{\mathrm{A}}(\nu, t) = 0. \qquad (4.6)$$

The sum of Regge poles (4.4) is the asymptotic form for $A^{\mathrm{A}}(\nu, t)$. Suppose that it is already a good numerical approximation to $A^{\mathrm{A}}(\nu, t)$, or at least to its imaginary part, for $\nu > \bar{\nu}$, so that the integral (4.6) receives negligible contribution from values of ν greater than $\bar{\nu}$. That is, we may replace the infinite upper integration limit with $\bar{\nu}$. Inserting the definition (4.5) and performing the integration over the Regge-pole terms, we arrive at

$$\int_0^{\bar{\nu}} d\nu \ \mathrm{Im} \, A^{\mathrm{A}}(\nu, t) = -\tfrac{1}{2}\pi \sum_i \beta_i(t)(2\bar{\nu})^{\alpha_i(t)+1} / \Gamma(\alpha_i(t) + 2) \qquad (4.7)$$

where we have used the properties of the Γ-function in (2.14). The relation (4.7) is called a finite-energy sum rule[160–163].

For an amplitude $A^{\mathrm{S}}(\nu, t)$ that is even under crossing we can apply the same argument as above to $\nu A^{\mathrm{S}}(\nu, t)$ to give the finite-energy sum rule

$$\int_0^{\bar{\nu}} d\nu \ \nu \, \mathrm{Im} \, A^{\mathrm{S}}(\nu, t) = \tfrac{1}{4}\pi \sum_i \beta_i(t)(2\bar{\nu})^{\alpha_i(t)+2} / \Gamma(\alpha_i(t) + 3). \qquad (4.8)$$

Provided that the convergence criteria are satisfied we can apply the same argument to $\nu^{2n} A^{\mathrm{A}}(\nu, t)$ and to $\nu^{2n+1} A^{\mathrm{S}}(\nu, t)$, where n is an integer, to obtain sum rules of different moments. The above derivation only gives finite-energy sum rules for crossing-odd and crossing-even parts of the physical amplitude $A(\nu, t)$. For a general amplitude one must consider both odd and even-moment sum rules[164].

The finite-energy sum rules relate the Regge pole parameters $\beta_i(t)$ and $\alpha_i(t)$ to the low energy amplitudes. For amplitudes which are dominated at low energy by resonances this leads to a relation between Regge poles and resonances which is known as duality.

4.2 Duality

Equations (4.7) and (4.8) can be used as phenomenological tools for obtaining information about Regge poles by using phase shifts in the low-energy integral. Finite-energy sum rules were first used in this manner and gave very valuable information on the Regge pole parameters[165,166]. Usually only the differential cross section and limited polarisation measurements were available at the higher energies and so it was impossible to obtain the individual helicity amplitudes. This led to ambiguities in Regge-pole fits to high energy data. However at low energies phase-shift analysis determines the decomposition into individual amplitudes, the extra information coming from the assumption that only a few partial waves contribute. As the finite-energy sum rules are written for amplitudes, we can use our knowledge of the low-energy amplitudes to obtain information about the Regge-pole amplitudes.

The role of the upper integration limit $\bar{\nu}$ is crucial in this procedure. Strictly speaking, equations (4.7) and (4.8) are valid only for $\bar{\nu}$ such that the Regge-pole expression for the amplitude is a good numerical approximation for $\nu > \bar{\nu}$. However phase-shift analysis generally stops short of this and in practice $\bar{\nu}$ has to be taken to be the upper limit of the phase-shift analysis, corresponding typically to values of $\sqrt{s}$ close to 2 GeV. The significance of this can be seen if we write the finite-energy sum rule (4.6) as

$$\int_0^{\bar{\nu}} d\nu \ \mathrm{Im}\left(A^{\mathrm{A}}(\nu, t) - A^{\mathrm{A}}_{\mathrm{Regge}}(\nu, t)\right) = 0. \tag{4.9}$$

This states that $A^{\mathrm{A}}_{\mathrm{Regge}}(\nu, t)$ describes the amplitude at low energy on the average. If we cut off the integral at low energies then we are assuming that this averaging takes place even over intervals smaller than $[0, \bar{\nu}]$ and this seems to happen in practice. Figure 4.1, after the original[162], illustrates this point. It compares

$$p_{\mathrm{Lab}}\left(\sigma^{\mathrm{Tot}}(\pi^- p) - \sigma^{\mathrm{Tot}}(\pi^+ p)\right) \tag{4.10}$$

with the Regge fit to high energy data, in this case the exchange of the ρ trajectory, extrapolated to low energies. This averaging over small energy ranges is called semi-local duality and it is this which enables us to obtain information from the seemingly empty (4.9) by taking $\sqrt{\bar{\nu}} \approx 2$ GeV.

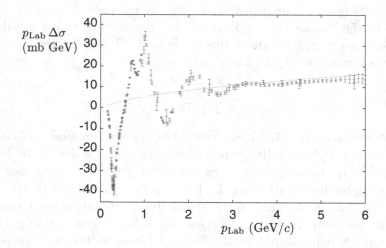

Figure 4.1. Comparison of $p_{\mathrm{Lab}}(\sigma^{\mathrm{Tot}}(\pi^- p) - \sigma^{\mathrm{Tot}}(\pi^+ p))$ with the Regge fit to high energy data

4.3 Two-component duality and exchange degeneracy

The $\pi^- p$ and $\pi^+ p$ high-energy elastic scattering amplitudes receive equal contributions from pomeron exchange, so that pomeron exchange cancels in the difference (4.10) between the two total cross sections. So figure 4.1 shows that the non-pomeron reggeon t-channel exchanges are dual to the s-channel resonances. This is assumed to be so in the case of $\pi^- p$ and $\pi^+ p$ scattering separately. Figure 4.2 shows that the extrapolations to low energy of the Regge fits to the high-energy total cross sections in each case give good descriptions to the average low-energy cross sections. For each, the resonances sit on a non-resonance background. So we see that assuming that the non-pomeron t-channel exchanges are dual to the s-channel resonances leads us to assume also that pomeron exchange is dual to the low-energy s-channel non-resonance background. This is two-component duality.

Although Regge theory sums t-channel exchanges, we have seen in the previous section that s-channel effects provide constraints through finite-energy sum rules. Consider $K^- p$ and $K^+ p$ scattering, which have the same t-channel quantum numbers but very different s-channel quantum numbers. In terms of the reggeon exchanges

$$
\begin{aligned}
A(K^- p) &= A_{I\!P} + A_{f_2} + A_\omega + A_{a_2} + A_\rho \\
A(K^+ p) &= A_{I\!P} + A_{f_2} - A_\omega + A_{a_2} - A_\rho.
\end{aligned}
\tag{4.11}
$$

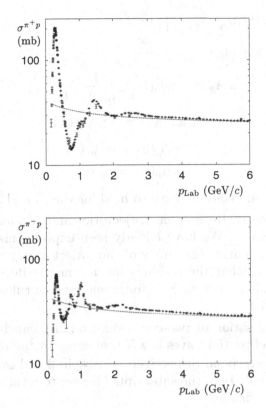

Figure 4.2. Comparison of the low-energy $\pi^+ p$ and $\pi^- p$ total cross sections with the Regge fits to high energy data

At sufficiently high energy, when pomeron exchange dominates, the two cross sections come together. However at lower energies they are very different: as we saw in figure 3.1 the $K^- p$ cross section falls steeply with increasing energy and the $K^+ p$ cross section is almost flat. This implies that the reggeon exchanges collectively have almost decoupled from the $K^+ p$ amplitude. A similar situation holds for $K^- n$ and $K^+ n$ scattering, with the reggeon exchanges collectively decoupling from the latter.

The way to understand these effects from a t-channel point of view is to say that at $t = 0$ the imaginary parts of the ρ and a_2 exchanges must cancel, and so must those of the f_2 and ω exchanges. It is necessary to take the cancellation between poles of the same isospin to achieve the effect in both $K^+ p$ and $K^+ n$ scattering. For the a_2 exchange,

$$\operatorname{Im} A_{a_2} = -\pi \operatorname{Im} \beta_{a_2}(t) \frac{1}{\Gamma(\alpha_{a_2}(t) + 1) \sin(\pi \alpha_{a_2}(t))} \left(1 + e^{-i\pi\alpha_{a_2}(t)}\right) (2\nu)^{\alpha_{a_2}(t)}$$

$$= \pi \beta_{a_2}(t) \frac{1}{\Gamma(\alpha_{a_2}(t) + 1)} (2\nu)^{\alpha_{a_2}(t)} \tag{4.12}$$

and for the ρ exchange

$$\text{Im} \, A_\rho = -\pi \beta_\rho(t) \frac{1}{\Gamma(\alpha_\rho(t) + 1)} (2\nu)^{\alpha_\rho(t)}. \tag{4.13}$$

So to achieve the cancellation at $t = 0$ we require

$$\beta_\rho(0) = -\beta_{a_2}(0)$$
$$\alpha_\rho(0) = \alpha_{a_2}(0). \tag{4.14}$$

Obviously similar relations have to hold for the f_2 and ω exchanges.

This degeneracy of Regge-pole trajectories and residues is known as exchange degeneracy. We have already seen experimental evidence for the approximate exchange degeneracy of the trajectories, not just at $t = 0$, and there is evidence that the residues too are nearly degenerate not just at $t = 0$. This evidence comes from finite-energy sum rules which, as in (4.9), are valid for all t.

An explicit realisation of two-component duality can be seen in the individual s-channel partial waves in πN scattering by taking combinations $f_{l\pm}^0$ and $f_{l\pm}^1$ that correspond respectively to isospin-0 and isospin-1 exchange in the t-channel[166]. Here the subscripts $l\pm$ refer to total angular momentum $J = l \pm \frac{1}{2}$. Specifically,

$$f_{l\pm}^0 = \frac{1}{3}(f_{l\pm}^{\frac{1}{2}} + 2 f_{l\pm}^{\frac{3}{2}})$$
$$f_{l\pm}^1 = \frac{1}{3}(f_{l\pm}^{\frac{1}{2}} - f_{l\pm}^{\frac{3}{2}}) \tag{4.15}$$

where the superscripts $\frac{1}{2}$ and $\frac{3}{2}$ refer to the s-channel isospin. The dimensionless partial wave amplitudes f_l are related to the amplitudes A_l of (1.46) by

$$f_l = \rho(s) A_l = \frac{1}{2i}(\eta_l(s) e^{2i\delta_l(s)} - 1). \tag{4.16}$$

As $\eta_l \leq 1$, f_l must lie inside or on a circle of centre $\frac{1}{2}i$ and radius $\frac{1}{2}$, the unitarity circle. In the neighbourhood of a resonance, a partial-wave amplitude can be approximated by the sum of a resonance term, typically a Breit-Wigner form, and a complex background which is usually slowly varying with energy. For a resonance below the inelastic threshold, where the only decay of the resonance is back to the initial state,

$$f_l = \frac{m_R \Gamma_l e^{i\phi_l}}{m_R^2 - s - i m_R \Gamma_l} + B_l \tag{4.17}$$

where m_R is the resonance mass and Γ_l its width. If the resonance is narrow, then near the pole the first term swamps the second and the amplitude lies on the unitarity circle. As the energy is increased the amplitude moves round the circle in an anticlockwise direction and the resonance energy occurs at the top of the circle. When inelastic decays of the resonance must be included, (4.17) becomes

$$f_l \approx \frac{m_R \Gamma_l^{\text{El}} e^{i\phi_l}}{m_R^2 - s - i m_R \Gamma_l^{\text{Tot}}} + B_l \tag{4.18}$$

where ϕ_l is some phase. The behaviour of the partial-wave amplitude is then similar, but the curve described by the amplitude is wholly inside the unitarity circle and is rotated by the phase ϕ_l. If the background is important, then the resonance circle can be displaced to any part within the unitarity circle.

As the pomeron does not contribute to the t-channel $(I = 1)$-exchange amplitude, two-component duality predicts that the $f_{l\pm}^1$ should be given entirely by s-channel resonance states. That is, we should find that a simple sum of resonances accounts for the entire amplitude in every partial wave. However, the $f_{l\pm}^0$ should not be accounted for by s-channel resonances alone. There should be a predominantly-imaginary smooth background term representing the pomeron contribution on which the s-channel resonances are superimposed. Pion-nucleon partial-wave analyses do indeed show that the t-channel $I = 1$ amplitudes, with only the s-channel S-wave as an exception, are represented by clear resonance circles in the complex phase-shift plane, with very little background. In contrast the resonance circles in the t-channel $I = 0$ amplitudes are superimposed on a large and predominantly-imaginary background.

The exchange degeneracy of the residues is less exact than that of the trajectories. In K^+p and K^+n scattering there are small but significant reggeon contributions, so the cancellation of the residues at $t = 0$ is not exact. The fit to the pp total cross section in figure 3.1 has a rather large reggeon contribution, so the notion of duality for this reaction is not supported, although it does indicate why the reggeon contribution to σ^{pp} is smaller than that to $\sigma^{\bar{p}p}$. This failure of duality is not understood.

4.4 The Veneziano model

Veneziano constructed a very simple model[167] for the non-pomeron part of the scattering amplitude which is explicitly crossing symmetric and analytic, has Regge behaviour, satisfies the finite-energy sum rules and exhibits duality. Although it does not provide a precise quantitative description of the real physical situation in hadron physics, it provides an explicit realisation of the above features and it initiated extremely important developments, and led to string theory.

The model contains strictly-linear trajectories, zero-width resonances and an infinite set of daughter trajectories. It was initially applied to the scattering of identical neutral scalar particles of mass m. For a reaction which is identical in all three channels, for example $\pi^0\pi^0$ scattering, the amplitude has the form

$$A(s,t) = \bar{\beta}\left[\mathrm{B}(-\alpha(s),-\alpha(t)) + \mathrm{B}(-\alpha(s),-\alpha(u)) + \mathrm{B}(-\alpha(t),-\alpha(u))\right]$$
$$(4.19)$$

which is explicitly $s \leftrightarrow t$, $s \leftrightarrow u$, $t \leftrightarrow u$ crossing-symmetric. In (4.19), $\bar{\beta}$ is a constant, α is a real linear trajectory

$$\alpha(s) = \alpha(0) + \alpha' s, \tag{4.20}$$

and $\mathrm{B}(x,y)$ is the Euler Beta-function

$$\mathrm{B}(x,y) = \frac{\Gamma(x)\Gamma(y)}{\Gamma(x+y)}. \tag{4.21}$$

Reactions which do not have resonances in all three channels do not have all the terms in (4.19).

$\mathrm{B}(x,y)$ may be written purely as a sum of poles in either variable:

$$\mathrm{B}(x,y) = \sum_{n=0}^{\infty} \frac{(-1)^n}{n!} \frac{1}{x+n} \frac{\Gamma(y)}{\Gamma(y-n)} = \sum_{n'=0}^{\infty} \frac{(-1)^{n'}}{n'!} \frac{1}{y+n'} \frac{\Gamma(x)}{\Gamma(x-n')}. \tag{4.22}$$

Thus for the first term of the amplitude (4.19) we obtain

$$\mathrm{B}\left(-\alpha(s),-\alpha(t)\right) = \sum_{n=0}^{\infty} \frac{R_n(t)}{n-\alpha(s)} = \sum_{n'=0}^{\infty} \frac{R'_n(s)}{n'-\alpha(t)} \tag{4.23}$$

with R_n a polynomial of degree n:

$$R_n(t) = \frac{1}{n!}\left(\alpha(t)+1\right)\left(\alpha(t)+2\right)\cdots\left(\alpha(t)+n\right) \tag{4.24}$$

and similarly for the second and third terms. Equation (4.23) shows that the Veneziano amplitude fulfils the duality requirement that the amplitude can be represented as a sum of poles either in the s or the t channel. The positions of the poles give the resonance masses as

$$M_n^2 = \frac{n-\alpha(0)}{\alpha'}. \tag{4.25}$$

The angular momenta of the resonances at position M_n can be obtained by expanding the numerator of the corresponding pole in Legendre polynomials. The residues of the s-channel poles of the full amplitude $A(s,t)$ are polynomials in t and u:

$$A(s,t) = \bar{\beta}\left(\sum_{n=0}^{\infty} \frac{R_n(t) + R_n(u)}{n - \alpha(s)} + \mathrm{B}(-\alpha(t), -\alpha(u)) \right). \quad (4.26)$$

Since we can write the scattering angle in the s-channel centre-of-mass system as

$$\cos\theta_s = \frac{t - u}{s - 4m^2}$$

the residue $R_n(t) + R_n(u)$ can be expanded in Legendre polynomials with argument $\cos\theta_s$:

$$R_n(t) + R_n(u) = \sum_{l=0}^{n} c_l P_l(\cos\theta_s). \quad (4.27)$$

The residue is symmetric under the interchange $s \leftrightarrow u$ and hence under $\cos\theta_s \leftrightarrow -\cos\theta_s$, in conformity with Bose statistics. Hence only even values of l contribute to the sum. So if n is even there is a resonance of spin n, together with daughters (see section 2.7 and figure 2.11) with the same mass M_n and with spins $n-2$, $n-4$, $\ldots$, 2, 0. If n is odd, the spins are $n-1$, $n-3$, $\ldots$, 1.

To show that the amplitude (4.19) exhibits Regge behaviour for large s we use Stirling's formula to obtain

$$\frac{\Gamma(x + \mu)}{\Gamma(x + \nu)} = x^{\mu - \nu}\left(1 + O(x^{-1}) \right). \quad (4.28)$$

Thus from (4.21) and (4.28), and identifying x as $-\alpha's$, we have for fixed t and s large in any direction in the complex plane except along the real axis,

$$\mathrm{B}(-\alpha(s), -\alpha(t)) = \Gamma(-\alpha(t))\,(-\alpha' s)^{\alpha(t)}\,(1 + O(s^{-1})). \quad (4.29)$$

and in the limit of large s with t fixed we obtain the result

$$A(s,t) \sim \bar{\beta}\,\Gamma(-\alpha(t))\,(1 + e^{-i\pi\alpha(t)})\,(\alpha' s)^{\alpha(t)} \quad (4.30)$$

which is the Regge behaviour (2.17) for an even-signatured trajectory.

Note that strictly we cannot take the limit $s \to \infty$ for s on the real axis where the poles lie. These poles would be moved into the lower half of the complex s-plane like proper resonance poles if α were given an imaginary part. A point of interest is that in the Veneziano model choosing the scale factor s_0 of (2.17) to be $1/\alpha'$ makes the coupling function $\beta(t)$ constant.

The Veneziano amplitude is important because it demonstrates that simple functions exist which satisfy the theoretical requirements of analyticity, crossing and duality. However it also demonstrates that there is no unique function. The $B(-\alpha(s), -\alpha(t))$ of (4.20) can be replaced by

$$B(m - \alpha(s), n - \alpha(t)) \tag{4.31}$$

for any integers m, $n \geq 1$, and similarly for the s, u and t, u terms. Any function which can be represented as a sum of terms like (4.31) satisfies the finite-energy sum rules, so there is no constraint on the resonance parameters unless additional assumptions are made. A specific example is provided by $\pi\pi$ scattering. In the case of $\pi^+\pi^-$ scattering, isospin 0 and 1 are allowed in the s-channel, so for example it contains the ρ and the f_2, but in $\pi^0\pi^0$ scattering the ρ is excluded.

Veneziano models have been constructed also for multiparticle amplitudes: see the book by Frampton[168] or the review by Mandelstam[169], for example.

4.5 Pion-nucleon scattering

We have seen in section 3.1 that three Regge poles, the pomeron, the f_2 and the ρ, are required to explain the $\pi^\pm p$ total-cross-section data. These poles contribute to the three charge states in the linear combinations

$$A(\pi^- p \to \pi^- p) = A_{I\!P} + A_{f_2} + A_\rho$$
$$A(\pi^+ p \to \pi^+ p) = A_{I\!P} + A_{f_2} - A_\rho$$
$$A(\pi^- p \to \pi^0 n) = (\sqrt{2}/3) A_\rho. \tag{4.32}$$

Fits to the $\pi^- p$ and $\pi^+ p$ total cross sections imply that the f_2 contribution to the $C = +1$ amplitude is appreciably larger than the contribution of the ρ to the $C = -1$ amplitude. The dominance of pomeron and f_2 exchange is also evident in the near equality of the $\pi^- p$ and $\pi^+ p$ differential cross sections, and from the relative smallness of the charge-exchange differential cross section. There are two independent s-channel helicity amplitudes in pion-nucleon scattering: the no-helicity-flip amplitude A_{++} and the helicity-flip amplitude A_{+-}. As the pion has zero spin it does not have a helicity label and we have also omitted the $\frac{1}{2}$ for the nucleon helicity, retaining only the sign. In the language of section 2.9, A_{++} is an $n = 0$ amplitude whereas A_{+-} corresponds to $n = 1$. With the amplitudes

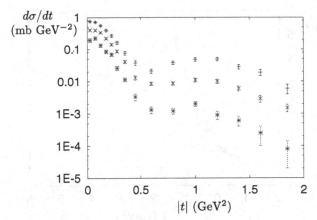

Figure 4.3. Differential cross section for $\pi^- p \to \pi^0 n$ at, from top to bottom, $p_{\text{Lab}} = 3.67$, 5.85 and 13.3 GeV/c.

normalised in the same way as for the unpolarised case, as in (1.19), the differential cross section is given by

$$\frac{d\sigma}{dt} = \frac{1}{64\pi|\mathbf{p}_1|^2 s}\left(|A_{++}|^2 + |A_{+-}|^2\right) \tag{4.33}$$

and the polarised-target asymmetry P by

$$P\frac{d\sigma}{dt} = -\frac{1}{32\pi|\mathbf{p}_1|^2 s} \,\text{Im}\left(A_{++}^* A_{+-}\right). \tag{4.34}$$

Consider charge-exchange scattering first. The differential cross section data are shown in figure 4.3 and exhibit three notable features:

- shrinkage of the forward peak with increasing s

- a turnover near $t = 0$

- a dip at $t \approx -0.55$ GeV^2.

The first feature is a remarkable success for Regge theory as only one leading pole, the ρ, contributes. This can be seen even more clearly if we plot the effective trajectory $\alpha_{\text{eff}}(t)$ as a function of t, with $\alpha_{\text{eff}}(t)$ defined by

$$\frac{d\sigma}{dt} = |f(t)|^2 \, s^{2\alpha_{\text{eff}}(t)-2} \tag{4.35}$$

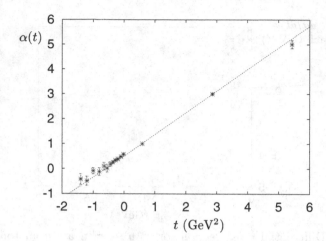

Figure 4.4. Data for $\alpha_\rho(t)$ obtained at $t < 0$ from $\pi^- p$ charge-exchange data and the physical states at $t > 0$. The line is $\alpha(t) = 0.5 + 0.9t$ as in the fit of figure 2.8.

where $f(t)$ is some function of t only. The results are shown in figure 4.4 together with the resonances lying on the ρ trajectory and the same straight-line fit to it as in figure 2.8.

The turnover at small t in figure 4.3 is reproduced most naturally by taking the s-channel helicity-non-flip amplitude A_{++} to be appreciably smaller than the helicity flip amplitude A_{+-}. As the latter is $n = 1$, then (2.75) tells us that near $t = 0$

$$A_{+-} \sim \sin \tfrac{1}{2}\theta_s \qquad (4.36)$$

so that it vanishes at $t = 0$ and the turnover is obtained.

The dip around $t = -0.55$ GeV^2 is explained by the vanishing of the dominant A_{+-} amplitude as $t \approx -0.55$ GeV^2 corresponds to the wrong-signature point $\alpha(t) = 0$ for the ρ trajectory and so, from (2.18), the amplitude is expected to be zero there. The fact that the cross section does not vanish completely here is attributed to A_{++} receiving a contribution either from cuts or from daughter trajectories.

Pion-nucleon charge exchange was considered a great success of simple Regge theory until the polarised-target asymmetry P was measured. In a one-pole model this is predicted to be zero as A_{++} and A_{+-} have the phase given by the signature factor of the pole and so from (4.34) P vanishes. However the polarisation is non-zero and appreciable, again requiring either cuts or additional low-lying poles.

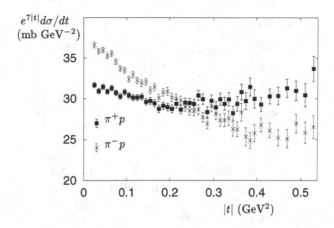

Figure 4.5. $\pi^- p$ and $\pi^+ p$ differential cross sections, multiplied by $e^{b|t|}$ with $b = 7$ GeV^{-2} to make the plots clearer, at $\sqrt{s} = 10.4$ GeV

For this reaction finite-energy sum rules give useful information[165,166]. These confirm the dominance of the helicity-flip amplitude and the zero in A_{+-} at $t \approx -0.55$ GeV^2. They also suggest a zero in A_{++} at $t \approx -0.2$ GeV^2. Because of the dominance of the flip amplitude this does not produce a dip in $d\sigma/dt$ for $\pi^- p \rightarrow \pi^0 n$, but is associated with a cross-over effect in the elastic differential cross sections. The simplest interpretation of the $I = 1$ t-channel exchanges is in terms of the ρ and the daughter ρ' trajectory[165], although the latter cannot be reliably separated from cut effects. However these are certainly small.

The cross-over is a noticeable feature in the $\pi^\pm p$ differential cross section data at moderate energies, $\pi^+ p$ being smaller than $\pi^- p$ for $|t| < 0.2$ GeV^2 and being greater at larger values of $|t|$. This is illustrated in figure 4.5 which compares the $\pi^- p$ and $\pi^+ p$ differential cross sections[170] at $\sqrt{s} = 10.4$ GeV. Applying finite-energy sum rules to $\pi^\pm p$ elastic scattering[165,166] shows that much the largest $C = +1$ exchange amplitude is A_{++}; that is, s-channel helicity is conserved in $C = +1$ exchange. In terms of the amplitudes for the three exchanges the difference in the differential cross sections is given by

$$\frac{d\sigma}{dt}(\pi^+ p) - \frac{d\sigma}{dt}(\pi^- p) =$$

$$\frac{1}{32\pi |\mathbf{p_1}|^2 s} \; \text{Re}\left((A_{++}^{I\!P} + A_{++}^{f_2})^* A_{++}^{\rho} + (A_{+-}^{I\!P} + A_{+-}^{f_2})^* A_{+-}^{\rho} \right).$$

$$(4.37)$$

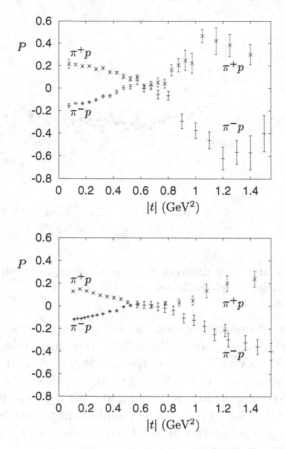

Figure 4.6. $\pi^- p$ and $\pi^+ p$ polarisation at $p_{\text{Lab}} = 6$ GeV/c (upper figure) and 10 GeV/c (lower figure)

If we assume that both the pomeron and the f_2 conserve s-channel helicity then the second term on the right-hand side of (4.37) vanishes and the first term must change sign to produce the cross-over effect. As the pomeron is predominantly imaginary we can conclude directly from the data that Im A^ρ_{++} has a zero at the cross-over point. This conclusion is supported by detailed fits to the data and, as mentioned earlier, is in agreement with finite-energy-sum-rule calculations. At sufficiently high energy the cross-over will disappear and the differential cross section will be dominated by pomeron exchange, as we saw in figure 3.7.

Finally we discuss the $\pi^\pm p$ polarised-target asymmetry data which are shown in figure 4.6, from which it can be seen that they exhibit a striking mirror symmetry: $P(\pi^+ p) \approx -P(\pi^- p)$. This immediately implies that pomeron and f_2 interference effects in the polarisation are very small, as

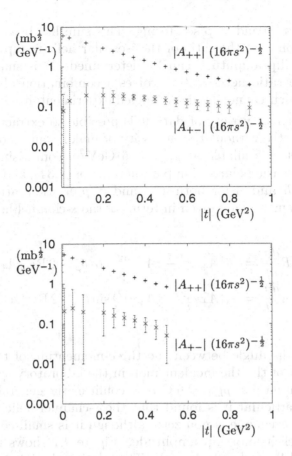

Figure 4.7. The moduli of the non-flip and flip helicity amplitudes in $\pi^\pm p$ scattering at $p_{\text{Lab}} = 6$ GeV/c (upper figure) and 16 GeV/c (lower figure)

they would contribute to $P(\pi^\pm p)$ with the same sign. It follows that neither can contribute significantly to the flip amplitude, supporting the belief that pomeron exchange conserves s-channel helicity to a good approximation. Polarisation in πp elastic scattering at high energy is thus mostly due to interference of the pomeron helicity-non-flip amplitude with the ρ helicity-flip amplitude. The polarisation should decrease with energy approximately as $s^{\alpha_\rho(t)-\alpha_{I\!P}(t)}$, and the data are compatible with this. The double zero in the polarisation at $t \approx -0.55$ GeV^2 is explained by the zero in the helicity-flip amplitude at this point.

Although pomeron-ρ interference is the dominant contribution to the polarisation it cannot be the whole story as the data are not exactly mirror-symmetric. The isoscalar-isovector interference terms cancel in the sum of

$P\,d\sigma/dt$ for π^+p and π^-p scattering. This sum has been studied[61], with the conclusion that the ratio of the isoscalar helicity-flip amplitude to the helicity non-flip amplitude can be determined. It is small, but non-zero. However the ratio necessarily involves a combination of f_2 and pomeron exchange which cannot be disentangled experimentally.

Given a sufficient variety of data it is possible to extract the helicity amplitudes directly without the necessity of making any model assumptions. One such set is available at $p_{\mathrm{Lab}} = 6$ GeV/c, comprising the differential cross sections, the polarisation parameter P of (4.34), and the spin-rotation parameters R and A for both π^+p and π^-p elastic scattering. The spin-rotation parameters are given in terms of the s-channel helicity amplitudes by

$$64\pi|\mathbf{p}_1|^2 s R \frac{d\sigma}{dt} = -(|A_{++}|^2 - |A_{+-}|^2)\cos\theta_p - 2\,\mathrm{Re}\,(A_{++}A_{+-}^*)\sin\theta_p$$

$$64\pi|\mathbf{p}_1|^2 s A \frac{d\sigma}{dt} = -(|A_{++}|^2 - |A_{+-}|^2)\sin\theta_p - 2\,\mathrm{Re}\,(A_{++}A_{+-}^*)\cos\theta_p$$

$$(4.38)$$

where θ_p is the angle between the three-momentum of the scattered proton and that of the the incident pion in the laboratory system. An analysis[171] of the data at $p_{\mathrm{Lab}} = 6$ GeV/c confirms the general picture outlined above. In particular it is found that the s-channel helicity-flip amplitude for isoscalar exchange is non-zero, although it is small compared with the dominant helicity-non-flip amplitude. Figure 4.7 shows the result of this analysis for $|A_{++}(t)|$ and $|A_{+-}(t)|$. The relative phase between the two amplitudes is found to vary slowly with t and to be close to $180°$ throughout. Although there is not such a complete set of data the analysis was repeated at $p_{\mathrm{Lab}} = 16$ GeV/c, with similar results[171].

In summary, pion-nucleon scattering conforms rather well to the straightforward Regge picture. The isovector t-channel is dominated by ρ exchange, with a contribution from the daughter ρ'. Both have zeros at the relevant wrong-signature point. The isoscalar t-channel is dominated by the pomeron and f_2 which exhibit no untoward features. The finite-energy sum rules suggest a contribution from the f_2 daughter trajectory, which is small and could also be attributed in whole or in part to cut effects. In general, cut effects are small.

5

Photon-induced processes

The Regge theory we have described in the previous chapters applies equally well when either or both of the initial hadrons is replaced with a current, for example the electromagnetic current. In particular, we may replace one of the hadrons with a real photon. This is obviously true if we may use the vector-meson dominance model, in which the photon is assumed to behave just like an on-shell ρ or another vector particle. But the vector-dominance assumption is not necessary: the applicability of Regge theory does not depend on it. Most reactions discussed in this chapter are ones involving pomeron exchange. However in section 5.6 we discuss pion photoproduction with two objectives: one is to compare and contrast π^0 photoproduction with $\pi^- p$ charge exchange; and the other is to look at the role of pion exchange in charged pion photoproduction.

5.1 Photon-proton and photon-photon total cross sections

The photon-proton total cross section should be fitted by forms similar to the hadron-hadron total cross sections in figures 3.1 and 3.2. This is verified in figure 5.1. Extending the fit to low energies confirms the two-component duality hypothesis, as is seen in figure 5.2. We discuss in section 7.5 whether the fit should include an additional component, the hard pomeron.

The $\gamma\gamma$ total cross section can be predicted from the known pp, $\bar{p}p$, pn, $\bar{p}n$ and γp total cross sections. We need to determine the $C = +1$ exchange contributions. We can do this because, as we explained in section 3.1, the data show that the f_2 trajectory couples much more strongly to the proton than does the a_2. The γp and γn cross sections are found to be almost equal at low energies, so f_2 exchange dominates over a_2 exchange here also. Thus

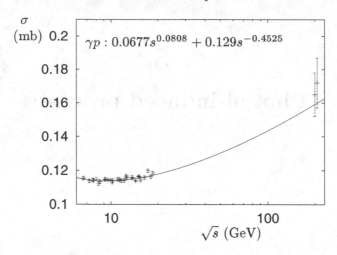

Figure 5.1. γp total cross section; the curve takes account of the exchange of the soft pomeron, f_2 and a_2

we may extract the f_2 and pomeron couplings by applying the factorisation (2.25) separately to the reggeon and pomeron terms in the γp and pp total cross sections of figures 5.1 and 3.1. The result is shown in figure 5.3.

The total hadronic $\gamma\gamma$ cross section was measured at LEP in the ranges 10 GeV $\leq \sqrt{s} \leq$ 110 GeV by the OPAL collaboration[172], and 5 GeV $\leq \sqrt{s} \leq$ 145 GeV by the L3 collaboration[173]. The final cross sections are rather sensitive to the Monte Carlo model used for the unfolding of detector effects, different Monte Carlo simulations producing different results. The resulting uncertainty is included in the errors on the OPAL data. The L3 data are shown with the use of two Monte Carlo simulations, the errors corresponding to the statistical and systematic errors being combined in quadrature. It is clear in figure 5.3 that the result of factorisation is compatible with the OPAL data, but the energy dependence is not sufficiently strong to match that of the L3 data. This may be an indication that a more rapidly-rising component must be added.

5.2 Vector-meson-dominance model

A striking feature of many photon-induced reactions is their similarity to hadron-induced reactions. This finds its simplest realisation in the vector-meson dominance model[174–176], in which the electromagnetic interactions are mediated by vector mesons V, as indicated in figure 5.4. The amplitude for the reaction $\gamma^*A \to B$, where A and B are any hadronic

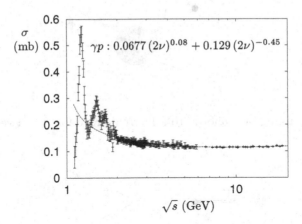

Figure 5.2. γp total cross section at low energy with the high-energy fit of figure 5.1

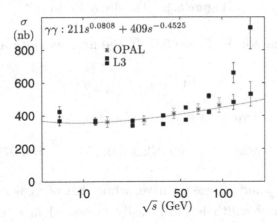

Figure 5.3. $\gamma\gamma$ total cross section[172,173] with the prediction obtained from factorisation

states, for transverse photons whose four-momentum q satisfies $q^2 = -Q^2$, can be written as

$$A^T_{\gamma^* A \to B}(Q^2, s, t, \ldots) = \sum_V \left(\frac{e}{\gamma_V}\right) \frac{m^2_V}{m^2_V + Q^2} A^T_{V A \to B}(s, t, \ldots) \qquad (5.1)$$

where $A^T_{V A \to B}(s, t, \ldots)$ is the on-shell transverse vector-meson amplitude. The coupling constants e/γ_V are in principle measurable in the $e^+ e^-$ decay

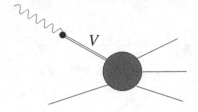

Figure 5.4.   Coupling of a photon to a hadronic reaction, via a vector meson V

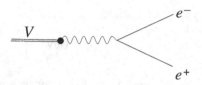

Figure 5.5.   The decay $V \rightarrow e^+ e^-$

of the vector meson V, figure 5.5. In the narrow-width approximation

$$\Gamma_{V \rightarrow e^+ e^-} = \frac{\alpha^2}{3} \frac{4\pi}{\gamma_V^2} m_V. \tag{5.2}$$

The values of $4\pi/\gamma_V$ for the ρ, ω and ϕ are

$$\frac{4\pi}{\gamma_\rho^2} = 0.496 \pm 0.023 \quad \frac{4\pi}{\gamma_\omega^2} = 0.042 \pm 0.0015 \quad \frac{4\pi}{\gamma_\phi^2} = 0.0716 \pm 0.0017 . \tag{5.3}$$

The signs of γ_ρ and γ_ω are positive, while that of γ_ϕ is negative[177].

The $e^+ e^-$ decay width is most easily measured in $e^+ e^-$ annihilation to hadronic final states. For any final state f the total cross section for the reaction $e^+ e^- \rightarrow V \rightarrow f$ via a single vector meson V is given by

$$\sigma_{e^+ e^- \rightarrow f}(s) = 4\pi\alpha^2 \frac{4\pi}{\gamma_V^2} \left(\frac{m_V^2}{s} \right) \frac{m_V \Gamma_V B_V^f}{(s - m_V^2)^2 + m_V^2 \Gamma_V^2} \tag{5.4}$$

where Γ_V is the total width of the meson V and B_V^f is its branching ratio to the final state f. However, there is a problem as in (5.2) and (5.4) the photon is timelike and the vector meson is on-shell, while in (5.1) the photon is spacelike and the vector meson is off-shell. The resulting continuation in Q^2 which is required can introduce unknown form factors.

The amplitude for longitudinal photons must vanish at $Q^2 = 0$ and this behaviour has to be imposed explicitly. This is most simply achieved by arbitrarily including a factor Q^2/m_V^2 into (5.1) to give

$$A^L_{\gamma^* A \to B}(Q^2, s, t, \ldots) = \sum_V \left(\frac{e}{\gamma_V} \right) \left(\frac{Q^2}{m_V^2} \right) \frac{m_V^2}{m_V^2 + Q^2} A^L_{VA \to B}(s, t, \ldots) \quad (5.5)$$

but this might not be the correct prescription.

So far we have not indicated to which amplitudes the vector-dominance prescription should be applied. Since (5.1) and (5.5) can be regarded as pole approximations to a dispersion relation in Q^2 it is most natural to apply the model to invariant amplitudes free of kinematical singularities. This is complicated for particles with high spin, so it is more usual instead to use the s-channel helicity amplitudes which have a direct relation to experimental data. The applications we shall make are not affected by this approximation.

In many applications the vector mesons in the summation (5.1) are restricted to the lowest-lying ones, namely the ρ, ω and ϕ. However higher-mass vector mesons, for example the ρ', ω' and ϕ' states which are known to exist, can be considered and, at sufficiently high mass, replaced by a continuum. Models of this type were first used to interpret nucleon form-factor data, which require the inclusion of the higher-mass states. In contrast the pion-form factor data are reasonably compatible with ρ dominance in both the spacelike and timelike regions for $|Q^2| \leq 1.5$ GeV^2, as discussed in section 3.2.

The inclusion of more higher-mass states leads to less and less predictability, as the electromagnetic couplings are poorly known and off-shell effects are increasingly important, so the connection between the vector-meson on-shell coupling and that required in the application of the model becomes increasingly tenuous. An additional complication is the occurrence of "off-diagonal" effects. The general vector-dominated form of the forward amplitude for $\gamma^* p$ scattering for transverse photons is

$$A^T_{\gamma^* p} = \sum_{m,n} \frac{e}{\gamma_m} \frac{m_m^2}{m_m^2 + Q^2} A^T_{mn} \frac{e}{\gamma_n} \frac{m_n^2}{m_n^2 + Q^2} \quad (5.6)$$

where A^T_{mn} is the transverse forward scattering amplitude for $V_m p \to V_n p$. The diagonal terms $m = n$ correspond to elastic scattering. The off-diagonal processes that have been measured are found to be small compared with the elastic cross sections and so they are frequently neglected. Even in the resulting diagonal approximation there is still considerable freedom. From the optical theorem (1.25), applied to (5.6), the total cross section for transverse photons becomes

$$\sigma^T_{\gamma^*p}(s, Q^2) = 4\pi\alpha \sum_n \frac{1}{\gamma_n^2} \left(\frac{m_n^2}{m_n^2 + Q^2}\right)^2 \sigma_n^T(s) \tag{5.7}$$

where σ_n^T is the total transverse cross section for $V_n p$ scattering.

Unfortunately the electromagnetic couplings of only the ρ, ω and ϕ are known with any degree of accuracy, and for the higher mass vector mesons there is much freedom in the choice of the couplings. Further, little is known about the cross sections σ_n^T as a function of m_n^2, so obviously there is considerable flexibility in the application of (5.7).

Equation (5.7) can be approximated as an integral over the vector-meson continuum:

$$\sigma^T_{\gamma^*p}(s, Q^2) = \int dm^2 \frac{m^4 \rho^T(s, m^2)}{(m^2 + Q^2)^2} \tag{5.8}$$

with

$$\rho^T(s, m^2) = \frac{1}{4\pi^2\alpha} \sigma_{e^+e^-}(m^2) \sigma^T(s, m^2). \tag{5.9}$$

The physical picture is that the photon dissociates into a $q\bar{q}$ pair of mass m, which might or might not be bound. It then scatters on the proton with a cross section $\sigma^{\text{Tot}}(s, m^2)$. The equivalence of (5.7) and (5.8) is easily seen by replacing $\sigma_{e^+e^-}(m^2)$ in (5.8) with a sum of vector-meson poles, as in (5.4), to give

$$\sigma^T_{\gamma^*p}(s, Q^2) = \frac{1}{\pi} \sum_n \left(\frac{e}{\gamma_n}\right)^2 \int dm^2 \frac{m_n^2 m^2}{(m^2 + Q^2)^2} \sigma_n(s, m^2)$$

$$\times \frac{m_n \Gamma_n}{(m^2 - m_n^2)^2 + m_n^2 \Gamma_n^2}$$

$$= 4\pi\alpha \sum_n \frac{1}{\gamma_n^2} \frac{1}{2\pi i} \int dm^2 \frac{m_n^2 m^2}{(m^2 + Q^2)^2} \sigma_n(s, m^2)$$

$$\times \left(\frac{1}{m^2 - m_n^2 - im_n\Gamma_n} - \frac{1}{m^2 - m_n^2 + im_n\Gamma_n}\right). \tag{5.10}$$

In the narrow-width approximation, $\Gamma_n \to 0$, the denominators $m^2 - m_n^2 \mp im_n\Gamma_n \to m^2 - m_n^2 \mp i\epsilon$ so the last term in (5.10) is simply $2\pi i\delta(m^2 - m_n^2)$, giving the result (5.8).

In applications it is usual to separate out the lowest-lying poles in (5.9) and start the continuum integral (5.8) from some threshold m_0^2.

In general the vector-meson-dominance model is descriptive rather than predictive but, despite its many limitations, when it is used with care it

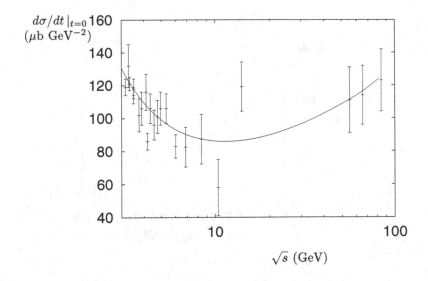

Figure 5.6. Forward differential cross section for exclusive ρ photoproduction. The curve is the result from naive vector-meson dominance scaled by a factor 0.84.

provides a good qualitative guide and in some circumstances a quantitative description. This includes the photoproduction of the light vector mesons, which we describe in the next section. However, the model does not take into account the fact that the coupling of the photon to quarks, unlike that of vector mesons, is pointlike, and sometimes this is important. We saw an example in section 2.8: amplitudes having one or more photons as external legs may have fixed poles in the complex-l plane, while purely-hadronic amplitudes cannot. We take the matter up again in section 9.2.

5.3 Vector-meson photoproduction

We may calculate the cross section for exclusive ρ^0 photoproduction, $\gamma p \to \rho^0 p$, with essentially no free parameters, by assuming vector-meson dominance and the additive-quark model. In the simplest version of vector-meson dominance, the forward cross section is given by

$$\frac{d\sigma}{dt}(\gamma p \to \rho^0 p)\bigg|_{t=0} = \frac{4\pi\alpha}{\gamma_\rho^2}\frac{d\sigma}{dt}(\rho^0 p \to \rho^0 p)\bigg|_{t=0}. \tag{5.11}$$

For the amplitude $\rho^0 p \to \rho^0 p$ the exchange can only be $C = +1$, so the leading trajectories are the pomeron and the reggeons f_2, a_2. According to the additive-quark model, the forward amplitude is simply the average of

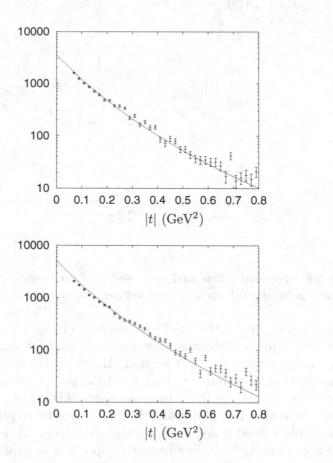

Figure 5.7.   Data[178] for $\gamma p \rightarrow \rho p$ at $\sqrt{s} = 6.86$ GeV (upper figure) and 10.8 GeV (lower figure) with Regge curves. The data are events per 0.02 GeV^{-2}.

the forward amplitudes for $\pi^+ p \rightarrow \pi^+ p$ and $\pi^- p \rightarrow \pi^- p$. The imaginary parts of these can be deduced from the total cross sections by the optical theorem (1.25), and the pomeron and reggeon contributions are known separately. The real parts are then given from the known signature factors, so the complete amplitude is determined at $t = 0$. The coupling e/γ_ρ can be obtained from the known width of the $\rho \rightarrow e^+ e^-$ decay, using (5.4).

It turns out that that the energy dependence of this prediction is correct but the normalisation is too high[179,180]. Multiplying by a factor of 0.84 provides an excellent description of all the data. This is illustrated in figure 5.6. The factor of 0.84 can easily be attributed[181] to corrections

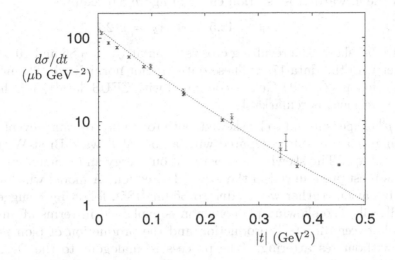

Figure 5.8. ZEUS data[183] for $\gamma p \to \rho p$ with the standard Regge-model prediction. The lower t data are at $\sqrt{s} = 71.7$ GeV and the higher-t data at 94 GeV.

to e/γ_ρ because of the finite width of the ρ, from the effects of closed channels associated with nearby thresholds[182] or from the inclusion of higher resonances.

To calculate the non-forward differential cross section for $\gamma p \to \rho^0 p$ we need the Regge trajectories for the pomeron and the f_2, a_2, which we know from hadron scattering. We also need the mass scale s_0 by which we must divide s before raising it to the Regge power. As in the case of hadronic scattering we use the Veneziano-model prescription and take it to be the inverse of the trajectory slope. As we have seen in chapter 3, it is well established that the trajectories couple to the proton through the Dirac electric form factor (3.16). Wherever it can be experimentally checked the differential cross section for ρ^0 photoproduction is found to have the same slope at small t as the $\pi^\pm p$ elastic differential cross section[178,184]. Within the context of the pomeron and f_2, a_2 exchange, the inference is that the form factor of the ρ is the same as that of the pion, $F_\rho(t) = F_\pi(t)$ with $F_\pi(t)$ given in (3.22). The amplitude for $\gamma p \to \rho^0 p$ is then

$$T(s,t) = iF_1(t)F_\rho(t)\Big(A_{I\!P}\,(\alpha'_{I\!P}s)^{\alpha_{I\!P}(t)-1}e^{-\frac{1}{2}i\pi(\alpha_{I\!P}(t)-1)}$$

$$+ A_{I\!R}\,(\alpha'_{I\!R}s)^{\alpha_{I\!R}(t)-1}e^{-\frac{1}{2}i\pi(\alpha_{I\!R}(t)-1)}\Big). \qquad (5.12)$$

Normalising the amplitude such that $d\sigma/dt = |T(s,t)|^2/(16\pi)$ in μb GeV^{-2}, then the forward cross section curve in figure 5.6 requires

$$A_{I\!P} = 42.5 \qquad A_{I\!R} = 112.7 . \tag{5.13}$$

Figure 5.7 shows the resulting cross section at $\sqrt{s} = 6.86$ and 10.4 GeV, together with the data[178]. These data are not normalised. The normalised prediction at $\sqrt{s} = 94$ GeV is compared with ZEUS data[183] in figure 5.8. The agreement is remarkable.

The ρ^0 shape is found to be skewed, with too many low-mass events and too few high-mass events compared with a normal P-wave Breit-Wigner resonance shape. The skewing is dependent on energy and momentum transfer, and is most pronounced in the forward direction. A model which describes this behaviour rather well is due to Söding[185], following a suggestion by Drell[186]. The skewing phenomenon is explained in terms of an interference between direct ρ^0 production and the production of pion pairs with and without rescattering. The process is analogous to the Deck mechanism discussed in section 3.9, the ρ in figures 3.29 and 3.30 being replaced with a pion and the a_1 with a ρ. The interference term changes sign from positive to negative as $m_{\pi\pi}$ passes through the ρ mass (see [137] for a simple explanation of this effect). It is primarily this which gives rise to the skewing, although the interference term contributes little to the integrated cross section. The Drell-Söding model also provides a good description[187] of the t dependence of the cross section and of the angular correlation of the photoproduced pion pairs. An application of the Drell-Söding model to ρ photoproduction and electroproduction at high energy can be found in [188]. A simple phenomenological parametrisation of the skewing effect was proposed by Ross and Stodolsky[189], which entails multiplying a normal P-wave Breit-Wigner resonance by a factor $(m_\rho/m_{\pi\pi})^k$, with k an energy and t-dependent parameter. A slowly-varying background is sometimes added. This provides a good description[356,348] of the skewing effect.

Exclusive ω photoproduction parallels that of the ρ^0, apart from the cross section being an order of magnitude smaller because of the much smaller width for the $\omega \to e + e^-$ decay and therefore the much smaller value of the $\omega \to \gamma$ coupling e/γ_ω. A more interesting case is provided by exclusive ϕ photoproduction. Because of the Okubo-Zweig-Iizuka (OZI) rule[190–193] this is dominated by pomeron exchange alone, so the cross section should rise as $s^{2\alpha_{I\!P}(0)-2}/b(s)$, where $b(s)$ is the near-forward t-slope. Figure 5.9 shows a comparison with the data in the approximation that $b(s)$ is a constant, and the normalisation is chosen to fit the data. The fit is not affected significantly if the forward peak is allowed to shrink in the canonical way, that is $b(s) = b_0 + 2\alpha'_{I\!P} \log(s/s_0)$.

The normalisation does not agree with a simple application of vector-meson

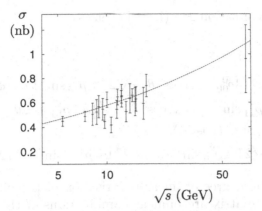

Figure 5.9. Total cross section for $\gamma p \to \phi p$ with pomeron-exchange fit

dominance and the additive-quark model. The pomeron-exchange contribution to the $K^{\pm}p$ total cross section is 87% as large as for the $\pi^{\pm}p$ total cross section. Thus the effective coupling of the pomeron to strange quarks is only 73% of its coupling to the light quarks. The square of this then appears in ϕ photoproduction. Using this and the known width of the decay $\phi \to e^{+}e^{-}$ predicts a cross section twice as large as that observed. There are two possible explanations of this, and probably both are involved. One is that the vector dominance model becomes less reliable as the mass of the vector meson increases. The other is that the discrepancy arises from wave-function effects, which are associated with the ϕ having a comparatively small radius and which should disappear at large Q^2.

However, the cross section for the process $\gamma p \to J/\psi\, p$ rises much more rapidly with increasing energy than can be explained by soft-pomeron exchange alone. We explain in section 7.8 that the data are consistent[194] with the presence of a second pomeron.

5.4 Spin effects in vector-meson photoproduction

In the application of vector-meson dominance to ρ^0 photoproduction we have implicitly assumed that spin effects are unimportant, in particular when we equate the $\rho^0 p$ amplitude with the average of the π^+p and π^-p amplitudes. We have seen that the hypothesis of s-channel helicity conservation appears to work well in purely hadronic interactions. Exclusive ρ photoproduction provides another test via ρ^0-decay angular distributions.

The full information on the ρ^0-decay angular distribution is contained in independent density-matrix elements ρ_{ij}^k which are discussed in appendix E for the more general case of ρ^0 electroproduction[195]. For real photons with linear polarisation P_γ, (E.3) becomes

$$\frac{4\pi}{3} T(\cos\theta, \phi, \psi) =$$

$$\tfrac{1}{2}(1 - \rho_{00}^0) + \tfrac{1}{2}(3\rho_{00}^0 - 1)\cos^2\theta - \sqrt{2}\operatorname{Re}\rho_{10}^0 \sin 2\theta \cos\phi - \rho_{1-1}^0 \sin^2\theta \cos 2\phi$$

$$- P_\gamma \cos 2\psi \, (\rho_{11}^1 \sin^2\theta + \rho_{00}^1 \cos^2\theta - \sqrt{2}\operatorname{Re}\rho_{10}^1 \sin 2\theta \cos\phi$$

$$\qquad - \rho_{1-1}^1 \sin^2\theta \cos 2\phi)$$

$$- P_\gamma \sin 2\psi \, (\sqrt{2}\operatorname{Im}\rho_{10}^2 \sin 2\theta \sin\phi + \operatorname{Im}\rho_{1-1}^2 \sin^2\theta \sin 2\psi) \qquad (5.14)$$

where P_γ is the degree of linear polarisation of the photon. The elements of the density matrix are bilinear combinations of the helicity amplitudes $T_{\lambda_\rho\lambda_{N'},\lambda_\gamma\lambda_N}$, where λ_γ, λ_ρ are the helicities of the initial photon and the ρ, and λ_N, $\lambda_{N'}$ the helicities of the initial and final proton. These bilinear combinations are summed over the nucleon and photon helicities (see (E.8)), so the ρ helicities appear explicitly as the lower indices of the matrix elements. The upper indices refer to the contribution from different photon polarisation states, in this case 0 for an unpolarised beam and 1 or 2 for a linearly polarised beam.

The decay angular distributions also allow a separation of the natural and unnatural-parity t-channel-exchange amplitudes[196]. As expected, ρ photoproduction is found to be completely dominated by natural-parity exchange. At low energies there is a small contribution from unnatural-parity exchange, about 5%, which is consistent with one-pion exchange.

If we restrict ourselves to natural-parity exchange, there are only three independent helicity amplitudes: the non-flip amplitude T_{11}, the single-flip amplitude T_{01}, and the double-flip amplitude T_{1-1}. If we take the further step of neglecting the double-flip amplitude then the non-zero density matrix elements are

$$\rho_{00}^0 = \frac{|T_{01}|^2}{|T_{11}|^2 + |T_{01}|^2}$$

$$\operatorname{Re}\rho_{10}^0 = \frac{2\operatorname{Re}(T_{11}T_{01}^*)}{2(|T_{11}|^2 + |T_{01}|^2)}$$

$$\rho_{00}^1 = -\rho_{00}^0$$

$$\operatorname{Re}\rho_{10}^1 = -\operatorname{Re}\rho_{10}^0$$

$$\rho_{1-1}^1 = \frac{|T_{11}|^2}{|T_{11}|^2 + |T_{01}|^2}$$

$$\operatorname{Im}\rho_{10}^2 = \operatorname{Re}\rho_{10}^0$$

$$\operatorname{Im}\rho_{10}^2 = -\rho_{1-1}^1 \, . \qquad (5.15)$$

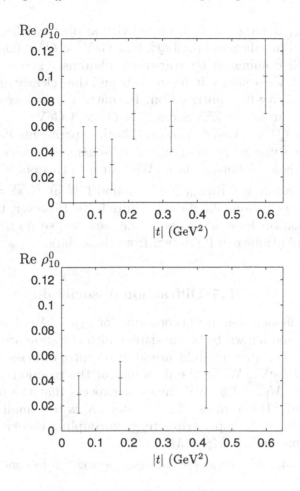

Figure 5.10. ρ density-matrix element $\mathrm{Re}\,\rho^0_{10}$ at $\sqrt{s} = 4.3$ GeV for real photons (upper figure)[197] and at 45 GeV for $\langle Q^2 \rangle = 0.41$ GeV^2 (lower figure)[198]

We see from (5.14) that s-channel helicity conservation requires that the data show a $\sin^2\theta\cos^2\psi$ correlation, which is indeed observed at all energies. However a detailed study of the density matrix elements at $\sqrt{s} = 4.3$ GeV^2 with a plane-polarised beam[197] shows that the interference term $\mathrm{Re}\,\rho^0_{10}$ between the non-flip and single-flip amplitudes departs significantly from zero in the range 0.2 GeV$^2 < |t| < 0.8$ GeV^2. This is shown in figure 5.10. The dominant ρ^1_{1-1} and $\mathrm{Im}\,\rho^2_{1-1}$ are close to $\frac{1}{2}$ and $-\frac{1}{2}$ respectively, as expected. At this energy there is still a significant contribution from f_2, a_2 exchange which cannot be separated from pomeron exchange.

The ρ density matrices have not been measured at HERA energies for real photons, but they have been for virtual photons[198,199]. The ZEUS collaboration has data at small Q^2, 0.25 GeV$^2 < Q^2 < 0.85$ GeV^2, a region which is still dominated by transverse photons. The data for $\mathrm{Re}\,\rho_{10}^0$ at $\langle\sqrt{s}\rangle = 45$ GeV are shown in figure 5.10, and they clearly imply the presence of a small helicity-flip contribution. In contrast a non-zero value for $\mathrm{Re}\,\rho_{10}^0$ is not apparent in the ZEUS data for $Q^2 > 3$ GeV^2, nor in the H1 data which are at $Q^2 > 1$ GeV^2. However, both experiments have unambiguous evidence for some non-conservation of s-channel helicity for longitudinal photons in these higher-Q^2 data. We return to this point in section 7.8.

A non-zero result for $\mathrm{Re}\,\rho_{10}^0$ is also found[178] at $\langle\sqrt{s}\rangle = 8.5$ GeV, with the opposite sign from the data in figure 5.10. However, this could be due to contamination by e^+e^- pairs in the data set, so no firm conclusion on minor contributions can be drawn from these data.

5.5 Diffraction dissociation

Diffraction dissociation in photoproduction, $\gamma p \to Xp$, has been studied at HERA[200] and shown to be consistent with the standard Regge theory of section 3.4. The present data are at two centre-of-mass energies, $\langle\sqrt{s}\rangle = 187$ and 231 GeV. While the data are for the reaction $\gamma p \to XY$, if one requires that $M_Y < 1.6$ GeV then one expects that the proton final state will dominate. If the mass of the system X is too small then resonances will dominate, so to apply triple-reggeon-coupling theory a lower cutoff of 4 GeV^2 is imposed on $M^2 = M_X^2$.

Following (3.47), the cross section may be written in general as

$$\frac{d\sigma}{dt\,dM^2} = \sum_{i,j,k} g_k^{ij}(t)\Big(\frac{M^2}{s}\Big)^{1-\alpha_i(t)-\alpha_j(t)} \Big(\frac{M^2}{s_0}\Big)^{\alpha_k(0)-1} e^{(\phi_i(t)-\phi_j(t))} \quad (5.16)$$

where

$$g_k^{ij}(t) = f_i^p(t) f_j^p(t) f_k^\gamma(0) G_k^{ij}(t). \quad (5.17)$$

Here, $f_i^p(t)$ and $f_j^p(t)$ are the couplings of the reggeons i, j to the proton, $f_k^\gamma(t)$ is the coupling of reggeon k to the photon and $G_k^{ij}(t)$ is the appropriate three-reggeon coupling. In photoproduction the exchanges i and j must have the same signature so the phase difference $\phi_i(t) - \phi_j(t) = -\frac{1}{2}\pi(\alpha_i(t) - \alpha_j(t))$.

Three different sets of terms are distinguished.

- Diffractive: $\quad\displaystyle \frac{ij}{k} = \frac{I\!P\,I\!P}{I\!P} + \frac{I\!P\,I\!P}{I\!R}$

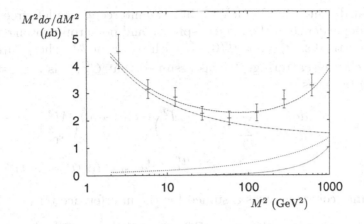

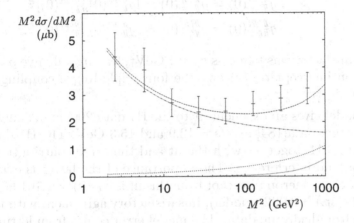

Figure 5.11. Photoproduction diffraction-dissociation cross section[200] at $\langle\sqrt{s}\rangle = 187$ GeV (upper figure) and 231 GeV (lower figure). The lines, from the bottom upwards, are the reggeon-exchange contribution, its interference with pomeron exchange, pomeron exchange, and the total.

- Non-diffractive: $\dfrac{ij}{k} = \dfrac{I\!R I\!R}{I\!P} + \dfrac{I\!R I\!R}{I\!R}$

- Interference: $\dfrac{ij}{k} = \dfrac{I\!P I\!R}{I\!P} + \dfrac{I\!R I\!P}{I\!P} + \dfrac{I\!P I\!R}{I\!R} + \dfrac{I\!R I\!P}{I\!R}$

The data were fitted[200] assuming linear degenerate meson trajectories (see figure 2.8) with intercepts and slopes fixed at the typical hadronic values of 0.55 and 0.9 GeV^{-2} respectively. The pomeron slope was also fixed at the standard value of 0.25 GeV^{-2}, but the intercept was left free. At small $|t|$ the t dependence of the reggeon-proton and pomeron-proton couplings were parametrised as $f_i^p(t) = f_i^p(0)e^{b_i^p t}$ with $i = I\!\!P$ or $I\!\!R$. The f_i^p are determined from elastic scattering. It was assumed that $G_k^{ij}(t)$ is a constant, so that (5.16) becomes

$$\frac{d\sigma}{dt\,dM^2} = \sum_{i,j,k} g_k^{ij}(0)\Big(\frac{M^2}{s}\Big)^{1-\alpha_i(t)-\alpha_j(t)}\Big(\frac{M^2}{s_0}\Big)^{\alpha_k(0)-1}$$

$$\times\ e^{(b_i^p+b_j^p)t}\cos(\tfrac{1}{2}\pi(\alpha_i(t)-\alpha_j(t))). \qquad (5.18)$$

Maximal coherence was assumed for the interference terms:

$$g_{I\!\!P}^{I\!\!P I\!\!R}(0) = g_{I\!\!P}^{I\!\!R I\!\!P}(0) = |g_{I\!\!P}^{I\!\!P I\!\!P}(0)g_{I\!\!P}^{I\!\!R I\!\!R}(0)|^{\frac{1}{2}}$$

$$g_{I\!\!R}^{I\!\!P I\!\!R}(0) = g_{I\!\!R}^{I\!\!R I\!\!P}(0) = |g_{I\!\!R}^{I\!\!P I\!\!P}(0)g_{I\!\!R}^{I\!\!R I\!\!R}(0)|^{\frac{1}{2}}. \qquad (5.19)$$

The scale factor was taken as $s_0 = 1 \text{ GeV}^2$, leaving only five parameters: the pomeron intercept $\alpha_{I\!\!P}(0)$ and the four triple-Regge couplings $g_{I\!\!P}^{I\!\!P I\!\!P}$, $g_{I\!\!R}^{I\!\!P I\!\!P}$, $g_{I\!\!P}^{I\!\!R I\!\!R}$ and $g_{I\!\!R}^{I\!\!R I\!\!R}$.

The model gives an excellent fit to the H1 data[200] and, simultaneously, to fixed-target data[184] at $\sqrt{s} = 12.9$ and 15.3 GeV. The H1 data are shown in figure 5.11, together with the fit and the three contributions: pomeron, reggeon and interference. The importance of the latter is evident. The result for the pomeron intercept from this fit is $\alpha_{I\!\!P}(0) = 1.101\pm0.010$ (stat.) $\pm$ 0.022 (syst.) ±0.022 (model), in satisfactory agreement with the values obtained from hadronic data. The model error comes from letting the "fixed" parameters vary between the limits set by hadronic data. Theoretical understanding of the interference is weak and there is no reason to impose (5.19). For example, an equally satisfactory fit can be obtained without interference but with an additional contribution from isovector exchange. This gives $\alpha_{I\!\!P}(0) = 1.071 \pm 0.024$ (stat.) ± 0.021 (syst.) ± 0.018 (model), which is again compatible with the hadronic value.

5.6 Pion photoproduction

Pion photoproduction provides another example of the application of finite-energy sum rules and, in the case of charged pion photoproduction, of the problems associated with pion exchange. Pion photoproduction differs from pion-nucleon scattering, described in section 4.5, in that there are many features of the data which cannot be explained by a simple Regge-pole model, that is one dominated by exchange-degenerate trajectories whose contributions vanish where the signature factors (2.18) are zero. Regge cuts are certainly required, and there are important contributions from daughter trajectories. A detailed discussion of pion photoproduction can be found in [201].

Pion photoproduction can be discussed either in terms of the four independent s-channel helicity amplitudes $T^{\lambda_\gamma}_{\lambda_N \lambda'_N}$, where λ_γ is the photon helicity and λ_N, λ'_N are the helicities of the initial and final nucleons, or in terms of four independent t-channel amplitudes F_i. These amplitudes, the relations between them, and the definition of cross sections are given in appendix D. Here we summarise the salient features.

The s-channel helicity amplitudes satisfy the conditions

$$T^1_{+-} = T^{-1}_{-+} \qquad T^1_{--} = T^{-1}_{++}$$
$$T^1_{++} = T^{-1}_{--} \qquad T^1_{-+} = -T^{-1}_{+-}. \tag{5.20}$$

We have used the usual convention of denoting the nucleon helicities only by their sign, as there is no ambiguity. Because of the relations (5.20) it is also conventional to use the $\lambda_\gamma = 1$ helicity amplitudes and omit the photon-helicity label. As the pion has zero helicity there is automatic helicity flip at the photon-pion-reggeon vertex, so the net helicity flip is defined by the nucleon helicities. Then T_{-+} is a non-flip amplitude, T_{++} and T_{--} are both single-flip amplitudes and T_{+-} is double-flip.

The amplitudes F_i have definite parity in the t-channel. F_1 and F_2 are respectively natural and unnatural-parity t-channel amplitudes; and at large s F_3 and F_4 are respectively natural and unnatural-parity t-channel amplitudes. Further, the amplitudes F_2 and F_3 satisfy the constraint $F_3 = 2m_p F_2$ at $t = 0$, where m_p is the nucleon mass.

In pion photoproduction it is possible in principle to make a complete set of measurements[203] which determine the amplitudes up to an overall phase. However, unlike pion-nucleon scattering, such a complete set of measurements has not yet been made and the use of finite-energy sum rules to provide supplementary information has been essential. The cross section for a polarised photon incident on a polarised target is

$$\frac{d\sigma}{dt}(\mathbf{P}, P_T, \phi, P_\odot) = \frac{d\sigma}{dt}\bigg|_{\text{unpolarised}} \bigg(1 - P_T \cos 2\phi\ \Sigma$$
$$+ P_x(-P_T \sin 2\phi\ H + P_\odot F) - P_y(-T + P_T \cos 2\phi\ R)$$
$$- P_z(-P_T \sin 2\phi\ G + P_\odot E)\bigg). \tag{5.21}$$

Here Σ, R, T, E, F, G and H are functions of s and t, and are defined in terms of the helicity amplitudes in appendix D. The other quantities

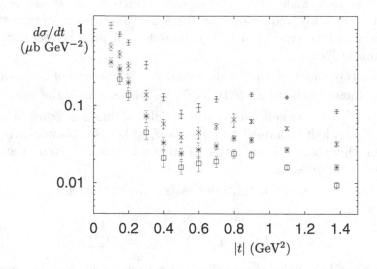

Figure 5.12. π^0 photoproduction differential cross section[202] at, from top to bottom, $E_\gamma = 6$, 9, 12 and 16 GeV

describe the polarisation state of the proton and the photon: **P** is the polarisation vector of the proton, P_T is the transverse polarisation of the photon, $P_\odot$ is the right circular polarisation of the photon, and ϕ is the angle between the plane of photon polarisation and the production plane. The data which are available for both charged and neutral pion photoproduction are the unpolarised differential cross section; the polarised beam asymmetry Σ for linearly polarised photons; and the polarised target asymmetry T. In addition for π^0 photoproduction there are data on R, the recoil proton asymmetry, and on the quantities G and H which require simultaneous polarisation of beam and target.

We start with $\gamma p \rightarrow \pi^0 p$, for which the principal reggeon exchange is expected to be the ω. A significant contribution from ρ exchange is not expected as the γ-π-ω coupling is about three times the γ-π-ρ coupling, and we have seen in (3.8) that the coupling of the ω to the nucleon is about four times the coupling of the ρ. C-parity excludes any contribution from f_2 or a_2 exchange.

The unpolarised differential cross sections[202] for $\gamma p \rightarrow \pi^0 p$ at $E_\gamma = 6$, 9, 12 and 16 GeV are shown in figure 5.12. They look very similar to the differential cross sections for $\pi^- p \rightarrow \pi^0 n$. Just as the latter reaction is dominated by the ρ-exchange single-flip amplitude, π^0 photoproduction is dominated by the ω-exchange single-flip amplitudes. The dominance of

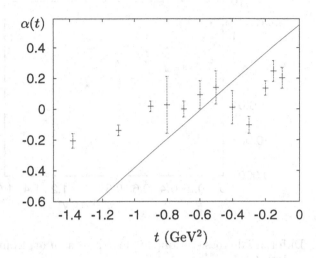

Figure 5.13. Effective ω trajectory obtained from the π^0-photoproduction differential cross section. The line is $\alpha(t) = 0.44 + 0.92t$ as in figure 2.13.

helicity flip can be inferred from the turnover in the cross section near $t = 0$. The origin of the minimum in the cross section at $t \approx -0.55$ GeV^2 is the zero in ω exchange due to the vanishing of the signature factor there, just as for ρ exchange in $\pi^- p$ charge exchange. However π^0 photoproduction is more complicated[204] than $\pi^- p$ charge exchange, as can be seen from the effective trajectory in figure 5.13. The line is the extrapolated fit to the physical states on the ω trajectory. This figure should be compared with the corresponding one for ρ exchange, figure 4.4. It is clear that for π^0 photoproduction a lower-lying trajectory or a cut is required, or possibly both.

The relevant lower-lying trajectory is the one associated with b_1 exchange, but the possibility of their being a significant contribution from it can be ruled out. As the b_1 trajectory is lower-lying than the ω trajectory, we would expect the dip to deepen with increasing energy. This is in contradiction to the data. Further, b_1 exchange cannot produce an effective trajectory lying above the ω trajectory as is clearly required from figure 5.13. So cuts are an essential and important part of the reaction.

The high-energy data alone do not determine the amplitudes, so it is essential to use finite-energy sum rules. The two principal results of a combined fit[205,204] to the data and the sum rules are the following: to a good approximation the two helicity-flip amplitudes T_{++} and T_{--} are equal (which

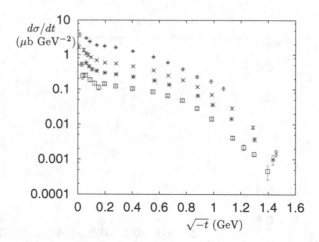

Figure 5.14.　Differential cross section[206] for $\gamma p \to \pi^+ n$ at, from top to bottom, $E_\gamma = 5$, 8, 11 and 16 GeV

is a consequence of the constraint between F_2 and F_3 (D.21) and of (D.24)), and dominate the cross section; both $\operatorname{Im} T_{++}$ and $\operatorname{Im} T_{--}$ have zeros in the region of $t \approx -0.5$, -0.6 GeV^2, but the zeros in the real parts of these amplitudes occur close to $t = -0.2$ GeV^2 so that the cut contribution is primarily real.

These results confirm our qualitative expectations for ω exchange, but the cut contributions do not follow the conventional picture and are not understood. Thus π^0 photoproduction provides a considerable contrast to $\pi^- p$ charge exchange. Not only are cut effects important in determining the structure of the flip amplitudes, but also it is clear that no simple cut model can account for all the structure observed. In general π^0 photoproduction seems to set rather difficult problems for high-energy models, even though the basic structure reflects the expected behaviour.

The differential cross section data[206] for $\gamma p \to \pi^+ n$ at $E_\gamma = 5$, 8, 11 and 16 GeV are shown in figure 5.14. Note that $d\sigma/dt$ is plotted as a function of $\sqrt{-t}$ to highlight the forward peak, which is an obvious feature. What is not so obvious at first sight is the energy dependence: the cross section decreases as approximately s^{-2}. This is evident in the effective trajectory in figure 5.15, which shows that $\alpha_{\text{eff}} \approx 0$ over the whole t range. Another obvious feature is that there is no dip at $t \approx -0.55$ GeV^2.

The forward peak, in which the differential cross section rises by more than a factor of 2 between $t = -m_\pi^2$ and $t = t_{\min}$, is naturally associated with pion exchange. Pion exchange is unique among reggeon exchanges.

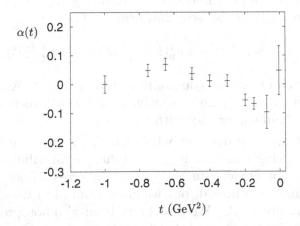

Figure 5.15. Effective trajectory for $\gamma p \to \pi^+ n$

The $\Gamma(-\alpha(t))$ in (2.15) has a pole near the physical region, so for small t one can approximate it by the pion-pole term. The pion-exchange term vanishes in the forward direction, even though it occurs in the non-flip amplitude T_{+-} as well as in the double-flip amplitude T_{-+}, and so by itself it cannot produce a forward peak. The simplest prescription to cure this is to put a pion pole directly into the s-channel helicity amplitudes with no kinematic factors that are not required by s-channel angular-momentum conservation[207,208]. In pion photoproduction this means replacing the factorising form $t/(t - m_\pi^2)$ by $m_\pi^2/(t - m_\pi^2)$ in the non-flip amplitude. As we can rewrite $m_\pi^2/(t - m_\pi^2)$ as $t/(t - m_\pi^2) - 1$ we see that we are simply adding a specific smooth background which interferes destructively with the pole and which presumably should be associated with a cut. The pion-exchange contribution to the double-flip amplitude is unaltered as there s-channel angular momentum conservation requires that the t dependence is unchanged. This simple prescription describes all features of the near-forward charged-pion-photoproduction data. It also has the advantage that it can be applied to all pion-exchange reactions, not just photoproduction. It has the disadvantage of being completely obscure theoretically, although the effect can be reproduced in cut models albeit in a complicated way.

Pion exchange contributes only to the t-channel amplitude F_2, which we have seen must satisfy the constraint $F_3 = 2m_p F_2$ at $t = 0$. This constraint is satisfied trivially by pion exchange as it vanishes at $t = 0$. However this is not the case for the cut. The constraint equation automatically requires a cut in F_2 to be accompanied by a cut in F_3, that is the cut occurs both in the unnatural and natural-parity amplitudes. This immediately predicts the behaviour of the polarised beam asymmetry $\Sigma = (\sigma_\perp - \sigma_\parallel)/(\sigma_\perp - \sigma_\parallel)$ at small t. From (D.27) we see that, near $t = 0$,

$$\Sigma \sim \left(|F_3|^2 - 4m_p^2|F_2|^2\right)\Big/\left(|F_3|^2 + 4m_p^2|F_2|^2\right) \qquad (5.22)$$

so $\Sigma = 0$ at $t = 0$ from the constraint equation (D.21). As $|t|$ increases F_2 decreases rapidly but F_3 varies slowly. Hence Σ will increase rapidly to a value close to 1, in conformity with the data.

The fact that $\alpha_{\text{eff}} \approx 0$ over the whole t range is evidence for strong cuts. It is not surprising that the reggeon-exchange contributions, other than the pion, are not obvious in the differential-cross-section data. Both ρ and a_2 exchange are allowed, but we know from (3.8) that their couplings to the nucleon are weak. We also know from π^0 photoproduction that ρ exchange is not important and that b_1 exchange does not make a significant contribution. However, finite-energy sum rules[209] show that ρ and a_2 exchange, and their associated cuts, cannot be completely excluded. Their presence is not seen directly in the data but through interference with the strong pion cut in the natural-parity amplitude. The absence of a dip in the differential cross section could be taken to imply that the ρ pole does not have a zero there. However, the presence of the strong cuts, evident from the energy dependence of the cross section, invalidates that argument. The evidence from sum rules is that the zero is required[209].

There are two principal conclusions to be drawn from pion photoproduction. The first is that cuts can sometimes be very important, even dominant. The second is that there is no simple prescription for calculating the cuts. The data provide a salutary reminder that the application of Regge theory is not always straightforward. Fortunately, often it is, as we have seen in chapter 3.

6

QCD: perturbative and nonperturbative

In this chapter we describe briefly the basic physical principles of QCD and give an overview of methods used and results obtained in the application of QCD to diffractive scattering.

6.1 Basics of QCD

Today we think that we know the basic degrees of freedom, quarks and gluons, in terms of which we should be able to describe all strong interaction phenomena. The fundamental Lagrangian is

$$\mathcal{L}_{\text{QCD}}(x) = -\tfrac{1}{4}F^a_{\lambda\rho}(x)F^{\lambda\rho a}(x) + \sum_q \bar{q}(x)(i\gamma^\lambda D_\lambda - m_q)q(x). \qquad (6.1)$$

Here $q(x)$ are the quark fields for the various quark flavours $q = u, d, s, c, b, t$ with masses m_q. We denote the gluon potentials by $A^a_\lambda(x)$ $(a = 1, \ldots, 8)$ and the gluon field strengths by

$$F^a_{\lambda\rho}(x) = \partial_\lambda A^a_\rho(x) - \partial_\rho A^a_\lambda(x) - g_0 f_{abc} A^b_\lambda(x) A^c_\rho(x) \qquad (6.2)$$

where g_0 is the bare strong coupling constant and f_{abc} are the structure constants of $SU(3)$. The covariant derivative of the quark fields is

$$D_\lambda q(x) = (\partial_\lambda + ig_0 A^a_\lambda \mathbf{t}_a)q(x). \qquad (6.3)$$

Here $\mathbf{t}_a = \tfrac{1}{2}\lambda_a$, where λ_a are the Gell-Mann matrices of the $SU(3)$ group; see appendix B. The gluon potential and field-strength matrices are defined as

$$\mathbf{A}_\lambda(x) = A^a_\lambda(x)\,\mathbf{t}_a$$
$$\mathbf{F}_{\lambda\rho}(x) = F^a_{\lambda\rho}(x)\,\mathbf{t}_a. \qquad (6.4)$$

The Lagrangian (6.1) is invariant under $SU(3)$ gauge transformations. Let $\mathbf{U}(x)$ be an arbitrary matrix function, where for fixed x the $\mathbf{U}(x)$ are $SU(3)$ matrices:

$$\mathbf{U}(x)\mathbf{U}^\dagger(x) = 1$$
$$\det \mathbf{U}(x) = 1. \tag{6.5}$$

With the transformation laws

$$q(x) \to \mathbf{U}(x)q(x)$$

$$\mathbf{A}_\lambda(x) \to \mathbf{U}(x)\mathbf{A}_\lambda(x)\mathbf{U}^\dagger(x) - \frac{i}{g_0}\mathbf{U}(x)\partial_\lambda\mathbf{U}^\dagger(x) \tag{6.6}$$

we find

$$\mathbf{F}_{\lambda\rho}(x) \to \mathbf{U}(x)\mathbf{F}_{\lambda\rho}(x)\mathbf{U}^\dagger(x) \tag{6.7}$$

and invariance of $\mathcal{L}_{\mathrm{QCD}}$:

$$\mathcal{L}_{\mathrm{QCD}}(x) \to \mathcal{L}_{\mathrm{QCD}}(x). \tag{6.8}$$

So far we have considered the Lagrangian at the level of classical fields. The transition to quantum field theory is most conveniently done using the path-integral formalism.

This approach to quantisation by functional integration[210,211] has turned out to be the most suitable one for non-Abelian gauge theories; it is also the basis of most genuine nonperturbative treatments of QCD. For definiteness we sketch here some essential features for the simple case of a real scalar field $\phi(x)$. Let us consider the vacuum expectation value of the time-ordered product of a functional of the scalar field $\langle 0|\mathrm{T}\,F[\phi]|0\rangle$. In the functional approach it is given by the average over all possible field configurations with a weight determined by the classical action. This is formally written as

$$\langle 0|\mathrm{T}\,F[\phi]|0\rangle = \langle F[\phi]\rangle = \frac{1}{Z}\int \mathcal{D}\phi\,F[\phi]\,e^{iS[\phi]} \tag{6.9}$$

with

$$Z = \int \mathcal{D}\phi\,e^{iS[\phi]} \quad \text{and} \quad S[\phi] = \int d^4x\,\mathcal{L}(\phi, \partial_\mu\phi). \tag{6.10}$$

In order to give this integral a meaning we first go to a finite volume in space-time and discretise it, for instance by a hypercubic lattice. The action $S[\phi]$ and the functional $F[\phi]$ are expressed as functions of the field ϕ at the discrete lattice points and the integration variables are the field values at these points. As an intermediate step we have to form the high-dimensional integral

$$\frac{1}{Z} \int \prod_{i=1}^{n} d\phi_i \, F(\phi_1 \ldots \phi_n) \, \exp\left(iS(\phi_1 \ldots \phi_n)\right) \tag{6.11}$$

where $\phi_i = \phi(x_i)$ and x_i are the lattice points. The oscillatory integrals are made convergent by introducing a small imaginary part, for instance by giving the the masses a small imaginary part, $m^2 \to m^2 - i\epsilon$, so that the exponential acquires a term $-\epsilon \sum_i \phi_i^2$ which damps the integrand for large values of the fields.

The functional integral (6.10) is defined as the limit of (6.11) for vanishing lattice spacing and infinite volume. We shall not dwell on the enormous mathematical and technical problems of this limit, since later we shall deal with the only class of functional integrals which are under control analytically, namely Gaussian ones for which the action $S[\phi]$ is quadratic in ϕ. That is

$$S[\phi] = \int d^4x \, d^4y \, \phi(x) K(x-y) \phi(y). \tag{6.12}$$

In this case we can perform the functional integration. We define the generating functional $W[J]$ from which the vacuum expectation values of any power of ϕ can be obtained by functional differentiation.

$$W[J] = \langle 0 | \mathrm{T} \exp\left(i \int d^4x \, J(x)\phi(x)\right) | 0 \rangle$$

$$= \frac{1}{Z} \int \mathcal{D}\phi \, \exp\left(i \int d^4x \, J(x)\phi(x)\right) \exp\left(i \int d^4x \, d^4y \, \phi(x)K(x-y)\phi(y)\right)$$

$$= \exp\left(-\tfrac{1}{4}i \int d^4x \, d^4y \, J(x) \, K^{-1}(x-y) J(y)\right) \tag{6.13}$$

where the inverse kernel $K^{-1}(x-y)$ is defined by

$$\int d^4y \, K^{-1}(x-y)K(y-z) = \delta^4(x-z). \tag{6.14}$$

We see that the inverse kernel $K^{-1}(x-y)$ is proportional to the vacuum expectation value of the product of two field operators:

$$\langle 0 | \mathrm{T}\phi(x)\phi(y) | 0 \rangle = -\frac{\delta}{\delta J(x)} \frac{\delta}{\delta J(y)} W[J] \Big|_{J=0} = \tfrac{1}{2} i K^{-1}(x-y). \tag{6.15}$$

For the free field, for instance, we have a quadratic action with the kernel $K(x-y) = -\tfrac{1}{2}(\partial^2 + m^2 - i\epsilon) \delta^4(x-y)$, so that $\tfrac{1}{2} i K^{-1}(x-y)$ is the familiar scalar propagator in position space.

It is evident that the integral (6.10) is much more suitable for numerical and rigorous mathematical treatment if the weight $\exp(iS[\phi])$ does not oscillate but is strongly exponentially damped. This can be achieved by going to a Euclidean field theory by the substitution

$$x^0 \to -ix^4, \quad x^4 \text{ real.} \tag{6.16}$$

This may be achieved by a suitable analytic continuation. After the expectation value has been obtained in Euclidean field theory the resulting expectation values are brought back to the real Minkowski world by the inverse substitution $x^4 \to ix^0$. Lattice QCD calculations are done in Euclidean space.

The transition to Euclidean field theory does not affect only the space-time continuum. It is convenient to transform all four-vectors in order to keep the usual relations among them. The zero component of the four potential A_0 is replaced with iA_4. As a consequence the relations between the field strength and the potential preserve the conventional form and the Euclidean action is positive.

With these methods one can develop the perturbation theory to give a power series in the coupling parameter g_0 and derive the corresponding Feynman rules, which are summarised in appendix C for unrenormalised amplitudes and Green's functions. However, when calculating higher-order effects naively from these rules we encounter infinities which are treated properly in renormalisation theory.

The basic procedure in the renormalisation theory of QCD is as follows. When we replace the classical fields in the Lagrangian (6.1) with field operators, we encounter products of two or more field operators at the same space-time point. Such objects are highly singular and must be treated in a limiting procedure. Thus as a first step we must introduce some regularisation procedure which makes all quantities finite. The two most popular schemes for QCD are dimensional regularisation[212,213], where the number of space-time dimensions is changed from $d = 4$, and lattice regularisation[214], where the (Euclidean) space-time continuum is replaced by a grid.

The second step is to rescale the fields by suitable factors, the square roots of the so-called wave-function renormalisation constants, and to choose new "renormalised" coupling and mass parameters instead of the "bare" parameters of the original Lagrangian. The last step is then to remove the regularisation keeping the renormalised parameters fixed. In this way we get the final answers of the theory, to be compared with experiment.

In QCD the renormalised coupling constant $g(M)$ can, for instance, be defined from the value of the renormalised three-gluon vertex function at a convenient Euclidean momentum scale M. Clearly, M can be chosen freely and $g(M)$ will vary with M in a given theory defined by definite values for the bare parameters. The equation governing this dependence of $g(M)$ on M is the renormalisation-group or Callan-Symanzik equation. In the widely-used $\overline{\text{MS}}$ scheme[215] the mass scale M for defining $g(M)$ enters in a more formal way[27]. The Callan-Symanzik equation in this scheme is[27]

$$M\frac{dg(M)}{dM} = \beta(g(M))\,. \tag{6.17}$$

The function β can be calculated in perturbation theory. The expansion of β in powers of g starts with the term of order g^3:

$$\beta(g) = -\frac{g}{2}\left(\frac{\beta_0}{2\pi}\alpha_s + \frac{\beta_1}{4\pi^2}\alpha_s^2 + \frac{\beta_2}{64\pi^3}\alpha_s^3 + \cdots\right)$$

$$\alpha_s = \frac{g^2}{4\pi} \tag{6.18}$$

where

$$\beta_0 = 11 - \tfrac{2}{3}n_f \tag{6.19}$$

and n_f is the number of quark flavours. The terms up to order g^9 have been calculated[27,216].

For $n_f \le 16$ we have $\beta_0 > 0$ and the solution of the differential equation (6.17) for small g is

$$\alpha_s(M) = \frac{g^2(M)}{4\pi} = \frac{4\pi}{\beta_0 t}\left(1 - \frac{2\beta_1}{\beta_0^2}\frac{\log t}{t} + O\left(\frac{1}{t^2}\right)\right)$$

$$t = 2\log\left(M/\Lambda^{(n_f)}\right) \tag{6.20}$$

where $\Lambda^{(n_f)}$ is an integration constant, which is understood to be defined in the $\overline{\text{MS}}$ scheme unless specified otherwise. This shows that $g(M) \to 0$ as $M \to \infty$. To be more precise, this limit holds if there exists a value M_0 for which the coupling strength is small enough for the above expansions to be applicable. The effective coupling strength goes to zero as the momentum or energy scale in the interaction of quarks and gluons becomes very large. This is the famous property of asymptotic freedom of QCD[217,218]. The Λ parameter in (6.20) in essence sets the scale above which the coupling strength becomes small, so that perturbative calculations have a chance of being reliable. Note that, instead of the original dimensionless bare coupling parameter g_0, the coupling strength in the renormalised theory, that is the function $g(M)$, is governed by the dimensioned parameter Λ. This phenomenon is called "dimensional transmutation"[219].

approximate energy range M (GeV)	number of active quark flavours n_f	$\Lambda^{(n_f)}$ (MeV)
> 200	6	88 ± 11
$10 - 200$	5	208 ± 25
$3 - 10$	4	288 ± 30
< 3	3	326 ± 30

Table 6.1. The number of active quark flavours and $\Lambda^{(n_f)}$ in the $\overline{\text{MS}}$ scheme for various energy ranges

Today six quark flavours have been identified experimentally, so in principle we should set $n_f = 6$. However, when performing calculations in perturbation theory using the $\overline{\text{MS}}$ scheme with $n_f = 6$ we encounter problems. Typically the perturbation series is ill-behaved unless the parameter M, which in principle is arbitrary, is chosen suitably and unless the effects of quarks with mass $m_q \gg M$ are included using an appropriate resummation technique. In the calculation of some quantity, M is in general the typical energy scale involved. However, when a process involves more than one energy scale, it is often very tricky and even controversial how to choose an appropriate value for M. The resummation technique amounts to using as the expansion parameter in the perturbation series $g(M)$ calculated for n_f equal to the number of active quark flavours, that is the number of quark flavours with masses $m_q < M$. When M crosses a threshold where a new quark flavour comes into play n_f changes, and $\Lambda^{(n_f)}$ and $\Lambda^{(n_f+1)}$ are related by matching conditions[220,221]. From comparisons of theoretical QCD predictions with experiment one finds [27] a value for $\Lambda^{(5)}$ as quoted in table 6.1. The values for $n_f = 3, 4$ and 6 are calculated using the matching conditions[220,27].

Even equipped with all these methods and tools we still face big problems if we want to derive results for observable quantities from the Lagrangian (6.1). The most notable is that $\mathcal{L}_{\text{QCD}}$ is expressed in terms of quark and gluon fields whose quanta have not been observed as free particles. In the real world we observe only hadrons, that is colourless objects, with the quarks and gluons permanently confined. Nevertheless it has been possible in some cases to derive first-principles results which can be compared with experiment, starting from $\mathcal{L}_{\text{QCD}}$ of (6.1). These are found in the following contexts.

(1) Short-distance phenomena in which all relevant momentum scales are much bigger than the QCD scale-parameter Λ. Because of asymptotic free-

dom (recall (6.20)) the QCD coupling parameter becomes small in this regime and one can make reliable perturbative calculations. Perturbation series in QCD are most likely to be only asymptotic series; see section 6.7. Examples of quantities governed by short-distance interactions are the total cross section for electron-positron annihilation into hadrons and the total hadronic decay rate of the Z-boson; see for instance [222].

(2) Long-distance phenomena in which all momentum scales are at most a few times Λ. Here one is in the nonperturbative regime of QCD and one has to use numerical methods to obtain first-principles results from $\mathcal{L}_{\mathrm{QCD}}$, or rather from the lattice version of $\mathcal{L}_{\mathrm{QCD}}$ introduced by Wilson[214]. Typical quantities one can calculate in this way are hadron masses and other low-energy hadron properties. For reviews of these methods see [223,224].

6.2 Semi-hard collisions

There is a third regime of hadronic phenomena, high-energy hadron-hadron collisions, which involve both long and short-distance physics. Thus, none of the above-mentioned theoretical methods applies directly. Traditionally high-energy hadron-hadron collisions are classified as hard or soft ones, though often a so-called hard collision is really only semi-hard.

A typical semi-hard reaction is the Drell-Yan process, for example

$$
\begin{aligned}
\pi^- + N &\to \gamma^* + X \\
&\hookrightarrow l^+ l^-
\end{aligned}
\tag{6.21}
$$

where $l = e$ or μ. The invariant mass of the lepton pair is assumed to be large and provides a hard scale. However, the masses of the π^- and N in the initial state stay fixed and thus we are not dealing with a purely short-distance phenomenon.

In the reaction (6.21) we see in action the partons, the fundamental quanta of the theory. These are the quarks and gluons; see for example figure 6.1. In the usual theoretical framework for hard reactions, the QCD-improved parton model[225,222], one describes the reaction of the partons by perturbation theory. In the Drell-Yan case this is $q\bar{q}$ annihilation into a virtual photon. Since the parton process involves only high energies and high momentum transfers, this should be reliable. All the long-distance physics due to the bound state nature of the hadrons is then lumped into parton distribution functions of the participating hadrons. The theoretical basis for this procedure is the factorisation hypothesis, which after early investigations of soft initial- and final-state interactions[226,114] was formulated and studied in low orders of QCD perturbation theory[227–233]. Subsequently,

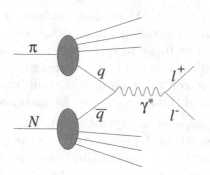

Figure 6.1. Lowest-order diagram for the Drell-Yan reaction (6.21) in the QCD-improved parton model

great theoretical effort has gone into proving factorisation to all orders in the framework of QCD perturbation theory. The result seems to be that factorisation is applicable in many but not all circumstances. For reviews see [234,235]. However, it is legitimate to ask whether factorisation is respected also by nonperturbative effects. To the best of our knowledge this question was first asked in [236–238]. There is some experimental evidence for a breakdown of factorisation in the Drell-Yan reaction[239,240]. In [238] and [241] it is argued that this should be expected due to QCD vacuum effects.

Factorisation relates the two quark-emission amplitudes in figure 6.1 to similar amplitudes that are relevant for other hard reactions, for example the deep-inelastic lepton scattering discussed in the next chapter. The figure may be used to calculate $d\sigma/dm_{ll}$, where m_{ll} is the invariant mass of the lepton pair. This quantity sums over all the possible hadronic states that accompany the lepton pair.

If, however, one is interested in the details of these accompanying hadronic states, things become more complicated: one has to take into account additional initial and final-state interactions. These may or may not lead to the production of additional particles; examples are shown in figure 6.2, where the zigzag lines represent pomerons. To calculate a cross section, one has to square figure 6.1 and the contributions exemplified by each of the figures 6.2a and 6.2b and perform an appropriate summation over the possible configurations of the final-state hadrons. In addition, there is the interference between figures 6.1 and 6.2a. By using Mueller's generalised optical theorem, which we introduced in section 3.4, it is possible[226,114] to show that when the summation is over all possible hadronic configura-

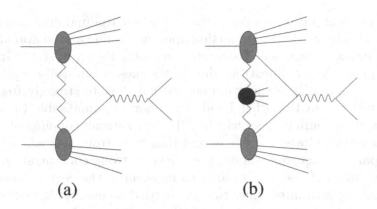

Figure 6.2. Pomeron-exchange corrections to figure 6.1: (a) produces no additional particles, but (b) does.

tions, as is necessary to calculate $d\sigma/dm_{ll}$, the interference term cancels the contributions from figure 6.2, so that $d\sigma/dm_{ll}$ is correctly calculated from the simple diagram of figure 6.1 alone. This cancellation is known as an AGK cancellation, after its authors[156] who considered various such cancellations. Note that the cancellation is an asymptotic one, at sufficiently high energy and large m_{ll}; we do not know how large these have to be for it to become effective. If one ignores this problem, as is usually done, the results generally seem to be reasonable.

6.3 Soft hadron-hadron collisions

Consider now soft high-energy collisions. A typical reaction in this class is proton-proton elastic scattering at large centre-of-mass energies and small momentum transfers. Here we have two scales, one becoming large, one staying finite: $\sqrt{s} \gg 1$ GeV, $|t| \lesssim 1$ GeV^2. Thus, none of the above calculational methods is directly applicable. Indeed, most theoretical papers dealing with reactions in this class develop and apply models which are partly older than QCD, partly QCD-motivated.

In 1952 Heisenberg[242] made some inspiring general field-theoretical considerations concerning high-energy scattering and particle production. He assumed that in a high-energy hadron-hadron collision in which the hadrons are highly-Lorentz-contracted a meson shock wave is formed. Then he analysed what type of particle spectrum one should expect to be produced from the evolution of this shock wave. For a free-meson theory one would just have to perform a Fourier transform of the meson wave to see its particle content. A highly-contracted wave in the longitudinal direction in ordinary space clearly will lead to a particle spectrum filling all the available longitudinal momentum space. For a renormalisable theory of strong interactions, he argued, the nonlinearities due to the meson-field self-coupling are in a sense weak, and the spectrum should not deviate strongly from the free-field result. On the other hand, for a non-renormalisable theory one has very strong nonlinearities which will lead to strong coupling of modes with high energy to low-energy ones and thus to a strong degradation of energy. The particle spectrum should then have a strong concentration at low absolute values of the longitudinal momentum in the centre-of-mass system. Heisenberg estimated that the mean particle energy $\bar{E}$ in the centre-of-mass system should grow only like log s. If we look today at the spectra of mesons produced in nucleon-nucleon collisions, we find that the whole available longitudinal momentum range is populated and that there is no spike at small momentum; see for instance [243]. Thus, using Heisenberg's argument, we can conclude that the theory of strong interactions should be of the renormalisable type. In fact, this conclusion could have been reached already in the 1960s, but before the advent of QCD the idea that field theory had anything to say on strong-interaction phenomena was not very popular.

In his article of 1952 Heisenberg also gave an argument, similar to the one used later by Froissart, Lukaszuk, and Martin to derive a rigorous bound, see section 1.7, that the inelastic total hadron cross sections should increase as $(\log s)^2$ at high energies if the interaction is "strong", that is of the non-renormalisable type. His reasoning was roughly as follows. Consider as an example a nucleon-nucleon collision at impact parameter b and centre-of-mass energy $\sqrt{s}$. Then energy $\sqrt{s}-2m_N$ is available for particle production, but in general only a fraction r of this will in fact be used up. He estimated that r should be determined by the overlap of the meson clouds of the nucleons calculated from the Yukawa potential. This led to the ansatz

$$r = \exp(-bm_\pi). \tag{6.22}$$

He estimated a minimal value $r_{\min}$ from the requirement that two mesons with mean particle energy $\bar{E}$ can be produced:

$$(\sqrt{s} - 2m_N)r_{\min} = 2\bar{E}. \tag{6.23}$$

But $r_{\min}$ corresponds through (6.22) to a maximal value of b

$$b_{\max} = -\frac{1}{m_\pi} \log r_{\min}. \tag{6.24}$$

Then the total inelastic cross section was estimated as

$$\sigma^{\text{Inel}} = 2\pi \int_0^{b_{\max}} b\,db = \pi b_{\max}^2$$

$$= \frac{\pi}{m_\pi^2}(\log r_{\min})^2$$

$$= \frac{\pi}{m_\pi^2}\left(\log \frac{\sqrt{s} - 2m_N}{2\bar{E}}\right)^2. \tag{6.25}$$

Thus having argued that the assumption of maximal strength gives $\bar{E} \propto \log s$, Heisenberg concluded that the finite range of the interaction and maximal strength make the cross section increase as $(\log s)^2$ for large s. Compare this with (1.68).

From the 1950s onward quite a number of approaches for describing hadron-hadron scattering at high energies have been proposed including the geometric, the eikonal, the valon, the additive-quark model, leading-log summations, topological expansions and strings, Regge poles, and two-gluon exchange. For a list of representative references to original papers on these approaches see [244]. Perturbative calculations for high-energy scattering amplitudes in quantum field theory are treated extensively in the classic book by Cheng and Wu[5].

We now argue that the theoretical description of measurable quantities of soft high-energy reactions such as the total cross sections should involve nonperturbative QCD in an essential way. Consider pure-gluon theory where $n_f = 0$, that is the term with the quark fields is absent in (6.1), and $\mathcal{L}_{\text{QCD}}$ contains no dimensioned parameter. Although the basic quanta are gluons, lattice calculations show quite conclusively that the asymptotic particles of the theory are massive hadrons, namely glueballs. For $n_f = 0$, the theory acquires one dimensioned parameter through the renormalisation procedure, the Λ parameter of (6.20). For dimensional reasons we must have $m_{\text{glueball}} \propto \Lambda^{(0)}$. By choosing a large renormalisation scale M, so that $g(M)$ is small and we may just keep the leading term in (6.20), this can be written as

$$m_{\text{glueball}} \propto Me^{-c/g^2(M)}, \qquad c = 8\pi^2/11. \tag{6.26}$$

Thus, masses in pure Yang-Mills theory, which contains no bare mass parameter, are a nonperturbative phenomenon, due to dimensional transmutation.

Scattering of glueball hadrons in massless pure-gluon theory should look very similar to scattering of hadrons in the real world, with finite total cross sections, amplitudes with analytic t dependence etc. At least, this would be our expectation. If the total cross section σ^{Tot} for glueball-glueball scattering had a finite limit as $s \to \infty$, from the same dimensional argument we would have

$$\lim_{s \to \infty} \sigma^{\mathrm{Tot}}(s) \propto M^{-2} e^{2c/g^2(M)}. \tag{6.27}$$

In this case, total cross sections in pure-gluon theory are also nonperturbative quantities. It is easy to see that this conclusion is not changed if $\sigma^{\mathrm{Tot}}(s)$ has a logarithmic behaviour for s large, for instance

$$\sigma^{\mathrm{Tot}}(s) \sim \sigma_0 (\log s/s_0)^2. \tag{6.28}$$

We must have again

$$\sigma_0 \propto \Lambda^{-2} \propto M^{-2} e^{2c/g^2(M)} \tag{6.29}$$

which shows that $\sigma^{\mathrm{Tot}}(s)$ cannot be expanded in powers of g. We would then expect in full QCD that total cross sections are also nonperturbative quantities, at least as far as hadrons made out of light quarks are concerned.

The arguments given above suggest trying to connect high energy soft scattering to other nonperturbative phenomena in QCD, such as the structure of the vacuum state, to which we turn next.

6.4 The QCD vacuum

According to current theoretical prejudice the vacuum state in QCD has a very complicated structure; for reviews see [247,248]. It has been noted[249] that by introducing a constant chromomagnetic field with strength B into the perturbative vacuum one can lower the vacuum-energy density $\varepsilon(B)$. The result of a one-loop calculation is

$$\varepsilon(B) = \tfrac{1}{2} B^2 + \frac{\beta_0 g^2(M)}{32\pi^2} B^2 \Big(\log \frac{B}{M^2} - \tfrac{1}{2}\Big) \tag{6.30}$$

where $g(M)$ is the strong coupling constant, M is again the renormalisation scale, and β_0 is given in (6.19). Thus, as long as we have asymptotic freedom, that is for $\beta_0 > 0$, the energy density $\varepsilon(B)$ looks as indicated schematically in figure 6.3 and has its minimum for $B = B_{\mathrm{vac}} \neq 0$. Therefore, we should expect the QCD vacuum to develop spontaneously a chromomagnetic field, the situation being similar to that in a ferromagnet below the Curie temperature where we have spontaneous magnetisation.

Of course, the vacuum state in QCD has to be relativistically invariant and cannot have a preferred direction in ordinary space and colour space. What has been considered are states composed of domains with random orientation of the gluon-field strength. This is analogous to Weiss domains in a

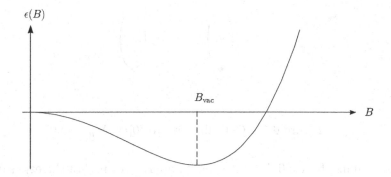

Figure 6.3.   Schematic behaviour of the vacuum energy density $\epsilon(B)$ as a function of a constant chromomagnetic field B according to (6.30)

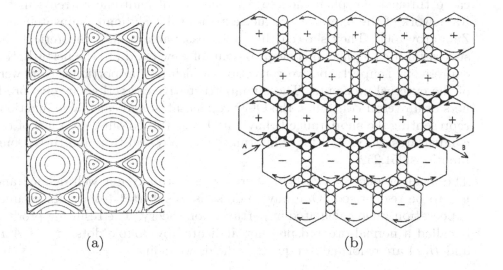

(a)   (b)

Figure 6.4.   (a) QCD vacuum according to Ambjørn and Olesen[245]; (b) the ether according to Maxwell[246]

ferromagnet. The vacuum state should then be a suitable linear superposition of states with various domains and orientation of the fields inside the domains. This implies that the orientation of the fields in the domains as well as the boundaries of the domains will fluctuate.

A very detailed picture for the QCD vacuum along these lines has been developed in [245]. It is amusing to compare this modern picture of the

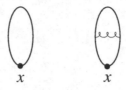

$$x \qquad\qquad x$$

Figure 6.5.   Contributions to $\langle 0|\bar{\psi}(x)\gamma_\mu\psi(x)|0\rangle$

QCD vacuum (figure 6.4a) with the "modern picture" of the ether developed by Maxwell more than 100 years ago[246] (figure 6.4b). The analogy is quite striking and suggests that we may also be able to find simpler views of the QCD vacuum. We should remember that Einstein made great progress by eliminating the ether from electrodynamics.

A breakthrough in the quantitative investigation of the vacuum properties came through the phenomenological analysis of hadronic currents in the framework of the operator-product expansion by Shifman, Vainshtein and Zakharov[250]. They showed that it is reasonable from a theoretical and successful from a phenomenological point of view to add to the perturbative expressions nonperturbative corrections which are proportional to powers of the external momenta. These nonperturbative corrections involve products of universal constants, the QCD condensates, and calculable functions representing the short-range behaviour. It has turned out that a few of the condensates can yield a lot of information on physical two- and three-point functions[251,252].

The local product of two field operators is normally not well defined and has to be regularised. One way to do so is to subtract from it its vacuum expectation value calculated in perturbation theory. The resulting product is called a normal-ordered product, indicated by double dots : : . If $A(x)$ and $B(x)$ are renormalised quantum fields we define

$$: A(x)\,B(x) : \; \equiv \lim_{\epsilon \to 0} \Big(A(x)\,B(x+\epsilon) - \langle 0|A(x)\,B(x+\epsilon)|0\rangle_{\text{pert}} \Big). \qquad (6.31)$$

with a space-like separation ϵ. Typical graphs contributing to the perturbative vacuum expectation value $\langle 0|\bar{\psi}(x)\psi(x)|0\rangle$, where ψ is a quark field, are shown in figure 6.5.

The simplest approach to the operator-product expansion[253] is to apply Wick's theorem to a product of composite operators in the interaction picture. For definiteness we consider the covariant version [254] of the time-ordered operator product of two vector currents

$$\Pi_{\rho\sigma}(x) = \text{T}^* : \bar{\psi}(x)\gamma_\rho\psi(x) : : \bar{\psi}(0)\gamma_\sigma\psi(0) : . \qquad (6.32)$$

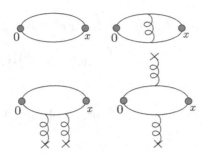

Figure 6.6.   Graphical representation of (6.33). The lines ending in a cross represent operators, the vacuum expectation values of which have to be taken in order to arrive at (6.36).

Summation over spin and colour indices is performed but not indicated explicitly. It can be shown to any order in perturbation theory that this product of operators at different positions can be expanded in a series of local operators to give

$$\Pi_{\rho\sigma}(x) = C^0_{\rho\sigma}(x,\mu)\,1 + \cdots + 2C^4_{\rho\sigma}(x,\mu) : \operatorname{tr} g^2 \mathbf{F}_{\mu\nu}(0)\mathbf{F}^{\mu\nu}(0) : + \cdots .$$
(6.33)

The coefficient functions $C^i(x^2,\mu)$ can be calculated in perturbation theory. They will depend in general on the renormalisation scale μ as do the remaining operator products. Here only the terms proportional to the unit operator and to the gauge and Lorentz-invariant operator : $\operatorname{tr} g^2 \mathbf{F}_{\mu\nu}(0)\mathbf{F}^{\mu\nu}(0)$: are shown, as they are the most interesting ones for us. Other terms involve operators like : $\bar{\psi}(0)\psi(0)$: or : $\operatorname{tr} g^3 \mathbf{F}_{\mu\nu}(0)\mathbf{F}^{\nu\rho}(0)\mathbf{F}^{\mu}_{\rho}(0)$: . Equation (6.33) is the operator-product expansion of the operator product $\Pi_{\rho\sigma}(x)$ of the vector currents. It is represented graphically in figure 6.6. The lines ending in a cross represent operators.

By definition of the normal-ordered product (6.31) the vacuum expectation value of $\Pi_{\rho\sigma}(x)$ calculated in perturbation theory reduces to $C^0_{\rho\sigma}(x,\mu)$:

$$\langle 0|\Pi_{\rho\sigma}(x)|0\rangle_{\text{pert}} = C^0_{\rho\sigma}(x,\mu) .$$
(6.34)

It is, however, conceivable that the expectation values of normal-ordered products with respect to the physical vacuum, representing the no-hadron state, do not vanish and genuine nonperturbative terms occur. For the Fourier-transformed expression, after some calculation one obtains for massless quarks

$$\langle 0|\tilde{\Pi}_{\rho\sigma}(q)|0\rangle = i \int d^4x \, e^{iq \cdot x} \langle 0|\Pi_{\rho\sigma}(x)|0\rangle$$
$$= \left(q_\rho q_\sigma - g_{\rho\sigma} q^2\right)\tilde{\Pi}(q^2) \tag{6.35}$$

where

$$\tilde{\Pi}(q^2) = \tilde{C}^0(q^2,\mu)\,1 + \cdots + 2\tilde{C}^4(q^2,\mu)\langle 0|\mathrm{tr}\, g^2 : \mathbf{F}_{\mu\nu}(0)\mathbf{F}^{\mu\nu}(0) : |0\rangle + \cdots \tag{6.36}$$

with

$$\tilde{C}^0(q^2,\mu) = -\frac{1}{8\pi^2}\left(1 + \frac{\alpha_s}{\pi}\right)\log\frac{-q^2}{\mu^2} + O(\alpha_s^2)$$
$$\tilde{C}^4(q^2,\mu) = \frac{1}{48\pi^2}\frac{1}{q^4}\left(1 + O(\alpha_s)\right). \tag{6.37}$$

By comparing the phenomenological expression for the correlator with (6.36) Shifman, Vainshtein and Zakharov[250] were able to determine the gluon condensate $\langle 0| : g^2 \,\mathrm{tr}\, \mathbf{F}_{\mu\nu}\mathbf{F}^{\mu\nu} : |0\rangle$ from charmonium spectra and predict correctly other resonances by applying an operator-product expansion like (6.36) to the product of currents interpolating these resonances. The accepted value for the gluon condensate currently is[255]

$$\langle g^2 FF\rangle \equiv 2\langle 0| : \mathrm{tr}\, g^2 \mathbf{F}_{\mu\nu}\mathbf{F}^{\mu\nu} : |0\rangle = (0.9 \pm 0.1\mathrm{GeV})^4. \tag{6.38}$$

The gauge-invariant two-gluon operator in pure gauge theory is related to the trace of the energy-momentum tensor

$$\theta_\mu^\mu \equiv \frac{\beta(g)}{g} : \mathrm{tr}\, \mathbf{F}_{\mu\nu}\mathbf{F}^{\mu\nu} : \tag{6.39}$$

where $\beta(g)$ is the β function defined in (6.18). From this it follows that $\langle g^2 FF\rangle$ is renormalisation-group invariant up to corrections of order α_s^2.

It is often convenient or even necessary to confine oneself to approximations in which internal fermion loops are not taken into account. This is the so-called quenched approximation. It can be realised in the limit $N_C \to \infty$. The number of colours N_C is increased to infinity, but the product $\alpha_s N_C$ is kept fixed. The value of the gluon condensate in a theory can be estimated from a low-energy theorem[256] relating the mass dependence of the gluon condensate to the condensate of light quarks, which is well known. It can be inferred that in quenched QCD the gluon condensate is about a factor 2 to 3 larger[256] than in full QCD:

$$\langle g^2 FF\rangle_{\text{quenched}} = (1.2 \pm 0.1 \ \mathrm{GeV})^4. \tag{6.40}$$

Though condensates are a well-established concept in the operator-product expansion it is hard to calculate them by genuinely nonperturbative methods like numerical lattice QCD. The problem lies just in the concept of normal ordering. One might assume that the short-distance behaviour in perturbative QCD is under complete control because of asymptotic freedom. However the Wilson coefficients are most probably not well defined since perturbation theory itself indicates that unsummable divergences lead to undetermined contributions which behave like an inverse power of q^2. This feature makes it impossible to subtract the perturbative part in a well-defined way. Since these ambiguities also play some role in the perturbative treatment of hard-scattering processes we devote section 6.7 to them. Needless to say, we are not prevented from applying our models by this difficulty. If we separate perturbative from nonperturbative contributions, we understand by perturbative contributions either only a few orders or a perturbative series where all internal momenta are greater than a certain cutoff. All uncertainties resulting from the ambiguous definition of normal ordering are absorbed in the models for nonperturbative effects. In the following double dots : : will indicate that we are considering the nonperturbative contribution only.

6.5 Nonlocal condensates

The introduction of condensates in QCD formed an important step away from the purely short-distance physics of perturbative QCD. In order to go to even larger distances and to handle confinement and related problems, at least in models, we have to go a step further and consider nonlocal condensates such as

$$\langle : \operatorname{tr} g^2 \mathbf{F}_{\mu\nu}(y)\mathbf{F}^{\mu\nu}(x) : \rangle . \tag{6.41}$$

In the following we treat the fields as classical ones, the quantisation coming through the functional averaging indicated by the brackets $\langle \rangle$.

Nonlocal condensates have advantages over local ones. They are well defined for $x \neq y$ and for large separations $z = x - y$ they are nonperturbative by definition. In a non-Abelian theory the matrix elements of the field-strength tensor are colour matrices. They are not locally gauge-invariant but transform under a local gauge transformation $\mathbf{U}(x)$ as

$$\mathbf{F}_{\mu\nu}(x) \rightarrow \mathbf{U}(x)\mathbf{F}_{\mu\nu}(x)\mathbf{U}^\dagger(x) \tag{6.42}$$

See (6.7). Therefore a nonlocal condensate such as (6.41) cannot be defined in a gauge-invariant way.

In order to overcome this problem we define a gauge-invariant gluon correlator (nonlocal condensate) in the following way. Let the connector or parallel transporter $\Phi(x, y)$ be the non-Abelian generalisation of the Schwinger string from point x to y:

$$\Phi(x, y) \equiv \mathrm{P}\exp\left(-ig\int_0^1 d\lambda\,(y - x)^\mu \mathbf{A}_\mu\,(x + \lambda\,(y - x))\right). \tag{6.43}$$

Here P denotes path ordering of the colour matrices in $\mathbf{A}_\mu$. Note that we are still on the classical level. For nonperturbative considerations it is most convenient to adopt the prescription from lattice regularisation and then go to the continuum limit:

$$\mathrm{P}\exp\left(-ig\int_0^1 d\lambda\,(y - x)^\mu \mathbf{A}_\mu\,(x + \lambda\,(y - x))\right) =$$

$$\lim_{N\to\infty}\prod_{n=1}^{N}\exp\left(-ig\frac{1}{N}(y - x)^\mu \mathbf{A}_\mu\left(x + \frac{n}{N}(y - x)\right)\right). \tag{6.44}$$

The ordering in a product $\prod_{i=1}^{N} a_i$ is defined as $a_N \ldots a_2\,a_1$.

Using the gauge transformation (6.6) of the colour potential and the definition of the path-ordered integral one can show that $\Phi(x, y)$ transforms under local gauge transformations as

$$\Phi(x, y) \to \mathbf{U}(y)\Phi(x, y)\mathbf{U}^\dagger(x), \tag{6.45}$$

and therefore the expression $:\,\mathrm{tr}\,\mathbf{F}_{\mu\nu}(y)\Phi(x, y)\mathbf{F}^{\mu\nu}(x)\Phi(y, x)\,:$ is gauge-invariant.

We can define the gauge-invariant nonlocal gluon condensate as

$$D_{\mu\nu\rho\sigma}(y - x) \equiv 2\langle:\,\mathrm{tr}\,g^2\mathbf{F}_{\mu\nu}(y)\Phi(x, y)\mathbf{F}_{\rho\sigma}(x)\Phi(y, x)\,:\rangle. \tag{6.46}$$

For comparison with results from lattice calculations and applications in nonperturbative approaches it is more convenient to consider first the correlator in Euclidean field theory. The transition to Minkowski metric will be discussed in sections 8.5 and 8.11.

Four-dimensional rotational invariance corresponding to Lorentz invariance restricts the tensor structure of $D_{\mu\nu\rho\sigma}(y - x)$ in such a way that it can be described by two invariant functions D and D_1:

$$D_{\mu\nu\rho\sigma}(z) = \frac{\langle g^2 FF\rangle}{12}\Bigg((\delta_{\mu\rho}\delta_{\nu\sigma} - \delta_{\mu\sigma}\delta_{\nu\rho})D(z^2)\kappa$$

$$+\Big(\tfrac{1}{2}\frac{\partial}{\partial z_\mu}(z_\rho\delta_{\nu\sigma} - z_\sigma\delta_{\nu\rho}) + \tfrac{1}{2}\frac{\partial}{\partial z_\nu}(z_\sigma\delta_{\nu\rho} - z_\rho\delta_{\nu\sigma})\Big)D_1(z^2)(1 - \kappa)\Bigg)$$

$$\tag{6.47}$$

where $z = y - x$. The constant κ can in principle take any value. The invariant functions are normalised to

$$D(0) = D_1(0) = 1.\qquad(6.48)$$

In an Abelian gauge theory such as QED the homogeneous Maxwell equations $\partial_\mu \epsilon^{\mu\nu\rho\sigma} F_{\rho\sigma} = 0$ imply that only the second tensor structure can appear which means that κ has to be zero. Hence the tensor multiplying $D(z^2)$ is typical of a non-Abelian theory.

The correlator (6.46) has been evaluated numerically[257] in Euclidean lattice QCD using a special method in order to enhance nonperturbative effects. In this way one has obtained $D_\parallel$ and $D_\perp$ defined through

$$D_\parallel(z^2) = \frac{1}{12}\langle g^2 FF\rangle\Big(D(z^2) + D_1(z^2)\Big)$$

$$D_\perp(z^2) = \frac{1}{12}\langle g^2 FF\rangle\Big(D_1(z^2) + z^2\frac{\partial}{\partial z^2}D_1(z^2)\Big)\qquad(6.49)$$

for $z^2 > 0$. It is found that the typically non-Abelian structure is dominant and the functions $D(z^2)$ and $D_1(z^2)$ fall off exponentially for $z^2 \to \infty$ with a correlation length $a \approx 1\,\mathrm{GeV}^{-1}$.

The correlator (6.46) can also be viewed as a two-point function of gauge-invariant currents in an extension of heavy-quark effective theory and it can be shown that confinement implies an exponential fall-off at large distances $|x - y|$. In this approach the correlation length has been calculated using QCD sum-rule techniques[258] and found to be consistent with the lattice results.

The special model forms[259]

$$D(z^2) = \int \frac{d^4k}{(2\pi)^4}e^{ikz}\frac{27\pi^4 k^2}{4a^2(k^2 + \frac{9\pi^2}{64a^2})^4}$$

$$= \frac{3\pi|z|}{8a}\left(K_1\left(\frac{3\pi|z|}{8a}\right) - \frac{1}{4}\frac{3\pi|z|}{8a}K_0\left(\frac{3\pi|z|}{8a}\right)\right)$$

$$D_1(z^2) = \int \frac{d^4k}{(2\pi)^4}e^{ikz}\frac{9\pi^4}{2a^2(k^2 + \frac{9\pi^2}{64a^2})^3}$$

$$= \frac{3\pi|z|}{8a}K_1\left(\frac{3\pi|z|}{8a}\right)\qquad(6.50)$$

are convenient for a continuation from Euclidean to Minkowski space since they have simple analytic properties in momentum space. The correlation length a is defined through

$$\int_0^\infty dz\, D(z^2) = a.\qquad(6.51)$$

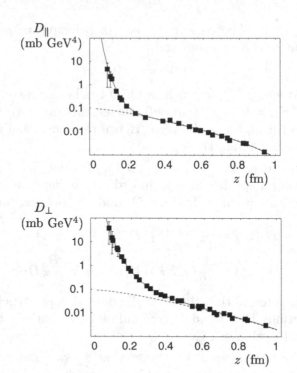

Figure 6.7.   Lattice data[257] for $D_\parallel$ and $D_\perp$ defined in (6.49). The fits use the functions (6.50) for the nonperturbative part and a singular term $1/z^4$ for the perturbative part. The solid line is the combined fit, and the dashed curve is the nonperturbative part alone.

These model forms are well adapted to later applications in high-energy scattering. Furthermore they give a good fit to the lattice data.   In [260] the results of the lattice calculations are fitted with the functions (6.50) for the nonperturbative part and a singular term $1/z^4$ for the perturbative contributions. In figure 6.7 the lattice data for quenched QCD as a function of distance $|z|$ are displayed together with the fit. The solid lines are the full fit, the dashed ones show it with only the nonperturbative part. For the nonperturbative part the following results were obtained in the quenched approximation:

$$a = 0.328 \pm 0.005 \,\text{fm} \quad \kappa = 0.89 \pm 0.02 \quad \langle g^2 FF \rangle = (1.22 \pm 0.02 \,\text{GeV})^4 \tag{6.52}$$

where the errors are the statistical ones. The results for the correlation length and κ are very stable. However the result for $\langle g^2 FF \rangle$ is plagued not only by numerical uncertainties but, due to the high mass dimension, also by the uncertainty in the physical scale.

6.6 Loops and the non-Abelian Stokes theorem

We generalise the connector $\mathbf{\Phi}(x, y)$ with a straight line defined in (6.43) to the line integral over an arbitrary continuous curve $C(x, y)$ from point x to point y:

$$\mathbf{V}[C(x, y)] \equiv \mathrm{P} \exp \left(-ig \int_{C(x,y)} dz^\mu \mathbf{A}_\mu(z) \right). \tag{6.53}$$

Again we consider first classical fields; the quantisation will come through the functional averaging, indicated by brackets $\langle\,\rangle$.

The line integral $\mathbf{V}[C(x, y)]$ has the same transformation property (6.44) as the straight line $\mathbf{\Phi}(x, y)$. It also satisfies the relation

$$\mathbf{V}[C(y, x)] = \mathbf{V}^\dagger[C(x, y)] = \mathbf{V}^{-1}[C(x, y)]. \tag{6.54}$$

If the curve $C(x, y)$ forms a closed loop, that is the points x and y coincide, and if the trace of $\mathbf{V}[C(x, x)]$ is taken, then the result is a gauge-invariant quantity, the Wilson loop. It is independent of the point x. In the following we shall denote this gauge-invariant quantity by $W[C]$:

$$W[C] \equiv \mathrm{tr}\, \mathbf{V}[C(x, x)] = \mathrm{tr}\, \mathrm{P} \exp \left(-ig \oint_{C(x,x)} dz^\mu \mathbf{A}_\mu(z) \right). \tag{6.55}$$

The loop together with the interpretation of links on the lattice as dynamical variables was introduced in Z_2 gauge theory by Wegner[261] and their importance in QCD was recognised by Wilson[214] in his classic paper on lattice QCD. In QCD the dependence of the loop on the geometrical properties of the curve C indicates whether there is confinement. If in Euclidean quenched QCD the expectation value $\langle W[C] \rangle$ of the loop falls off as the exponential of the minimal area enclosed by C we have confinement; if it depends on the perimeter there is no confinement. In Euclidean QCD there is a simple connection between rectangular loops and the potential of a static quark-antiquark pair. Let C_R be a rectangle with length T in the x_4 direction and width R in one space direction, say x_1. Then the static potential is given by

$$V_0(R) = -\lim_{T \to \infty} \frac{1}{T} \log \langle W[C_R] \rangle. \tag{6.56}$$

Thus the area law implies a linearly rising potential between a static quark-antiquark pair.

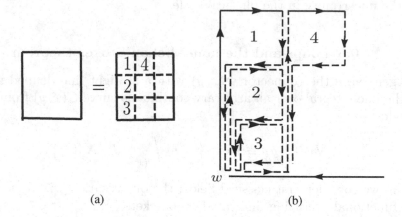

Figure 6.8.   (a) Illustration of the non-Abelian Stokes theorem. (b) Magnification of the right hand side of (a).

The Wilson loop $W[C]$ can be expressed through the field strength $\mathbf{F}_{\mu\nu}$ by applying the non-Abelian generalisation of the Stokes theorem[262–264]. This theorem can be derived in several ways; the simplest one relies on a geometrical representation. For that we introduce the plaquette, the loop of a small square $C_{\mu\nu}^d(x)$ in the plane defined by vectors in the μ and ν directions with small side-length d, and with lower left corner at the point x. If the potential in the exponent is expanded in a Taylor series about the point x one can express the plaquette $\mathbf{V}[C_{\mu\nu}^d(x)]$ through the field tensor $\mathbf{F}_{\mu\nu}(x)$:

$$\mathbf{V}[C_{\mu\nu}^d(x)] = \exp\left(-id^2 \, \mathbf{F}_{\mu\nu}(x) + O(d^3)\right). \qquad (6.57)$$

We can redraw the loop C_R of the simple rectangle at the left of figure 6.8a in a complicated way as in the figure. All inner line integrals (shown as dashed lines in figure 6.8b) of opposite direction cancel at the end because of the property (6.54). Therefore both curves, the simple and the complicated one, give the same expression. Applying (6.57) to the plaquettes 1, 2, ... we obtain

$$\mathrm{P}\exp\left(-ig\oint_{C_R(w)} dz^\mu \mathbf{A}_\mu\right) =$$
$$\prod_n \exp\left(-igd^2\mathbf{V}[C'(w,x_n)]\mathbf{F}_{\mu\nu}(x_n)\mathbf{V}[C'(x_n,w)] + O(d^3)\right) \quad (6.58)$$

where $C'(w,x_n)$ is the path from the lower left corner of C to the point x_n, as indicated explicitly in figure 6.8b. We introduce the parallel-transported field tensor $\mathbf{F}_{\mu\nu}[x; C(w, x)]$ by

$$\mathbf{F}_{\mu\nu}[x; C(w, x)] \equiv \mathbf{V}[C(w, x)]\mathbf{F}_{\mu\nu}(x)\mathbf{V}[C(x, w)]. \qquad (6.59)$$

In the limit of $d \to 0$ we can neglect the higher terms in d. Defining the surface-ordered integral as the limit of the right hand side of (6.58) we obtain

$$\mathrm{P} \exp\left(-ig \oint_C dz^\mu \, \mathbf{A}_\mu\right) = \mathrm{P} \exp\left(-ig \int_S d\sigma^{\mu\nu} \, \mathbf{F}_{\mu\nu}[x; C(w, x)]\right) \qquad (6.60)$$

where the surface integral $\int_S d\sigma^{\mu\nu}$ can be taken over any surface S with boundary C and w can be chosen anywhere on the surface. This is the non-Abelian Stokes theorem.

Under a local gauge transformation the parallel-transported field-strength tensor $\mathbf{F}_{\mu\nu}[x; C(w, x)]$ of (6.59) transforms as

$$\mathbf{F}_{\mu\nu}[(x; C(w, x)] \to \mathbf{U}(w)\mathbf{F}_{\mu\nu}[x; C(w, x)]\mathbf{U}^\dagger(w). \qquad (6.61)$$

The gauge-invariant correlator (6.46) can also be expressed through the transported field strength (6.59) as

$$D_{\mu\nu\rho\sigma}(y - x) = 2\langle : \mathrm{tr}\, g^2 \mathbf{F}_{\mu\nu}[y; C(w, y)]\mathbf{F}_{\rho\sigma}[x; C(w, x)] : \rangle \qquad (6.62)$$

where w is a point on the straight line connecting x and x'. Note that for a straight line the correlator is independent of w.

In the case of an Abelian gauge theory the parallel-transported field strength is identical to the untransported one since the gauge matrices $\mathbf{U}$ in equation (6.61) are complex numbers and cancel. Then (6.60) is the familiar Stokes theorem.

6.7 Stochastic-vacuum model

In this section we present a model which can be applied to evaluate functional integrals approximately[265,266]. It was originally formulated in Euclidean field theory, but it can easily be continued from Euclidean to Minkowski space and thus be applied to high-energy scattering. The essential assumption of its most radical and phenomenologically useful version is that the nonperturbative behaviour of QCD can be approximated by a Gaussian stochastic process (6.13) with the field strengths as stochastic variables, that is as integration variables in the functional integral. This assumption already yields one of the most striking features of QCD, namely confinement. A derivation of confinement in the framework of the model not only yields an important result, a relation between the confining potential of two heavy quarks and the correlators (6.62), but also is a good demonstration of the techniques necessary to apply the model to high-energy scattering.

We now state the assumptions of the model. For motivations and details we refer to reviews[267–269,244]. The principal assumption, that the nonperturbative behaviour of QCD can be approximated by a Gaussian process in the field strengths as in (6.13), can be expressed by the equation

$$\langle W[C]\rangle = \Big\langle \operatorname{tr} \mathrm{P} \exp\Big(-ig\int_{\mathcal{S}} \mathbf{F}_{\mu\nu}[x,C(w,x)]d\sigma^{\mu\nu}\Big)\Big\rangle$$

$$\approx \exp\Big(-\tfrac{1}{2}g^2\Big\langle\int_{\mathcal{S}} d\sigma^{\mu\nu}d\sigma'^{\rho\sigma}2\operatorname{tr}\mathbf{F}_{\mu\nu}[x,C(w,x)]\mathbf{F}_{\rho\sigma}[x',C(w,x')]\Big\rangle\Big).$$

(6.63)

To come from (6.13) to (6.63) one has to insert for ϕ the field strength $\mathbf{F}_{\mu\nu}$ and for $J(x)$ a distribution which vanishes everywhere except on the surface $\mathcal{S}$. The inverse kernel K^{-1} corresponds to the correlator of the field strengths.

In order to come from the representation (6.55) of the Wilson loop to the one containing the field strengths $\mathbf{F}_{\mu\nu}$ we have made use of the non-Abelian Stokes theorem (6.60). This introduces a path ordering, but since the trace is symmetric the path ordering of the integrals in the exponent of the second line of (6.63) plays no role. The assumption of a Gaussian process includes that all higher correlators can be expressed by the two-point correlator. If the points $x_1 \ldots x_4$ are ordered according to the surface ordering then for the four-point correlator, we have for example,

$$\langle\mathbf{F}(x_1)\mathbf{F}(x_2)\mathbf{F}(x_3)\mathbf{F}(x_4)\rangle = \langle\mathbf{F}(x_1)\mathbf{F}(x_2)\rangle\langle\mathbf{F}(x_3)\mathbf{F}(x_4)\rangle$$
$$+ \langle\mathbf{F}(x_1)\mathbf{F}(x_3)\rangle\langle\mathbf{F}(x_2)\mathbf{F}(x_4)\rangle$$
$$+ \langle\mathbf{F}(x_1)\mathbf{F}(x_4)\rangle\langle\mathbf{F}(x_2)\mathbf{F}(x_3)\rangle \quad (6.64)$$

with $\mathbf{F}(x_i) = \mathbf{F}_{\mu_i\nu_i}(x_i,C(w,x_i))$. The Gaussian approximation can be considered as the first step of the linked-cluster or cumulant expansion[270]. Higher cumulants describe the deviation from factorisation. Most of the general results of the model are valid beyond the Gaussian approximation only if the cumulant expansion converges.

Were it not for the path $C(w,x)$, which in general is not part of a straight line connecting x and x', we could directly insert (6.47) into (6.63). So we make the additional assumption that a reasonably-chosen reference point does not influence the results too strongly. A reasonable choice of the position is in the minimal plane spanned by the border C of the loop of (6.63)

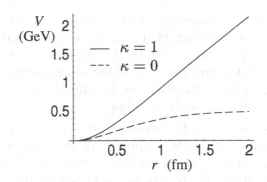

Figure 6.9. Contributions of the typically non-Abelian structure D ($\kappa = 1$) and of D_1 ($\kappa = 0$) to the nonperturbative part of the static quark-antiquark potential calculated from the stochastic-vacuum model

or, more generally, at a distance which is not great compared with the dimensions of the loop. For an unreasonable choice the Gaussian factorisation is no longer plausible.

If this strong assumption is made we can insert (6.47) into (6.63) and calculate the static potential using the relation (6.56) expressed through the field strength rather than through the potentials. Before we come to details we can show easily that the area law of the Wilson loop follows from (6.63) under the assumption of a finite correlation length of the correlator. Let the dimensions of the loop C be large compared with the correlation length. Performing the integration over $d\sigma'_{\rho\sigma}$ in the exponential of the second line of (6.63) one obtains $K\langle g^2 FF\rangle a^2$, where K is some number independent of x. The remaining integration over $d\sigma_{\mu\nu}$ then yields the area of the minimal surface whose border is the loop C.

A detailed analysis of (6.63) shows that the two different tensor structures in (6.47) give rise to very different behaviours of the potential. Only the term proportional to D, which is specific for a non-Abelian gauge theory, yields a linearly rising potential. The typical behaviours of the potential with $\kappa = 1$, to which only D contributes, and of the potential with $\kappa = 0$, to which only D_1 contributes, are given in figure 6.9.

We only quote the string tension σ_q, the asymptotic slope of the linear potential, resulting from that calculation:

$$\sigma_q = \frac{\kappa \langle g^2 FF\rangle}{48 N_C} \pi \int_0^\infty dz^2 \, D(z^2). \qquad (6.65)$$

The numerical values (6.52) from lattice calculations yield $\sigma_q = 0.17$ GeV^{-2}. Given the numerical uncertainties, especially of $\langle g^2 FF\rangle$, the agreement with the phenomenological value[271,272] 0.18 ± 0.02 GeV^2 is surprisingly good, and supports the assumption of Gaussian factorisation. Another strong point in favour of Gaussian factorisation is that in various representations of $SU(3)$ the expectation value of the Wilson loop is proportional to the eigenvalue of the Casimir operator of the representation[273,274]. Also, the spin structure of the potentials can be evaluated in the model and comes out to be in agreement with experiment.

Unfortunately Gaussian factorisation at the matrix level as implied by (6.64) is not sufficient for the evaluation of two loops which will be essential for high-energy scattering. Therefore in the following we make the even-stronger assumption of Gaussian factorisation for the colour components of the field strength tensor, which is our final input to the model:

$$\langle : F_1^{a_1} F_2^{a_2} F_3^{a_3} F_4^{a_4} : \rangle = \langle : F_1^{a_1} F_2^{a_2} : \rangle \langle : F_3^{a_3} F_4^{a_4} : \rangle + \langle : F_1^{a_1} F_3^{a_3} : \rangle \langle : F_2^{a_2} F_4^{a_4} : \rangle$$
$$+ \langle : F_1^{a_1} F_4^{a_4} : \rangle \langle : F_2^{a_2} F_3^{a_3} : \rangle \qquad (6.66)$$

where $F_i^{a_i}$ stands for $F_{\mu_i \nu_i}^{a_i}[x_{(i)}; C(w, x_{(i)})]$. Under the assumption that the dependence of the correlators on the path can be neglected,

$$\langle g^2 : F_{\mu\nu}^a(x; C(w, x)) F_{\rho\sigma}^b(x'; C(w, x')) : \rangle = \tfrac{1}{8}\delta^{ab} D_{\mu\nu\rho\sigma}(x - x') \qquad (6.67)$$

where $D_{\mu\nu\rho\sigma}$ is given by (6.47). The factorisation in the colour components leads to some technical difficulties in the evaluation of a single Wilson loop which have to be controlled by a technical prescription as explained in [275,244]. With this assumption one can calculate the gauge-invariant field distribution of a static quark-antiquark pair[275,276]. This is done by evaluating the product of two loops, a plaquette for the squared field strength and a large loop representing the static quark-antiquark pair. The results for the field-energy density due to the presence of a static quark-antiquark pair are displayed in figure 6.10; (a) is the pure Coulombic part, (b) includes the nonperturbative part. The formation of a nonperturbative string is clearly visible. The result in figure 6.10(b) compares favourably with numerical calculations[277] made in $SU(2)$ lattice QCD.

We summarise the main features of the model.

- In order to obtain the area law for the Wilson loop it is necessary to assume that the field strengths $\mathbf{F}_{\mu\nu}$ rather than the potentials $\mathbf{A}_\mu$ have Gaussian correlators.

- The model yields confinement only for a non-Abelian gauge theory like QCD but not for QED. In order to obtain confinement in an Abelian theory monopoles must condense. For an Abelian theory without monopoles the inhomogeneous Maxwell equations imply $\kappa = 0$ and hence a vanishing string tension according to (6.65).

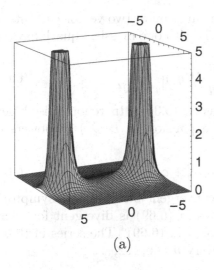 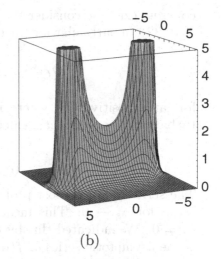

(a)   (b)

Figure 6.10.   Energy density in GeV fm^{-3} of a q$\bar{\text{q}}$ pair. The unit in the horizontal directions is the correlation length. The separation of the q$\bar{\text{q}}$ pair is 9 correlation lengths. (a) Colour-Coulomb contribution alone with $\alpha_s = 0.57$ and (b) including the nonperturbative contribution.

- In order to obtain the string tension (6.65) we have used the minimal surface bordered by the Wilson loop in evaluating the integral in (6.64). The assumption of vanishing higher cumulants is thus not independent of the choice of the surface.

- If the model is extended from a matrix factorisation to one in the colour components of the field strength (6.59) one can also calculate the product of two or more loops, and especially the density of the gauge invariant field strength squared $\sum_a (E_k^a)^2$, $\sum_a (B_k^a)^2$. One then obtains a string-like distribution of the field energy, figure 6.10(b). The total field energy obtained in this way is fully consistent with the potential and fixes $\alpha_s \approx 0.6$ in the nonperturbative region[275]. This implies that the scale which separates the perturbative from the nonperturbative regime is around 1 GeV.

6.8 Renormalons

In this section we discuss the relation between the perturbation series and some nonperturbative contributions, the so-called renormalon terms. This is also relevant for giving a more precise meaning to condensates. As a concrete example consider the correlator function of two vector currents as in (6.32), in particular for the case of QCD with one massless quark flavour,

$$D(Q^2) = -12\pi^2 q^2 \frac{d}{dq^2}\tilde{\Pi}(q^2)\Big|_{q^2=-Q^2} \tag{6.68}$$

for large positive Q^2, where $\tilde{\Pi}(q^2)$ is as in (6.35). In renormalised perturbation theory we can calculate the expansion of $D(Q^2)$ in powers of $\alpha_s = \alpha_s(Q)$:

$$D(Q^2) \equiv f(\alpha_s) \sim c_0 + c_1\alpha_s + c_2\alpha_s^2 + \cdots . \tag{6.69}$$

The series (6.69) is most probably not convergent but only an asymptotic series for $\alpha_s \to 0$. This means the series in (6.69) is divergent for every $\alpha_s \neq 0$. We indicated this by the symbol $\sim$ in (6.69). The series in (6.69) is the asymptotic series of $f(\alpha_s)$ if for every $n = 0, 1, 2, \ldots$

$$\lim_{\alpha_s \to 0} (\alpha_s)^{-n}[f(\alpha_s) - S_n(\alpha_s)] = 0 \tag{6.70}$$

where

$$S_n(\alpha_s) = \sum_{k=0}^{n} c_k\alpha_s^k \tag{6.71}$$

is the nth partial sum. The behaviour of c_k for large k is estimated[278] to be

$$c_k \propto k! . \tag{6.72}$$

For finite α_s the partial sums (6.71) typically give an increasingly better approximation of $f(\alpha_s)$ up to some value $\bar{n}$ which depends on α_s. For $n > \bar{n}$ the $S_n(\alpha_s)$ deviate more and more from $f(\alpha_s)$.

We note that $f(\alpha_s)$ and $f(\alpha_s) + Ce^{-c/\alpha_s}$, where $c > 0$, have the same asymptotic expansion for $\alpha_s \to 0$. Inserting the leading behaviour (6.20) of $\alpha_s(Q)$ for large Q we find

$$e^{-c/\alpha_s(Q)} \sim \left(\frac{\Lambda^2}{Q^2}\right)^{c\beta_0/(4\pi)} . \tag{6.73}$$

Thus, as we have stressed several times in this chapter, in principle perturbation theory can tell us nothing about power corrections, that is nonperturbative contributions.

Nevertheless, with the help of further hypotheses we can build a bridge between perturbative and some nonperturbative terms. For a review see [278]. Let us assume that $f(\alpha_s)$ can be represented as an integral

$$f(\alpha_s) = \frac{1}{\alpha_s}\int_0^\infty d\lambda\, \tilde{f}(\lambda)e^{-\lambda/\alpha_s}. \tag{6.74}$$

Figure 6.11.   Integration path in the Borel plane for (6.74)

Here $\tilde{f}(\lambda)$ is called the Borel transform of $f(\alpha_s)$ and the complex λ-plane is the Borel plane. To study the properties of the ansatz (6.74) we take a simple example which is encountered in the calculation below of $D(Q^2)$:

$$\tilde{f}_b(\lambda) = \frac{\Gamma(1+b)}{(1 - \lambda/\lambda_0)^{1+b}} \tag{6.75}$$

where $\lambda_0 > 0$ and b is a non-negative integer. We take the integration in (6.74) to run along the curve C as shown in figure 6.10, that is below the pole at $\lambda = \lambda_0$.

The function $\tilde{f}_b(\lambda)$ has a convergent power series expansion for $|\lambda| < \lambda_0$

$$\tilde{f}_b(\lambda) = \sum_{n=0}^{\infty} \lambda^n \frac{\Gamma(1+n+b)}{n!(\lambda_0)^n} . \tag{6.76}$$

On the other hand, inserting (6.75) into (6.74) and performing successive partial integrations gives

$$f_b(\alpha_s) = \sum_{k=0}^{n} \alpha_s^k \frac{\Gamma(1+k+b)}{(\lambda_0)^k} + R_n(\alpha_s) \tag{6.77}$$

$$R_n(\alpha_s) = (\alpha_s)^n \int_0^{\infty} d\lambda \, \tilde{f}_b^{(n+1)}(\lambda) e^{-\lambda/\alpha_s}. \tag{6.78}$$

From (6.77) we see easily that $f(\alpha_s)$ has the asymptotic expansion

$$f_b(\alpha_s) \sim \sum_{k=0}^{\infty} \alpha_s^k \frac{\Gamma(1+k+b)}{(\lambda_0)^k} \tag{6.79}$$

where the coefficients grow factorially. If we stop the expansion at the nth term, we have a remainder $R_n(\alpha_s)$ in (6.77). We can estimate $R_n(\alpha_s)$ from the contribution to the integral in (6.78) from the neighbourhood of the pole of $\tilde{f}_b^{(n+1)}(\lambda)$ as

$$R_n(\alpha_s) \propto e^{-\lambda_0/\alpha_s}. \tag{6.80}$$

We see from this simple example that the asymptotic series (6.79) leads us to expect a singularity at $\lambda = \lambda_0$ in the Borel plane which will yield a nonperturbative term in (6.77) proportional to $e^{-\lambda_0/\alpha_s}$. There will always

Figure 6.12.   Contributions to the renormalon in the Borel transform of the function D of (6.68)

be an ambiguity due to the arbitrary choice of the path shown in figure 6.11.

It is plausible that in QCD the situation is similar to what we found in our example. If in the two-loop contributions to the correlator function of two vector currents (6.68) the strong coupling g is replaced by the running coupling $g(k)$, where k^2 is the square of the momentum of the emerging gluon (see figure 6.12), one does indeed encounter[279] a singularity in the Borel plane at

$$\lambda_0 = 8\pi/\beta_0. \tag{6.81}$$

If the leading logarithmic contribution to the running coupling (6.20) is inserted,

$$\alpha_s(Q) = \frac{4\pi}{\beta_0} \frac{1}{\log(Q^2/\Lambda^2)} \tag{6.82}$$

the singularity is a pole. This pole on the positive real axis is called an infrared renormalon pole. Inserting the running coupling $\alpha_s(Q)$ and the value (6.81) for λ_0 into (6.80) one obtains a power correction proportional to Q^{-4}. It is clear from our example that this power correction corresponds to the nonperturbative term in the expansion (6.77) of $f(\alpha_s)$ and must in general depend on n. That is, the power-suppressed term depends on how many terms are included in the perturbative expansion. But we should stress that in general we have no guarantee that further nonperturbative terms, unrelated to the perturbation expansion, are absent. Since power suppressed terms are governed by condensates, see section 6.3, the contribution of resummed perturbative terms will influence the numerical values of these assumedly nonperturbative contributions. We have pointed out these ambiguities and the necessity of a truncation of the perturbative series in the discussion at the end of section 6.3.

The renormalon considerations give us confidence that the parametrisation of physical quantities with a partial sum from the perturbative expansion plus some power corrections makes sense. Such an analysis applies to quantities for which one can make an operator product expansion and thus one has a theory of power corrections, but also to other quantities. This is very useful in practice. Renormalon techniques have been used to calculate power suppressed corrections to high-energy scattering, for instance to the structure functions and to event shapes in jet production[280,281].

In QED one expects singularities on the negative λ-axis in the Borel plane (ultraviolet renormalons). In that case the integral in (6.74) can be performed without problems. This does not necessarily imply that the resulting finite expressions representing the summation of all orders of the renormalised perturbative expansion are sound and satisfactory from the physics point of view. This was investigated a long time ago in a soluble model field theory [282]. It was found that the summed renormalised perturbative expansion leads to a theory containing negative norm states which clearly is unacceptable. Simple arguments [283,284] and results of lattice simulations[285] suggest that also in pure QED the summed renormalised perturbative expansion will contain pathologies.

7

Hard processes

Modern high-energy physics may be said to have begun with the deep-inelastic electron-scattering experiments at the Stanford Linear Accelerator in the late 1960s. In these experiments, an electron is made to collide with a proton; the electron radiates a virtual photon γ^*, which strikes the proton and breaks it up, as is shown in figure 7.1. Similar experiments have since been performed, at progressively higher energies. Some of these instead have been with muon beams, and there have also been related experiments with neutrino beams. The highest energy achieved so far has been at the electron-proton collider HERA in Hamburg.

In this chapter we describe the phenomenology and the theory of deep-inelastic lepton scattering. The theory relates it to various other semi-hard processes and brings together perturbative QCD and Regge ideas.

7.1 Deep-inelastic lepton scattering

Figure 7.1 describes deep-inelastic electron or muon scattering. We see that, effectively, it explores the scattering of the virtual photon on the proton, so as to form a system X of hadrons. Factoring off a γ^* flux from figure 7.1, we may identify the remaining factor as the $\gamma^* p$ cross section. Figure 7.1 is well-defined but, as one may adopt different definitions of a γ^* flux, the precise definition of $\sigma^{\gamma^* p}$ is a matter of convention, except for the case of real photons; see (7.6). Whatever definition is used, the optical theorem relates the total $\gamma^* p$ cross section, summed over X, to the imaginary part of the forward virtual-Compton amplitude; see figure 7.2. It is conventional

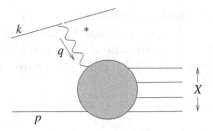

Figure 7.1. Electron-proton scattering: the electron radiates a virtual photon q which strikes the proton p and breaks it up, forming a system X of hadrons

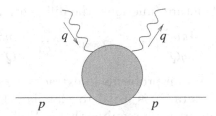

Figure 7.2. Forward virtual-Compton amplitude; see (7.3)

to introduce the variables

$$Q^2 = -q^2 \qquad \nu = p.q \qquad x = \frac{Q^2}{2\nu} \qquad y = \frac{p.q}{p.k}. \qquad (7.1)$$

Then $Q^2 \geq 0$ and $0 \leq x \leq 1$. The variable x was first introduced by Bjorken[286]. Many authors define instead $m\nu = p.q$, where m is the proton mass. The centre-of-mass $\gamma^* p$ energy is conventionally called W, so that W^2 plays the same role for $\gamma^* p$ scattering as s does for a purely hadronic process. In terms of the variables (7.1),

$$W^2 = Q^2 \left(\frac{1}{x} - 1 \right) + m^2. \qquad (7.2)$$

In most of the experiments so far, the proton is unpolarised, so the imaginary part of the $\gamma^* p$ forward amplitude must be averaged over the two possible proton polarisations r. It is conventional to define

$$W^{\mu\nu}(p,q) = \tfrac{1}{4} \sum_r \sum_X \langle p, r | J^\mu(0) | X \rangle \langle X | J^\nu(0) | p, r \rangle \, (2\pi)^3 \delta^4(p + q - P_X) \quad (7.3)$$

where J^μ is the electromagnetic current. This normalisation is such that the imaginary part of the forward Compton amplitude $T^{\mu\nu}(q, p; q, p)$ of

(2.63), averaged over proton polarisations and evaluated at $q^2 = 0$, is $2\pi W^{\mu\nu}(p, q)$. Because the electromagnetic interaction conserves parity, one finds[287,27] that $W^{\mu\nu}(p, q)$ may be decomposed into two independent amplitudes $F_1(x, Q^2)$ and $F_2(x, Q^2)$, called structure functions. They are conventionally defined by

$$W^{\mu\nu}(p, q) = - \left(g^{\mu\nu} + \frac{q^\mu q^\nu}{Q^2}\right) F_1(x, Q^2)$$

$$+ \frac{1}{\nu}\left(p^\mu + \frac{\nu}{Q^2} q^\mu\right)\left(p^\nu + \frac{\nu}{Q^2} q^\nu\right) F_2(x, Q^2). \qquad (7.4)$$

In terms of these structure functions, the differential cross section is

$$\frac{d^2\sigma}{dx\,dy} = \frac{4\pi\alpha^2}{xyQ^2}\left((1 - y)F_2 + xy^2 F_1 - \frac{m^2}{Q^2} x^2 y^2 F_2\right). \qquad (7.5)$$

Although there are certain problems in extracting $F_1(x, Q^2)$ and $F_2(x, Q^2)$ separately from the data, it is usual to plot data for $F_2(x, Q^2)$. Electromagnetic current conservation requires that $q_\mu W^{\mu\nu} = 0 = W^{\mu\nu} q_\nu$, with the consequence that, at fixed W, F_2 vanishes linearly with Q^2 when $Q^2 \to 0$. The real-photon cross section is

$$\sigma^{\gamma p} = \frac{4\pi^2\alpha}{Q^2} F_2 \bigg|_{Q^2=0}. \qquad (7.6)$$

For non-zero Q^2, the definition of the photon flux is ambiguous, but $\sigma^{\gamma^* p}$ is most simply defined by the same formula. Other definitions in common use[288,289] multiply (7.6) by functions of ν and Q^2 that are equal to 1 at $Q^2 = 0$. For non-zero Q^2, the photon may be polarised longitudinally as well as transversely. We may define cross sections $\sigma_T^{\gamma^* p}$ and $\sigma_L^{\gamma^* p}$ in terms of the transverse and longitudinal polarisation vectors, whose forms in the proton rest frame are given in (2.65). There is a similar ambiguity with the definitions of these cross sections, but the ambiguous virtual-photon flux cancels in their ratio

$$R = \frac{\sigma_L^{\gamma^* p}}{\sigma_T^{\gamma^* p}} = \frac{F_2(x, Q^2)}{F_1(x, Q^2)}\left(\frac{m^2}{\nu} + \frac{1}{2x}\right) - 1. \qquad (7.7)$$

With the convention of (7.6),

$$\sigma_T^{\gamma^* p} = \frac{4\pi^2\alpha}{\nu + 2xm^2} F_1$$

$$\sigma_L^{\gamma^* p} = \frac{4\pi^2\alpha}{2\nu x} F_L \qquad (7.8)$$

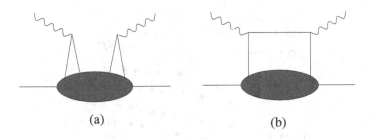

Figure 7.3. Parton-model description of figure 7.2. (b) is a part of (a), corresponding to the disconnected part of the lower amplitude. For large Q^2, the part (b) dominates.

where

$$F_L = F_2 - \frac{2\nu x}{\nu + 2xm^2} F_1. \tag{7.9}$$

The virtual photons in figure 7.2 couple to the hadronic component of the electromagnetic current, which is composed of contributions from each quark flavour and is given in (2.62). That is, each photon couples to a quark-antiquark pair as is shown in figure 7.3a. In the framework of the parton model, when Q^2 is sufficiently large the part of figure 7.3a that dominates is figure 7.3b, which involves just the disconnected part of the lower amplitude, so that both currents couple to the same quark flavour. This property survives perturbative-QCD corrections to the parton model. One writes

$$F_2(x, Q^2) = \tfrac{4}{9}(xu(x, Q^2) + x\bar{u}(x, Q^2)) + \tfrac{1}{9}(xd(x, Q^2) + x\bar{d}(x, Q^2))$$
$$+ \tfrac{1}{9}(xs(x, Q^2) + x\bar{s}(x, Q^2)) + \tfrac{4}{9}(xc(x, Q^2) + x\bar{c}(x, Q^2)) + \cdots. \tag{7.10}$$

The functions $u, \bar{u}, d, \bar{d}, \ldots$ are called quark and antiquark distribution functions. The factors x are conventionally included in their definitions for historical reasons: in the parton model the functions are interpreted as quark densities. The factors $\tfrac{4}{9}$ and $\tfrac{1}{9}$ are the squares of the corresponding quark and antiquark charges.

A striking discovery at HERA has been the rapid rise of $\sigma^{\gamma^* p}$ with W^2 at even quite small fixed values of Q^2; see figure 7.4. If one parametrises the rise as an effective power

$$\sigma(\gamma^* p) \sim F(Q^2)\, (W^2)^{\lambda(Q^2)} \tag{7.11}$$

then the power $\lambda(Q^2)$ is found to be significantly greater than the value that is familiar in purely hadronic collisions which, as we have seen in section 3.1,

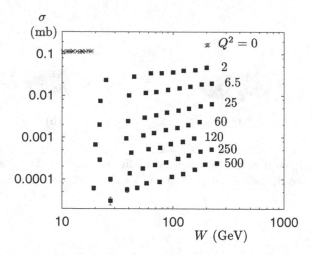

Figure 7.4. Data[290] for $\sigma^{\gamma^* p}$, defined by (7.6), at various values of Q^2, together with the real-photon data of figure 5.1

is just less than 0.1. The value of $\lambda(Q^2)$ has been extracted[291] by the H1 collaboration from their data and is shown in figure 7.5, from which it can be seen to increase with Q^2 and reach about 0.4 at the highest values that have been measured.

When $W^2 \gg Q^2$, x is small and, according to (7.2), $W^2 \sim Q^2/x$. Then the effective-power behaviour (7.11) corresponds to

$$F_2(x, Q^2) \sim f(Q^2)\, x^{-\lambda(Q^2)}. \tag{7.12}$$

When they extracted $\lambda(Q^2)$ from their data, to make the plot of figure 7.5, H1 assumed that the value of $\lambda(Q^2)$ at small x is independent of x at each Q^2. While the data are compatible with this assumption, they do not require it and it does not have theoretical justification. In particular, the two-pomeron model described in section 7.4 has significant variation of $\lambda(Q^2)$ with x, as is seen in figure 7.6.

Even before the HERA measurements, there were predictions[6,222] that $\lambda(Q^2)$ would be large at high values of Q^2. Such predictions arose from two different equations of perturbative QCD: the DGLAP equation and the BFKL equation. In the following we review the application of these equations and explain why it has since been realised that the predictions are not as clean as initially had been hoped.

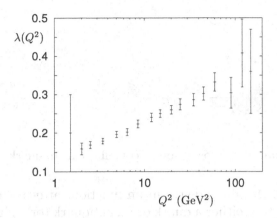

Figure 7.5. The effective power $\lambda(Q^2)$ of (7.11) extracted from H1 data[291]

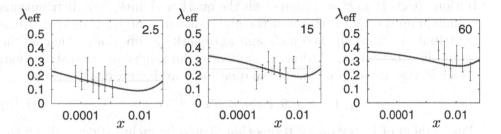

Figure 7.6. Data[291] for the effective power of $1/x$ at $Q^2 = 2.5$, 15 and 60 GeV^2. The horizontal lines correspond to the values plotted in figure 7.5 and the solid lines to the two-pomeron fit described in section 7.4.

7.2 The DGLAP equation

The DGLAP equation is named after its authors: Dokshitzer, Gribov, Lipatov, Altarelli and Parisi. A full account of it may be found in textbooks, for example [287,222]. The equation is derived from perturbative QCD and therefore is valid at sufficiently large Q^2. It enables one to calculate how the quark and antiquark distribution functions vary with Q^2 at each value of x, at least provided that x is not too small. Examples of lowest-order contributions to the Q^2 evolution of quark and gluon distributions are shown in figure 7.7. The first diagram describes a quark or an antiquark emerging from the proton and radiating a gluon before it absorbs the virtual photon. The second diagram shows a gluon emerging from the proton and changing into a quark-antiquark pair; one of the pair then absorbs the photon. This second type of process brings in another distribution function, the gluon distribution function $g(x, Q^2)$. The third diagram contributes to the

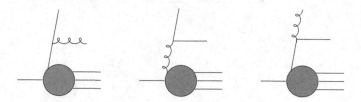

Figure 7.7.   Examples of lowest-order contributions to quark and gluon evolution

Q^2 variation of the gluon distribution function: after the emission from the proton of a parton, either a quark or an antiquark (or a gluon), it undergoes perturbative interactions and so transforms into the gluon under study.

When Q^2 is much larger than all the relevant quark masses, the gluon distribution affects the Q^2 variation of all the quark and antiquark distributions equally, and the Q^2 variation of the gluon distribution itself receives equal contributions from all the quark and antiquark distributions. That is, the gluon distribution's evolution is coupled to the singlet quark distribution, which is the sum of the sum of the quark and antiquark distributions:

$$q(x, Q^2) = u + \bar{u} + d + \bar{d} + s + \bar{s} + c + \bar{c} + \cdots. \qquad (7.13)$$

The number of heavy-quark terms that should be included depends on the value of W. So the relevant form of the DGLAP equation introduces the two-component quantity

$$\mathbf{u}(x, Q^2) = \begin{pmatrix} q(x, Q^2) \\ g(x, Q^2) \end{pmatrix}. \qquad (7.14)$$

It reads

$$Q^2 \frac{\partial}{\partial Q^2} \mathbf{u}(x, Q^2) = \int_x^1 dz\, \mathbf{P}(z, \alpha_s(Q^2))\, \mathbf{u}(x/z, Q^2). \qquad (7.15)$$

Here, $\mathbf{P}(z, \alpha_s(Q^2))$ is a 2×2 matrix, called the splitting matrix; it is calculated from perturbative QCD.

In order to apply the DGLAP equation, one should first choose some starting value Q_0^2 of Q^2 and fit the parton distributions there to experimental data, as functions of x. These functions cannot be calculated from perturbative QCD, but the DGLAP equation, together with assumptions about the form of the gluon distribution, determines how they change as Q^2 increases. The resulting agreement with experiment at higher values of Q^2 is impressive[292,293]. If[294] one starts at some fairly small value Q_0^2 of Q^2 with parton distributions that are rather flat in x at small x, then an application of the DGLAP equation results in distributions that rapidly become steeper as Q^2 increases, in good agreement with experiment. However, this application is controversial. It does not seem safe to apply a perturbative equation for choices of Q_0^2 as small as 1 GeV^2 or less, as is sometimes done. At the time of writing, the use of the DGLAP equation at small x is the subject of much discussion. It is necessary at present to expand the elements of the splitting matrix $\mathbf{P}(z, \alpha_s(Q^2))$ in powers of α_s and to truncate the expansions after two terms. The experimental data fit quite well[295] with the double-logarithmic rise

$$F_2(x, Q^2) \sim \exp\left(\sqrt{\frac{48}{\beta_0} \log \frac{1}{x} \log \log \frac{Q^2}{Q_0^2}}\right) \qquad (7.16)$$

calculated long ago[296] from the lowest-order approximation. Here, β_0 is the first coefficient of the QCD β-function; see (6.19). One can understand that (7.16) may be a good numerical approximation to $F_2(x, Q^2)$ when x is small but not too small. However, it is surely unsafe when x is very small, as we explain below.

7.3 The BFKL equation

The BFKL equation is named after its authors: Balitsky, Fadin, Kuraev and Lipatov[297,298]. A full account of it may be found in [6].

The BFKL equation is not applicable unless x is small and Q^2 large but not too large. Originally, it was hoped that with it one might use perturbative QCD to calculate the effective power $\lambda(Q^2)$ that parametrises the rise (7.11) or (7.12) of $F_2(x, Q^2)$ at small x. It gave values of $\lambda(Q^2)$ close to $\frac{1}{2}$, which is approximately what experiment finds at high Q^2. This led to talk of a second pomeron, the hard pomeron, with a $\lambda(Q^2)$ that might be calculated from perturbative QCD. However, it has since been realised that this calculation is almost certainly invalid. Apart from concerns that the approximations used to derive the BFKL equation do not take sufficient account of energy conservation[299], it is not correct to suppose that the equation enables $\lambda(Q^2)$ to be calculated from perturbative QCD alone[300]. While it is conceivable[301,302] that there is no such problem with purely-hard processes such as $\gamma^*\gamma^*$ collisions at very high Q^2, for semi-hard collisions such as γ^*p the BFKL equation inevitably receives important contributions from uncalculable nonperturbative effects.

The most primitive form of the BFKL equation is a lowest-order version, which also takes a fixed coupling α_s and ignores quarks. It reads

$$x\frac{\partial}{\partial x}g(x, Q^2) = x\frac{\partial}{\partial x}g^{(0)}(x, Q^2) + \alpha_s \int dk^2 \, K(Q^2, k^2)g(x, k^2). \qquad (7.17)$$

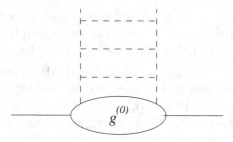

Figure 7.8. Generalised gluon ladders summed by the BFKL equation

Crudely speaking, the equation inputs a nonperturbative distribution $g^{(0)}$ and erects on top of it a sum of gluon ladders with $1, 2, 3, \ldots$ rungs: see figure 7.8. Its complete description is very much more complicated than this; for example, the vertical lines in figure 7.8 are actually not elementary gluons, but are rather Reggeised so as to sum all terms that have leading behaviour in $\alpha_s \log s$.

The function $K(Q^2, k^2)$ is called the BFKL kernel. Because it does not depend on any fixed mass scale its eigenfunctions are simple powers:

$$\int dk^2 \, K(Q^2, k^2)(k^2)^M = K(M)(Q^2)^M \qquad (7.18)$$

and hence (7.17) simplifies if we apply a Mellin transform to it. Define the double Mellin transform with respect to x and Q^2

$$g(N, M) = \int_0^1 dx \, x^{N-1} \int_0^\infty dQ^2 \, (Q^2)^{-M-1} g(x, Q^2) \qquad (7.19)$$

so that

$$g(x, Q^2) = \int \frac{dN}{2\pi i} x^{-N} \int \frac{dM}{2\pi i} (Q^2)^M g(N, M) \qquad (7.20)$$

where the N and M integrations are along lines in the complex planes of the two variables, in each case parallel to the imaginary axis. The initial locations of these lines must be chosen such that the x and k^2 integrations in (7.19) converge. For the M integration this depends on the details of the behaviour of $g(x, Q^2)$ for both small and large Q^2, but for the N integration one may initially take a line that has Re N large and positive; this is because, while the k^2 integration in (7.19) has no upper limit, x is integrated only up to 1. We cannot simply insert the representation (7.20) into the BFKL equation (7.17), because the BFKL equation is applicable only at small x while it is evident from (7.19) that the definition of $g(N, M)$ involves all values of x. So we must first separate $g(N, M)$ into a part described by the BFKL equation and a remainder that is important only when x is not small:

$$g(N, M) = g^{\mathrm{BFKL}}(N, M) + g^{\mathrm{remainder}}(N, M). \qquad (7.21)$$

In particular, the rightmost singularity in the complex N-plane, which gives the dominant behaviour of $g(x, Q^2)$ at small x, is present in $g^{\mathrm{BFKL}}(N, M)$ but not in $g^{\mathrm{remainder}}(N, M)$. Then

$$N g^{\mathrm{BFKL}}(N, M) = N g^{(0)}(N, M) - \alpha_s K(M) \, g^{\mathrm{BFKL}}(N, M). \qquad (7.22)$$

Evaluating $K(M)$ to lowest order in α_s gives[6]

$$K(M) = \frac{3}{\pi} \Big(2\psi(1) - \psi(M) - \psi(1 - M) \Big) \qquad (7.23)$$

where $\psi(M) = \Gamma'(M)/\Gamma(M)$. Unfortunately, this expression for $K(M)$ receives[303,304] a huge correction in next-to-leading order in α_s. It is not known beyond this order, but one must strongly suspect that it is illegal to expand it as a power series in α_s. Work is in progress to try to overcome this problem[305–307]. At the very least, some resummation is needed, but there is as yet no consensus about what is the right approach. If it were valid simply to ignore the corrections to (7.23), a simple calculation[6] would give for the effective power defined in (7.11)

$$\lambda(Q^2) = (12\alpha_s/\pi) \log 2. \qquad (7.24)$$

The calculation is too simple for a number of reasons[299], if only because it supposes that the coupling α_s does not run with Q^2. Nevertheless, with a reasonable value of α_s it gives a value of $\lambda(Q^2)$ between 0.4 and 0.5. As we explain in section 7.4, the data show the need for a contribution to $F_2(x, Q^2)$ corresponding to such a value, but because of all the problems with the BFKL equation this is probably a coincidence.

If one similarly ignores quarks in the DGLAP equation (7.15) and defines

$$P_{gg}(N, \alpha_s) = \int_0^1 dz \, z^N P_{gg}(z, \alpha_s) \qquad (7.25)$$

the equation becomes

$$M g(N, M) = P_{gg}(N, \alpha_s) \, g(N, M). \qquad (7.26)$$

While initially the N integration in (7.20) is to be taken along a line in the complex N-plane parallel to the imaginary axis with $\mathrm{Re}\,N$ large positive, we may move this line to the left until we encounter the rightmost singularity

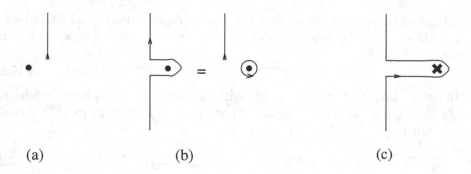

(a)   (b)   (c)

Figure 7.9. N-plane integration contour: (a) initially to the right of a pole; (b) moved past the pole; (c) moved past a branch point

of $g(N, M)$. If the singularity is at $N = \epsilon$ and is a simple pole, as in figure 7.9a, we may move it past the pole and pick up the pole residue, figure 7.9b. In this case, the leading behaviour of $g(x, Q^2)$ at small x is a simple power $x^{-\epsilon}$, multiplied by some function of Q^2. If the singularity is a branch point, moving the contour past it picks up a discontinuity across the branch cut attached to the branch point, figure 7.9c, and the power of x is divided by some function of $\log x$ whose precise form depends on the nature of the branch point. For values of M and N at which $g^{\mathrm{BFKL}}(N, M)$ diverges, the BFKL equation (7.22) tells us that

$$N = -\alpha_s K(M). \tag{7.27}$$

This is true if the divergence is generated by perturbation theory, so that $g^{(0)}$ is not divergent, and it does not matter whether it is a pole or a more complicated singularity. Because $g(M, N)$ diverges also at these values of M and N, the DGLAP equation (7.26) gives

$$M = P_{gg}(N, \alpha_s). \tag{7.28}$$

Putting these two conditions together, we find[308]

$$N = -\alpha_s K(P_{gg}(N, \alpha_s)). \tag{7.29}$$

With the form (7.23) for $K(M)$, this gives a splitting function $P_{gg}(N, \alpha_s)$ for which the first two terms of an expansion in powers of α_s are usually used in phenomenological applications[292–294].

But these two terms, and all subsequent terms in the expansion in powers of α_s, are singular at $N = 0$, and this apparently leads to the small-x dependence of $xg(x, Q^2)$ rapidly becoming very steep as Q^2 increases. However,

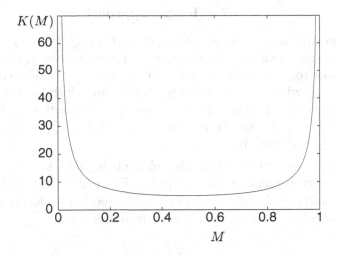

Figure 7.10.   The function $K(M)$ of (7.23)

the expansion is illegal near $N = 0$: it is fairly easy to establish that the solution of (7.29) gives a splitting function $P_{gg}(N, \alpha_s)$ which is not singular at $N = 0$. The function $K(M)$ is plotted in figure 7.10. To solve (7.29) we need the value(s) of M at which a line of height $-N/\alpha_s$ intersects the curve; when $N = 0$ this gives a pair of values that are complex, but finite. Mathematically, the solution has properties similar to those of the function

$$N - \sqrt{N^2 - \alpha_s} \qquad (7.30)$$

for which the expansion in powers of α_s is

$$\frac{\alpha_s}{2N} - \frac{\alpha_s^2}{8N^3} + \cdots . \qquad (7.31)$$

Every term in the expansion is singular at $N = 0$, but the expansion is valid only for $N^2 > \alpha_s$ and the complete function is not singular at $N = 0$. The lesson is that making the expansion is dangerous, and if possible one should use the complete solution[309] to the equation (7.29) for $P_{gg}(N, \alpha_s)$. However, (7.29) is valid only if the coupling α_s is taken to be fixed, when really it should be made to run with Q^2. More seriously, the expression (7.23) for $K(M)$ is correct only up to the first term of an expansion in powers of α_s, and this expansion too is dangerous because the next-to-leading term in the expansion is huge[303,304]. Work on these difficulties is continuing[310–312]. However, as we show in section 7.6, if one combines the DGLAP equation (7.26) with the Regge approach described in the next section, an expansion of $P_{gg}(N, \alpha_s)$ in powers of α_s does become useful for the calculation of the evolution at small x.

7.4 Regge approach

A powerful approach to the small-x data for $F_2(x, Q^2)$ is to extend the Regge phenomenology developed in chapter 3. Regge theory is not supposed to be a competitor with perturbative QCD, but to coexist with it, and to be applicable when W^2 is much greater than all the other variables, in particular when it is much greater than Q^2. According to (7.2), when x is small $W^2 \sim Q^2/x$, so the condition that $W^2 \gg Q^2$ is just that $x \ll 1$, however large Q^2 may be.

The simplest assumption is that the relevant singularities in the complex angular momentum plane are simple poles. From (2.3) and (7.19) the Mellin-transform variable N introduced in the last section is closely related to the complex angular momentum l: the correspondence is

$$N \leftrightarrow l - 1 \tag{7.32}$$

so that this assumption is equivalent to assuming that the relevant singularities in the complex N-plane are simple poles. This assumption may be not literally correct, but it turns out to give an excellent description of the small-x data[194]. Because of the kinematic factors involving ν that multiply F_2 in its definition (7.4) in terms of $W^{\mu\nu}$, it may be shown that a Regge trajectory $\alpha(t)$ contributes a power $(W^2)^{\alpha(0)-1}$ or $x^{1-\alpha(0)}$ to F_2. The argument is similar to that in section 2.8. So, according to the fits of section 3.1, we expect to find terms $x^{-0.08}$ and $x^{0.45}$ in the small-x behaviour of F_2, corresponding to soft-pomeron and reggeon exchange. But these are not sufficient to describe the rapid rise with $1/x$ seen in the data at small x and large Q^2. Another term $x^{-\epsilon_0}$ is needed, with $\epsilon_0 \approx 0.4$. We call this N-plane or l-plane singularity the hard pomeron. This does not explain what is its dynamical origin: maybe it is perturbative QCD, though, as we explained in section 7.3, initial hopes that it might be derived from the BFKL equation now seem unlikely to be realised.

So the simplest fit to the small-x data corresponds to

$$F_2(x, Q^2) \sim \sum_{i=0,1,2} f_i(Q^2)\, x^{-\epsilon_i}. \tag{7.33}$$

Here, the $i = 0$ term is hard-pomeron exchange, $i = 1$ is soft-pomeron exchange, and $i = 2$ is (f_2, a_2) exchange. Only even-C-parity exchanges are allowed. Regge theory gives no information about the form of the three coefficient functions $f_i(Q^2)$, beyond that they are analytic functions. Also we have seen that, because $\sigma^{\gamma p}$ is given by (7.6), at fixed W the function F_2 vanishes linearly with Q^2 at $Q^2 = 0$. This is guaranteed by gauge

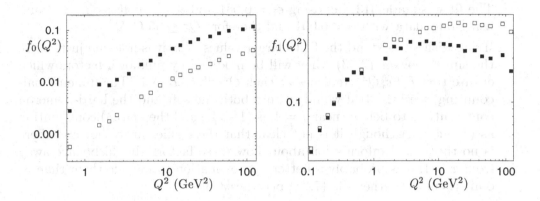

Figure 7.11. Fits to the coefficient functions $f_0(Q^2)$ and $f_1(Q^2)$ of (7.33) extracted from H1 and ZEUS data. The black points are for $\epsilon_0 = 0.36$ and the white points are for $\epsilon_0 = 0.5$.

invariance. If we assume that each term in (7.33) has this property, then

$$f_i(Q^2) \sim (Q^2)^{1+\epsilon_i} \tag{7.34}$$

near $Q^2 = 0$. One might hope that perturbative QCD will determine how the three coefficient functions behave at high Q^2, but we explain in section 7.6 that so far the theoretical difficulties described in the last section allow this only for the hard-pomeron coefficient function $f_0(Q^2)$. Figure 7.11 shows[313] how f_0 and f_1 vary with Q^2 if we fit (7.33) to the data at each Q^2, including values of x up to 0.02. This provides a guide on the likely functional forms of the coefficient functions. The $f_i(Q^2)$ can then be parametrised with suitable functions, whose shapes resemble the data points in the plots and which include a number of parameters. These parameters and the power ϵ_0 may then be varied[313] so as to give the best fit to all the small-x data for $F_2(x, Q^2)$ together with the data for $\sigma^{\gamma p}$. This results in the value

$$\epsilon_0 = 0.437 \tag{7.35}$$

with

$$f_0(Q^2) = 0.0015 \frac{(Q^2)^{1+\epsilon_0}}{(1 + Q^2/9.11)^{1+\frac{1}{2}\epsilon_0}} \tag{7.36}$$

and

$$f_1(Q^2) = 0.60 \left(\frac{Q^2}{1 + Q^2/0.59} \right)^{1+\epsilon_1}. \tag{7.37}$$

The fit was made[313] imposing $\epsilon_1 = 0.081$ and $\epsilon_2 = -0.45$ as in (3.6) and using only data with $x < 0.001$ and therefore $Q^2 \leq 35$ GeV^2.

If one wishes to extend the fit to larger values of x, it is not safe just to use the simple powers (7.33): they will be modified by unknown factors which ensure that $F_2(x, Q^2)$ vanishes at each Q^2 when $x \to 1$. The dimensional-counting rules[314,315] would require both the soft and the hard-pomeron contributions to behave near $x = 1$ as $(1-x)^7$, and the (f_2, a_2) contribution as $(1-x)^3$. Although it is not clear that these rules are valid, and there is no theoretical information about how these factors should behave away from $x = 1$, it is probably a better approximation to include them than to omit them altogether. So (7.33) becomes[313]

$$F_2(x, Q^2) = f_0(Q^2)\, x^{-\epsilon_0}(1-x)^7 + f_1(Q^2)\, x^{-\epsilon_1}(1-x)^7 + f_2(Q^2)\, x^{-\epsilon_2}(1-x^2)^3.$$
(7.38)

If we expand the factor $(1 - x^2)$ we find terms $x^{-(\epsilon_2 - 2)}$, $x^{-(\epsilon_2 - 4)}$, ... which are what we expect from meson daughter trajectories; see figure 2.13. Like-wise, expanding the factors $(1 - x)^7$ gives contributions which we interpret as resulting from daughters of the pomerons. Figure 7.12 shows that then, although the parameters were fixed using only data with $x < 0.001$ and $Q^2 \leq 35$ GeV^2, the fit is surprisingly good out to much larger x and all the way from $Q^2 = 0$ to 5000 GeV^2. The choice of $(1 - x^2)^3$ for the (f_2, a_2) term, rather than simply $(1 - x)^3$, is guided[313] by data[316] for the difference $F_2^p(x, Q^2) - F_2^n(x, Q^2)$ between the structure functions for a proton and a neutron target, which corresponds just to a_2 exchange.

Having concluded that the data for F_2 require a hard-pomeron component, it is necessary to test this with other data. The hard pomeron is seen clearly in the charm component of F_2. This describes events in which a D^* particle is produced, which are used to extract the contribution $F_2^c(x, Q^2)$ to the complete $F_2(x, Q^2)$ from events where the γ^* is absorbed by a charmed quark. That is, they measure the part $F_2^c(x, Q^2)$ of $F_2(x, Q^2)$ corresponding to the c and $\bar{c}$ terms of (7.10). The data[318] for $F_2^c(x, Q^2)$ must be treated with some caution because the experimentalists have to make a very large extrapolation to compensate for limited acceptance. Nevertheless the data have the striking property that, over a wide range of Q^2, they behave as a fixed power of x:

$$F_2^c(x, Q^2) = f_c(Q^2) x^{-\epsilon_0}$$
(7.39)

with $\epsilon_0 \approx 0.4$: see figure 7.13. It seems, therefore, that $F_2^c(x, Q^2)$ at small x receives a contribution from the hard pomeron and that the soft-pomeron contribution to it is negligibly small. This is a natural consequence of certain dipole models described in chapter 9. Even more surprisingly, the data are consistent with the hard-pomeron contribution to $F_2(x, Q^2)$ be-

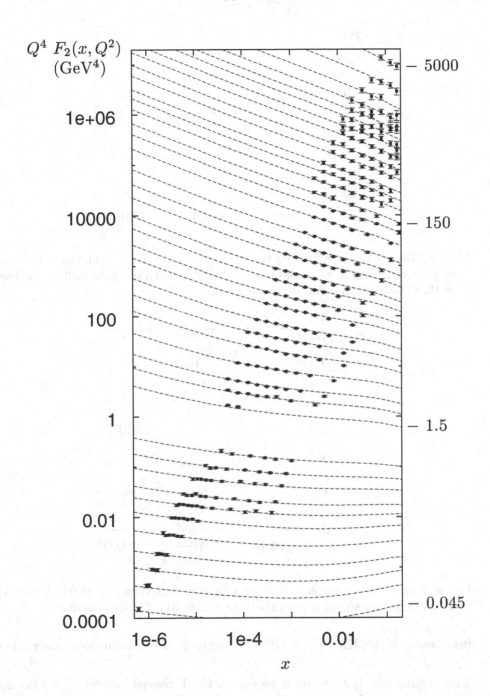

Figure 7.12. Regge fit to data for $F_2(x, Q^2)$ for Q^2 between 0.045 and 5000 GeV^2. The parameters were fixed using only data for $x < 0.001$ and therefore $Q^2 \leq 35$ GeV^2.

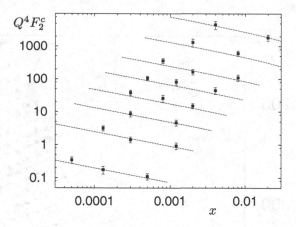

Figure 7.13. Data for $F_2^c(x, Q^2)$ from $Q^2 = 1.8$ to 130 GeV^2. The curves are the hard-pomeron component of the fit to $F_2(x, Q^2)$ shown in figure 7.12, normalised such that the hard pomeron is flavour-blind.

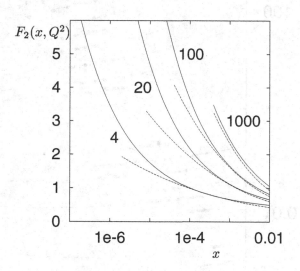

Figure 7.14. Two-pomeron fit to $F_2(x, Q^2)$ at various values of Q^2 (thick curves), with two-loop unresummed perturbative-QCD fit[317] (broken curves)

ing flavour-blind[313]: $F_2^c(x, Q^2)$ is close to $\frac{2}{5}$ of the hard-pomeron part of $F_2(x, Q^2)$.

The arguments that are used to derive the Froissart bound (1.68) do not apply to the real and virtual-Compton amplitudes. Nevertheless, many authors have invoked the bound to deduce that some sort of saturation must occur at small x, that is the DGLAP evolution equation (7.26) receives an

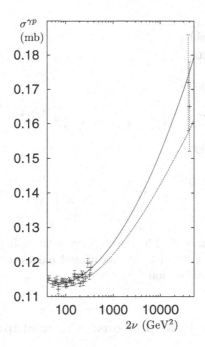

Figure 7.15. Data for $\sigma^{\gamma p}$. The upper curve is the extrapolation to $Q^2 = 0$ of the curves in figure 7.12; the lower curve omits the hard-pomeron term.

additional quadratic term[319], as we discuss in section 9.4. It is sometimes argued that there is a limit on how large the parton density can become within a finite radius of the proton, but we have already explained at the end of section 3.1 that this radius can grow without limit and, in connection with figure 3.3, that there is an ambiguity over what should be thought of as being "inside" the proton and what rather should be viewed as part of an exchange. So we maintain that, while saturation very possibly may occur, this is not a consequence of the Froissart bound, as we have explained in section 2.8, and there is no reason to suppose that the fits (7.38) and (7.39) break down at any experimentally-accessible values of x.

For the ranges of x and Q^2 where they overlap, the two-pomeron fit and conventional fits based on two-loop unresummed perturbative QCD agree with the data equally well. However, they no longer agree when they are extrapolated to smaller values of x, as is seen in figure 7.14.

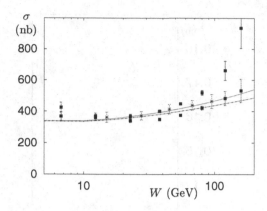

Figure 7.16. Data for $\sigma^{\gamma\gamma}$. The lower curve is the soft-pomeron/reggeon curve of figure 5.3; the upper curve has an additional hard-pomeron term. The data are explained at the end of section 5.1.

7.5 Real photons: a crucial question

Because the data of figure 7.12 include a point at extremely low Q^2, it should be reliable to extrapolate these data to $Q^2 = 0$. This extrapolation is the upper curve shown in figure 7.15 and, at $\sqrt{s} = 200$ GeV, is about 20 µb higher than the curve shown in figure 5.1. That is, at this energy, the fits with and without a hard-pomeron component differ by about 10%. The errors shown on the data in figure 7.12a are purely statistical; there is an additional systematic error of about 10% at the lowest Q^2. So at present it is not possible to decide whether the hard pomeron is present in the $\sigma^{\gamma p}$ data, though it seems likely.

There is also uncertainty with the data for $\sigma^{\gamma\gamma}$ from LEP (see section 7.7). We have seen figure 5.3 with data from the L3 and OPAL collaborations and the prediction obtained by applying factorisation to the $\sigma^{\gamma p}$ and σ^{pp} curves in figures 5.1 and 3.1, with no hard-pomeron component. The data from L3, particularly the upper set, may require such a component. Figure 7.16 shows an example of how adding it in might change the fit. The hard-pomeron component is the $Q^2 \to 0$ limit of a fit to the photon structure function F_2^γ described in section 7.7.

This is an important question. Is the hard pomeron already present at $Q^2 = 0$, or is it rather generated by perturbative evolution? If it is there already at $Q^2 = 0$ it presumably arises because the photon has a pointlike component, since if it is present in the $\bar{p}p$ total cross section it is very small.

7.6 Perturbative evolution

We saw in section 7.3 that the conventional approach to perturbative evolution of the proton structure function $F_2(x, Q^2)$ breaks down at small x. We also saw in section 7.4 that at small x Regge theory with the inclusion of a second pomeron, the hard pomeron, gives a very good description of $F_2(x, Q^2)$ and of its charm component $F_2^c(x, Q^2)$. The Regge approach and that of perturbative evolution are very different. We now show that perturbative evolution governs how the hard pomeron's contribution to the structure function increases with Q^2. It is found[320] that the parametrisation of the hard-pomeron coefficient function $f_0(Q^2)$ of (7.38) agrees very well with what is obtained from DGLAP evolution, over a large range of Q^2. As yet, we are not able to apply perturbative evolution to the soft pomeron.

First write the singlet DGLAP equation (7.15) as

$$\frac{\partial}{\partial t}\mathbf{u}(x, Q^2) = \int_x^1 dz\, \mathbf{P}(z, \alpha_s(Q^2))\, \mathbf{u}(x/z, Q^2) \tag{7.40}$$

where $\mathbf{P}$ is the splitting matrix as before, $t = \log(Q^2/\Lambda^2)$ and $\mathbf{u}(x, Q^2)$ is defined in (7.14). Write the Mellin transform with respect to x

$$\mathbf{u}(N, Q^2) = \int_0^1 dx\, x^{N-1}\mathbf{u}(x, Q^2) \tag{7.41}$$

and define

$$\mathbf{P}(N, \alpha_s(Q^2)) = \int_0 dz\, z^N \mathbf{P}(z, \alpha_s(Q^2)). \tag{7.42}$$

Then

$$\frac{\partial}{\partial t}\mathbf{u}(N, Q^2) = \mathbf{P}(N, \alpha_s(Q^2))\, \mathbf{u}(N, Q^2). \tag{7.43}$$

A power contribution $f(Q^2)x^{-\epsilon_0}$ to $F_2(x, Q^2)$ corresponds to a pole

$$\frac{\mathbf{f}(Q^2)}{N - \epsilon_0} \qquad \mathbf{f}(Q^2) = \begin{pmatrix} f_q(Q^2) \\ f_g(Q^2) \end{pmatrix} \tag{7.44}$$

in $\mathbf{u}(N, Q^2)$. With four active quark flavours and a flavour-blind hard pomeron, $f_q(Q^2) = \frac{18}{5}f_0(Q^2)$. We find[321], on taking the residue of the pole at $N = \epsilon_0$ on each side of the Mellin transform (7.43) of the DGLAP equation,

$$\frac{\partial}{\partial t}\mathbf{f}(Q^2) = \mathbf{P}(N = \epsilon_0, \alpha_s(Q^2))\, \mathbf{f}(Q^2). \tag{7.45}$$

If we include four flavours of quark and antiquark in the sum in (7.13), then at $Q^2 = 20\,\text{GeV}^2$ the singlet quark distribution $x\sum_f(q_f + \bar{q}_f) \sim 0.095x^{-\epsilon_0}$ at

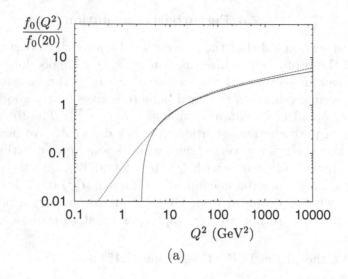

(a)

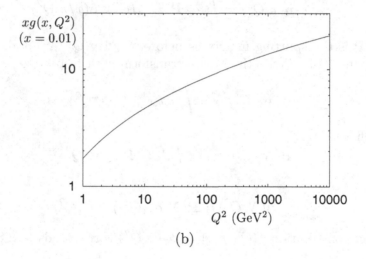

(b)

Figure 7.17. (a) Next-to-leading-order evolution with $\Lambda = 400$ MeV of the hard-pomeron coefficient $f_0(Q^2)$ (solid curve) and the fit (7.36) to the data (broken curve). (b) Evolution of the gluon structure function $xg(x, Q^2)$ at $x = 0.01$.

sufficiently small x. We showed in section 7.4 that the charmed-quark component F_2^c of F_2 is apparently governed almost entirely by hard-pomeron exchange at small x, even at small values of Q^2, and that, within the experimental errors, its magnitude is consistent with the hard pomeron being flavour-blind. According to perturbative QCD, the charmed quark origi-

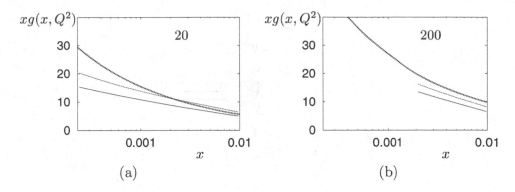

Figure 7.18. Gluon structure function $xg(x, Q^2)$ at (a) $Q^2 = 20$ and (b) 200 GeV^2. In each case the thick line is the distribution as described in the text and the thin lines are the limits extracted by conventional next-to-leading-order analysis of HERA data[328,324].

nates from a gluon in the proton, and[322] the two distributions are proportional to each other to a good approximation over a wide range of x and Q^2. This implies that the gluon distribution also is hard-pomeron dominated. The conventional approach is correct if x is not too small, because it does not then probe the splitting matrix at $z = 0$, which is where its expansion in powers of α_s is illegal. We assume that at $x = 0.01$ and $Q^2 = 20$ GeV^2 the value of $g(x, Q^2)$ extracted by the HERA experiments is reasonably close to the correct value. At $Q^2 = 20$ GeV^2 and $x = 0.01$, a next-to-leading-order fit[323,324] to the combined ZEUS and H1 data gives $xg(x, Q^2) = 5.7 \pm 0.7$. Other authors[325–327] find much the same value. This is 8 ± 1 times the hard-pomeron component of the singlet quark distribution.

Although an unresummed perturbation expansion of the splitting matrix $\mathbf{P}(N, \alpha_s)$ is not valid[321] for small values of N, we need $\mathbf{P}(N, \alpha_s)$ at $N = 0.437$, that is far from 0, so it is reasonable to hope that resummation is not needed. The numerical values of the elements of the splitting matrix $\mathbf{P}(N, \alpha_s)$ are known[222] in one and two-loop order, so it is straightforward to integrate (7.45). At the energies being considered it is necessary to use four flavours throughout as the charm contribution is active. The beauty contribution is so small that its omission has a negligible effect.

The result of integrating the differential equation (7.45) for the singlet quark distribution is shown in figure 7.17a, where the solid curve is the result of the two-loop-order perturbative QCD evolution according to (7.45) and the broken curve is the Regge fit to the data described in section 7.4. The ratio of the gluon distribution to the hard-pomeron component of the singlet quark distribution is taken to be 8.0 at $Q^2 = 20$ GeV^2. Figure 7.17b shows

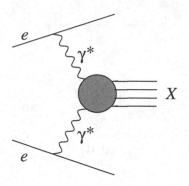

Figure 7.19. $\gamma^*\gamma^*$ scattering

how the gluon distribution

$$xg(x, Q^2) = f_g(Q^2)x^{-\epsilon_0} \tag{7.46}$$

evolves. Provided that one chooses Λ such that $\alpha_s(M_z^2) = 0.116$, which is the HERA value[323,324], there is little difference between the leading-order and next-to-leading-order results.

As we have seen in section 7.2, the conventional approach to evolution expands the splitting matrix $\mathbf{P}(N, \alpha_s)$ in powers of α_s. Because an unre-summed expansion that needs the splitting matrix at small N makes the splitting function larger than it really is, a gluon distribution of a given magnitude apparently gives stronger evolution than it really should. That is, the conventional approach will tend to under-estimate the magnitude of $xg(x, Q^2)$ in certain regions of (x, Q^2) space. This is verified by the results for the evolution of $xg(x, Q^2)$ obtained from integrating (7.45). Figure 7.18 shows the proton's gluon structure function at two values of Q^2, according to the solution of (7.45), which does not use the splitting matrix at small N, and compares it with what is extracted from the data by conventional means.

We cannot apply a similar analysis to the soft pomeron, because this would need the splitting matrix $\mathbf{P}(N, \alpha_s)$ at $N = 0.08$, which is too small for a expansion in powers of α_s to be meaningful. As we have explained in section 7.4, we do not yet know how to perform the necessary resummation.

7.7 Photon-photon interactions

In high-energy e^+e^- interactions some of the electrons and positrons do not annihilate or simply scatter electromagnetically, but are scattered by radiating virtual photons which then interact to produce a hadronic final state X as shown in figure 7.19. That is, the reaction is $e^+e^- \to e^+e^-X$ where X is some hadronic state. For convenience we consider three distinct cases:

- "untagged", when both the electron and the positron scatter through a small angle and neither is detected after the interaction; in nearly all events both photons are then almost-real

- "single-tag", when either the electron or the positron scatters through a sufficiently large angle to be detected; this corresponds to a highly virtual photon scattering on an almost-real photon

- "double-tag", when both the electron and the positron are detected; this leads to the interaction of two highly-virtual photons.

Of course these three cases are not distinct in any fundamental sense, and there is a smooth transition between them.

Let k_1 and k_2 be the four-momenta of the initial leptons and q_1 and q_2 the four-momenta of the two photons. In analogy with the definitions (7.1) for deep-inelastic ep scattering we define

$$y_1 = \frac{q_1.k_2}{k_1.k_2} \qquad y_2 = \frac{q_2.k_1}{k_1.k_2} \tag{7.47}$$

and

$$x_1 = \frac{Q_1^2}{2q_1.p_2} \qquad x_2 = \frac{Q_2^2}{2q_2.p_1} \tag{7.48}$$

where the photon virtualities are $Q_i^2 = -q_i^2$, $i = 1, 2$. The squared energy of the $\gamma^*\gamma^*$ pair in its centre-of-mass frame is $W^2 = (q_1 + q_2)^2 \approx y_1y_2s$, where $s = (k_1 + k_2)^2$ is the squared energy of the initial pair of leptons in their centre-of-mass frame. With this definition of the variables, $Q_i^2 \approx sx_iy_i$.

The cross section for $e^+e^- \to e^+e^-X$ depends on the angle $\tilde{\phi}$ between the lepton scattering planes in the centre-of-mass frame of the virtual photons. If we integrate over $\tilde{\phi}$ we obtain[329]

$$\frac{d^4\sigma}{dy_1 \, dy_2 \, dQ_1^2 \, dQ_2^2} = \left(\frac{\alpha}{2\pi}\right)^2 \left(P_{\gamma/e^-}^T(y_1, Q_1^2) \, P_{\gamma/e^+}^T(y_2, Q_2^2) \, \sigma^{TT}(Q_1^2, Q_2^2, W^2) \right.$$

$$+ P_{\gamma/e^-}^T(y_1, Q_1^2) \, P_{\gamma/e^+}^L(y_2) \, \sigma^{TL}(Q_1^2, Q_2^2, W^2)$$

$$+ P_{\gamma/e^-}^L(y_1) \, P_{\gamma/e^+}^T(y_2, Q_2^2) \, \sigma^{LT}(Q_1^2, Q_2^2, W^2)$$

$$\left. + P_{\gamma/e^-}^L(y_1) \, P_{\gamma/e^+}^L(y_2) \, \sigma^{LL}(Q_1^2, Q_2^2, W^2) \right) \frac{1}{Q_1^2 Q_2^2} \tag{7.49}$$

where

$$P_{\gamma/e}^T(y, Q^2) = \frac{1 + (1-y)^2}{y} - \frac{2m_e^2 y}{Q^2}$$

$$P_{\gamma/e}^L(y) = 2\frac{1-y}{y}. \tag{7.50}$$

The cross sections $\sigma^{ij}(Q_1^2, Q_2^2, W^2)$ may be defined to be the total cross sections for the process $\gamma^*(Q_1^2)\,\gamma^*(Q_2^2) \to$ hadrons; the indices $i, j = T, L$ denote the polarisations of the virtual photons as transverse or longitudinal in their centre-of-mass frame. The quantities

$$\frac{\alpha}{2\pi Q^2}P_{\gamma/e}^T(y, Q^2) \quad \text{and} \quad \frac{\alpha}{2\pi Q^2}P_{\gamma/e}^L(y) \tag{7.51}$$

may be regarded as the transverse and longitudinal photon-flux factors, though, just as with $\sigma^{\gamma^* p}$, the separation of each term of (7.49) into a flux factor times a cross section is not unique when the photons are not both real. Since real photons are only transversely polarised, the cross sections carrying an L index vanish in the limit $Q_i^2 \to 0$. When both photons are real, that is $Q_1^2 = Q_2^2 = 0$, only one term survives:

$$\sigma^{\gamma\gamma}(W^2) = \sigma^{TT}(0, 0, W^2). \tag{7.52}$$

In the case of one virtual and one real photon it is customary to rewrite the cross sections in terms of structure functions, just as in deep-inelastic lepton-nucleon scattering, and think of the real photon as the target particle whose structure is being probed by the virtual photon. If we take photon 1 as the probing photon and photon 2 as the target photon, then analogously to (7.3) it is conventional to define the tensor $W^{\mu\nu}$ as

$$W^{\mu\nu}(p, q) = \tfrac{1}{4}\sum_r \sum_X \langle q_2, \epsilon_r|J^\mu(0)|X\rangle\langle X|J^\nu(0)|q_2, \epsilon_r\rangle\,(2\pi)^3\delta^4(q_1 + q_2 - P_X) \tag{7.53}$$

where the sum is over the two possible transverse polarisations of the target photon. In analogy with (7.4), one defines photon structure functions $F_1^\gamma(x, Q_1^2)$ and $F_2^\gamma(x, Q_1^2)$ by

$$\frac{W^{\mu\nu}(x, Q_1^2)}{8\pi^2\alpha} = -\left(g^{\mu\nu} + \frac{q_1^\mu q_1^\nu}{Q_1^2}\right)F_1^\gamma(x, Q_1^2)$$

$$+ \frac{1}{\nu}\left(q_2^\mu + q_1^\mu\frac{\nu}{Q_1^2}\right)\left(q_2^\nu + q_1^\nu\frac{\nu}{Q_1^2}\right)F_2^\gamma(x, Q_1^2) \tag{7.54}$$

with $\nu = q_1.q_2$ and $x = Q_1^2/(2\nu)$. The factor $8\pi^2\alpha$ has been extracted in this definition, to take account of the fact that the target photon must first interact electromagnetically before the hadronic system X is created, so that then the structure functions describe the purely-hadronic part of the interaction. With the definition (7.51) of the virtual-photon flux factors,

$$F_1^\gamma = \frac{\nu}{4\pi^2\alpha}\sigma^{TT}(Q_1^2, 0, W^2)$$

$$F_2^\gamma = \frac{Q_1^2}{4\pi^2\alpha}(\sigma^{TT} + \sigma^{LT}). \qquad (7.55)$$

These definitions of the transverse and longitudinal cross sections are similar to those we gave for ep scattering, that is (7.8) and (7.6) generalised to $q^2 \neq 0$. When both photons are on shell

$$\sigma^{\gamma\gamma} = \frac{4\pi^2\alpha}{Q^2}F_2^\gamma\Big|_{Q^2=0} \qquad (7.56)$$

in direct analogy with (7.6).

In practice the target photon is not exactly real, as its parent lepton is scattered through a small finite angle. Although this angle is generally not measured, an upper limit $\theta_2^{\max}$ can be defined experimentally in the centre-of-mass frame of the initial lepton pair, and it is necessary to integrate up to this limit. When this is done the cross section for $e^+e^- \to e^+e^-X$ can be written as

$$\frac{d^3\sigma}{dx_1 dy_1}\Big|_{e^+e^-\to e^+e^-X} = \frac{4\pi\alpha^2}{x_1 y_1 Q_1^2}\Big((1-y_1)F_2^\gamma(x_1, Q_1^2)$$

$$+ x_1 y_1^2 F_1^\gamma(x_1, Q_1^2)\Big)f_{\gamma/e}(y_2, \theta_2^{\max})\,dy_2 \qquad (7.57)$$

with

$$\frac{\pi y_2}{\alpha}f_{\gamma/e}(y_2, \theta_2^{\max}) = (1 + (1-y_2)^2)\log\left(\frac{E_2(1-y_2)}{m_e y_2}\theta_2^{\max}\right) - 1 + y_2. \qquad (7.58)$$

The quantity $f_{\gamma/e}(y_2, \theta_2^{\max})\,dy_2$ is the number of quasi-real target photons, with $\theta_2^{\max}$ the maximum allowed scattering angle for the associated lepton and with y_2 in the range y_2 to $y_2 + dy_2$. As y_2 is not measured the experimental cross section is an integral of (7.57) over y_2, or equivalently over the $\gamma^*\gamma$ centre-of-mass energy W. In practice W can be determined by measuring the final-state hadronic energy in the reaction. For a given W we can rewrite (7.57) in terms of the $e\gamma \to eX$ cross section as

$$\frac{d^2\sigma}{dx\,dy}\Big|_{e\gamma\to eX} = \frac{4\pi\alpha^2}{xyQ^2}((1-y)F_2^\gamma + xy^2 F_1^\gamma) \qquad (7.59)$$

which is exactly similar to the form (7.5) for ep scattering.

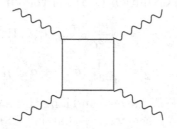

Figure 7.20.   Quark-box contribution to $\sigma^{\gamma^*\gamma^*}$

When both photons are nearly on mass shell the cross section becomes:

$$d^2\sigma\Big|_{e^+e^-\to e^+e^-X} = \sigma_{\gamma\gamma}(W^2)\, f_{\gamma/e}(y_1, \theta_1^{\max})\, f_{\gamma/e}(y_2, \theta_2^{\max})\, dy_1 dy_2. \quad (7.60)$$

The experimental cross section now involves an integral over y_1 and y_2, and the $\gamma\gamma$ centre-of-mass energy can again be obtained by measuring the final-state hadronic energy in the reaction.

The high-energy data for $\sigma^{\gamma\gamma}$ are shown in figure 5.3 together with the prediction assuming factorisation of the soft pomeron and of the $C = +1$ reggeon exchanges in $\sigma^{\gamma p}$ and σ^{pp}. As we have said, the figure suggests that the predicted energy dependence is insufficiently strong to match the data, indicating the possible need for the more rapidly-rising hard-pomeron component. We have seen in section 2.8 that photon cross sections can have contributions which are not present in hadron-hadron collisions, because photons do not appear in the intermediate states in the t-channel unitarity equations. Writing equations analogous to (2.28) for γp amplitudes and to (2.29) for $\gamma\gamma$ amplitudes we just have the two constraints

$$A_2^{\gamma p} = \frac{A_1^{\gamma p}}{1 + \rho(t) A_1^{pp}}$$

$$A_1^{\gamma\gamma} - A_2^{\gamma\gamma} = \rho(t) A_1^{\gamma p} A_2^{\gamma p} = \frac{\rho(t)(A_1^{\gamma p})^2}{1 + \rho(t) A_1^{pp}}. \quad (7.61)$$

Here, $\rho(t)$ is as defined in (2.27). The first equation shows that it is possible to have a singularity in both $A_1^{\gamma p}$ and $A_2^{\gamma p}$ which is not present in hadronic amplitudes. This could be the hard pomeron observed at HERA. The second equation allows two kinds of singularities. Firstly the γp singularity will result in a double $\gamma\gamma$ singularity, that is a γp cross section rising like a power of ν may be reflected by an additional logarithmic factor in the rise

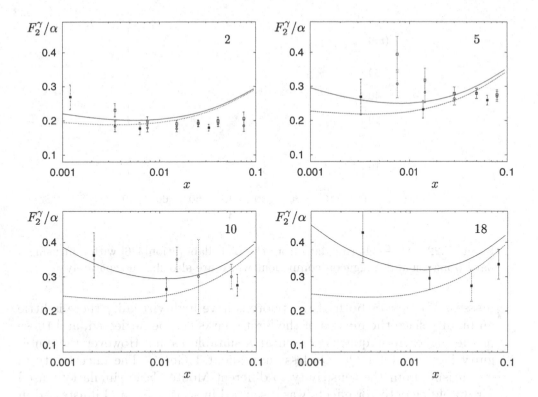

Figure 7.21. Data from OPAL[330–332] (black points), L3[333,334] (open points) and ALEPH[335] (crosses) for the photon structure function. The Q^2 values are approximately 2, 5, 10 and 18 GeV^2. The dotted lines are without the hard pomeron and the solid lines with a hard pomeron contribution included.

of the $\gamma\gamma$ cross section. The equations (7.61) allow a singularity for which both the position and the residue are independent of hadronic thresholds, and so may perhaps be calculable in perturbation theory. As this singularity cannot be a usual moving pole, because that would correspond to the existence of a bound state which would therefore after all be present in purely-hadronic amplitudes, it may be a candidate for the BFKL pomeron. However that is not the only possibility. The quark-box diagram of figure 7.20 is unique to $\gamma^*\gamma^*$, $\gamma^*\gamma$ and $\gamma\gamma$ interactions and corresponds to a fixed pole: see section 2.8 and compare with figure 2.14. Its contribution to the cross section decreases as W^{-2} for large W, up to a logarithmic factor. As a fixed pole, it is additional to reggeon exchange and both contributions must be included in a description of $\sigma^{\gamma\gamma}$. The quark-box contribution to the total $\gamma\gamma$ cross sections can be calculated exactly[329], but there is some uncertainty in applications because it depends on the choice of quark

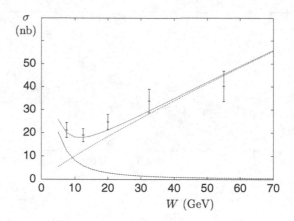

Figure 7.22. $\gamma\gamma \to$ charm: data from the L3 collaboration[336] with a fit using a hard pomeron and a reggeon component which are also shown separately

masses. If either or both of the photons have high virtuality there is little ambiguity since the masses of the light quarks can be neglected, and those of the higher-mass quarks chosen at a suitable value. However the ambiguity is relevant for Q_1^2, Q_2^2 less than about 1 GeV^2. The uncertainty in $\sigma^{\gamma\gamma}$ arising from the sensitivity to different Monte Carlo simulations used for unfolding detector effects was discussed in section 5.1 and illustrated in figure 5.3. Because of this it is not clear what the real energy dependence is.

As for the real $\gamma\gamma$ cross section, it is possible to calculate the photon structure function $F_2^\gamma(x, Q^2)$, using factorisation for the soft-pomeron and reggeon contributions to the γ^*p and pp cross sections and including the known box diagram. The result of this calculation is shown by the dashed lines in figure 7.21. It clearly lies below the data at small x and large Q^2, so it is natural to add a contribution from the hard pomeron. The simplest assumption one can make for the hard pomeron is that its contribution has a Q^2 dependence the same as for the proton structure function F_2, with the normalisation adjusted to fit the data. Typical results of the combined calculations are shown as the solid lines in figure 7.21. Clearly these data do not yet provide a sensitive test of the hard-pomeron contribution. Extrapolating the fit of figure 7.21 to $Q^2 = 0$ GeV^2 gives the result shown as the solid line in figure 7.16. We return to the two-pomeron description of $F_2^\gamma(x, Q^2)$ in section 9.5 in the context of the dipole formalism.

Strong energy dependence is seen in the cross section for charm production in $\gamma\gamma$ interactions at high energy. The L3 data[337] are shown in figure 7.22, with a fit using a hard pomeron and the contribution from the box diagram. An equally good description of these data can also be obtained[338] in perturbative QCD with a suitable choice of the photon's gluon distribution and of the mass of the charm quark. The two descriptions are indistinguishable over the present energy range. Note that the description of the photon structure function shown in figure 7.21, when extrapolated to $Q^2 = 0$ and assuming the hard pomeron is flavour blind, under-estimates the hard-pomeron contribution to charm production in $\gamma\gamma$ collisions by a factor of nearly 3.

Perturbative QCD also provides a good description[339] of the photon structure function F_2^γ. Analogously to (7.10), F_2^γ can be written as a sum over quark and antiquark distributions, and it satisfies an evolution equation similar to (7.15). The only substantive difference from the proton case is that full account must be taken of the box diagram. It is possible[339] to make a parameter-free QCD prediction of F_2^γ. The input parton distributions are separated into hadronic and pointlike contributions. Vector-meson dominance is used to relate the hadronic ones to the known[340,341] quark and gluon distributions of the pion. There is an intrinsic ambiguity in this as it is not clear whether one should take a simple incoherent sum or a coherent sum. In [339] a coherent sum was taken, thus maximising the u-quark component. It is assumed that the pointlike contribution for light quarks is generated entirely by the evolution equations and that the heavy quark contribution is via the box diagram and photon-gluon fusion, $\gamma^* g \to h\bar{h}$. The resulting parameter-free predictions are in good agreement with the data.

In contrast to F_2^γ at present energies, $\sigma^{\gamma^*\gamma^*}$ is expected to be sensitive to the hard pomeron. First consider the soft pomeron, for which factorisation gives

$$\sigma_1^{\gamma^*\gamma^*}(Q_1^2, Q_2^2, W^2) = \frac{\sigma_1^{\gamma^* p}(Q_1^2, W^2)\sigma_1^{\gamma^* p}(Q_2^2, W^2)}{\sigma_1^{pp}(W^2)} \tag{7.62}$$

where $\sigma_1^{\gamma^* p}$ and σ_1^{pp} are the soft-pomeron contributions to the $\gamma^* p$ and pp total cross sections. In the approximation of Bjorken scaling at large Q^2, the soft-pomeron contribution behaves as

$$\sigma_1^{\gamma^* p}(Q^2, W^2) \sim \frac{1}{Q^2}\Big(\frac{W^2}{Q^2}\Big)^{\epsilon_1}. \tag{7.63}$$

Also

$$\sigma_1^{pp} \sim \Big(\frac{W^2}{s_0}\Big)^{\epsilon_1} \tag{7.64}$$

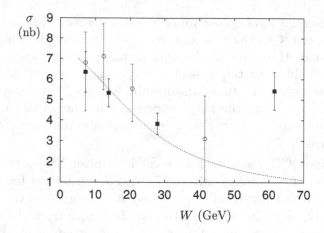

Figure 7.23. $\sigma^{\gamma^*\gamma^*}$ at $Q_1^2 \approx Q_2^2 \approx 16\,\mathrm{GeV}^2$ from L3 (black points[342]) and OPAL (open points[343]) with the calculation of the box diagram of figure 7.20

so that the factorisation (7.62) gives

$$\sigma_1^{\gamma^*\gamma^*}(Q_1^2, Q_2^2, W^2) \sim \left(\frac{s_0^2}{Q_1^2 Q_2^2}\right)^{1+\frac{1}{2}\epsilon}\left(\frac{W^2}{Q_1 Q_2}\right)^{\epsilon}. \tag{7.65}$$

Thus the characteristic feature of the soft-pomeron contribution to the total $\gamma^*\gamma^*$ cross section is its rapid decrease with increasing Q_1^2 or Q_2^2. The same is true of the reggeon contribution. On the other hand[302], for large virtualities the perturbative contribution to $\sigma^{\gamma^*\gamma^*}$ behaves like $1/Q^2$, where $Q^2 = \max\{Q_1^2, Q_2^2\}$, up to logarithmic terms. This follows from a simple dimensional argument. In perturbation theory with massless quarks, the only dimensioned quantities in forward scattering are W^2 and Q^2 as no internal scale appears. So for fixed Q^2/W^2 the cross section has to be proportional to $1/Q^2$.

The use of $\gamma^*\gamma^*$ interactions as a probe of the BFKL pomeron was first considered in [301,302]. Because the BFKL formalism is so uncertain, there have been estimates[344–347] based on various models of the hard component in γ^*p scattering. All agree, as expected, that the soft-pomeron and reggeon contributions are already small at even quite modest photon virtualities and that the $\gamma^*\gamma^*$ cross section is dominated by the box contribution over most of the available energy range. Data [343,342] are given in figure 7.23. It is only the highest-energy point that disagrees with the box contribution. We have explained in sections 7.3 and 7.4 that if a hard-pomeron component is indeed found in $\gamma^*\gamma^*$ collisions, it is very probably not associated with the BFKL equation.

7.8 Exclusive vector-meson production

Exclusive electroproduction of vector mesons, $\gamma^* p \to V p$, offers a variety of insights into the diffractive mechanism. Choosing different vector mesons, and varying the photon virtuality Q^2 and the momentum transfer t, allows one to move from the primarily-nonperturbative regime to the primarily-perturbative regime, and to explore kinematical regions where neither is dominant. Data are available for ρ, ω, ϕ, J/ψ, ψ' and Υ production. Further, simple two-body decays such as $\rho \to \pi^+ \pi^-$ and $\phi \to K^+ K^-$ allow a helicity decomposition of the amplitudes.

We saw in section 5.3 that the exclusive processes $\gamma p \to \rho p$ and $\gamma p \to \phi p$ are well described by conventional exchanges. But for $\gamma p \to J/\psi\, p$ this is not the case; in particular, as is seen in figure 7.24a, the total cross section rises significantly more rapidly than can be ascribed to soft-pomeron exchange. As for $\gamma p \to \phi p$, we expect from the OZI rule[190–193] that the exchanges ρ, ω, f_2, a_2 do not couple. If we include hard-pomeron exchange as well as the soft pomeron, the amplitude is

$$T(s,t) = i F_1(t)\, G_{J/\psi}(t) \sum_{i=0,1} A_{\mathbb{P}_i} (\alpha'_{\mathbb{P}_i} s)^{\alpha_{\mathbb{P}_i}(t)-1} e^{-\frac{1}{2} i \pi (\alpha_{\mathbb{P}_i}(t)-1)}. \qquad (7.66)$$

Here $F_1(t)$ is the proton's Dirac form factor, $G_{J/\psi}(t)$ is the coupling to the $\gamma J/\psi$ vertex, $\alpha_{\mathbb{P}_1}$ is the classical soft-pomeron trajectory and

$$\alpha_{\mathbb{P}_0}(t) = \alpha_{\mathbb{P}_0}(0) + \alpha'_{\mathbb{P}_0} t \qquad (7.67)$$

is the hard-pomeron trajectory. We expect $G_{J/\psi}(t)$ to be slowly varying over the range of t for which data are available, and it is sufficient to take it as a constant. The differential cross section, figure 7.24b, shows little or no shrinkage, so the slope of the hard-pomeron trajectory $\alpha_{\mathbb{P}_0}(t)$ is smaller than $\alpha'_{\mathbb{P}_1}$, assuming that it is linear in t. The fit to the H1 data[348] in figure 7.24 is with $\alpha_{\mathbb{P}_0}(0) = 1.44$ and $\alpha'_{\mathbb{P}_0} = 0.1$ GeV^{-2}. The coefficients in (7.66) are given by

$$A_{\mathbb{P}_0} = 0.016 \qquad\qquad A_{\mathbb{P}_1} = 0.17 \qquad (7.68)$$

with the cross section normalised such that $d\sigma/dt = |T(s,t)|^2$ in μb GeV^{-2}. Although the soft-pomeron contribution is appreciably smaller than the hard-pomeron term at the highest energy for which there are data, interference is still very important.

Given that the exclusive process $\gamma p \to J/\psi\, p$ includes contributions from hard-pomeron exchange, so also surely do $\gamma p \to \rho p$ and $\gamma p \to \phi p$. However, in the case of these lighter vector mesons, soft-pomeron exchange is much larger and is dominant.

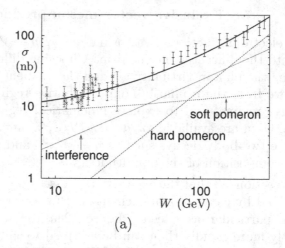

(a)

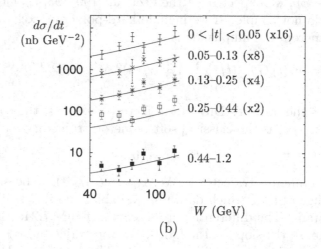

(b)

Figure 7.24. $\gamma p \to J/\psi\, p$: H1 data[348] for (a) the total cross section and (b) the differential cross section. The fits are (7.66) with $\alpha_{I\!P_0}(0) = 1.44$ and $\alpha'_{I\!P_0} = 0.1$ GeV^{-2}.

We have seen in section 7.4 that the contribution of the hard pomeron to $F_2(x, Q^2)$ becomes relatively more important as Q^2 increases. It is natural to expect that the same should be true for exclusive vector-meson production. The hard pomeron probably also becomes more important with increasing momentum transfer $|t|$. The exclusive production of ρ^0 mesons is the reaction for which most data exist. We have seen in section 5.3 that for real photons and for momentum transfers $|t| < 0.5$ GeV^2 the data are in remarkable agreement with Regge theory, when one includes f_2, a_2 and soft-pomeron exchange. However, the high-energy data for $|t| > 0.5$ GeV^2

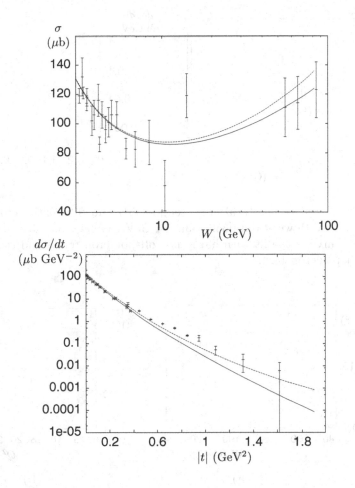

Figure 7.25. $\gamma p \to \rho p$: total cross section and differential cross section at $\sqrt{s} = 71.7$ GeV (lower-t data[349]) and 94 GeV (higher-t data[350]). The lower curves correspond to soft-pomeron and f_2, a_2 exchange; the upper curves include also hard-pomeron exchange.

require an additional contribution. The data are consistent with this being the exchange of the hard pomeron[180]. We have seen in section 7.4 that the coupling of the hard pomeron to quarks seems to be flavour-blind. So, in order to relate the strengths of the hard-pomeron couplings in $\gamma p \to J/\psi\, p$ and $\gamma p \to \rho^0 p$ we need just to include wave-function effects. We argued in section 7.5 that the hard pomeron couples mainly to the pointlike component of the photon. This in turn implies that in $\gamma p \to V p$ the strength of the hard-pomeron coupling depends on the magnitude of the vector-meson wave function at the origin and the relevant quark charges; therefore it is

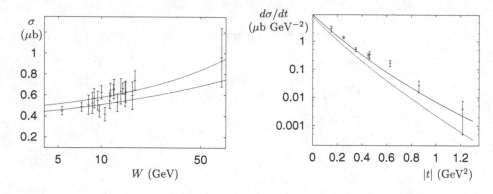

Figure 7.26. $\gamma p \to \phi p$: total cross section[351] and differential cross section at $\sqrt{s} = 71.7$ GeV (lower-t data[352]) and 94 GeV (higher-t data[350]). In each case the upper curve is the fit with hard and soft-pomeron terms and the lower curve is the soft-pomeron term.

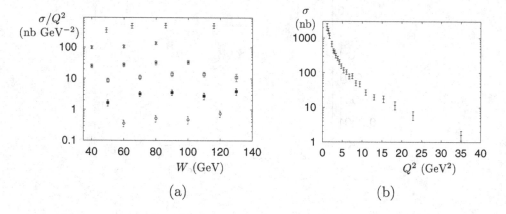

Figure 7.27. Cross section[353,348] for $\gamma^* p \to \rho p$ at (a), from top to bottom, $Q^2 = 2, 3, 5, 7, 11$ and 20 GeV^2 and (b) at $W = 75$ GeV

proportional to $\sqrt{\Gamma_{e^+e^-}/M_V}$. This implies that for $\gamma p \to \rho^0 p$ we should use $A_{\mathbb{P}_0} = 0.036$. Adding such a hard-pomeron term to the amplitude (5.12) gives[180] the upper curves in figures 7.25, from which it is clear that the hard-pomeron contribution is small at small t.

We may perform a similar analysis for ϕ photoproduction. As before, the flavour-blind coupling of the hard pomeron to quarks can be used to specify its contribution uniquely. For $\gamma p \to \phi p$ this gives $A_{\mathbb{P}_0} = 0.014$. We know that vector-meson dominance is not a good approximation for reactions involving the ϕ and that wave-function effects are important, so the nor-

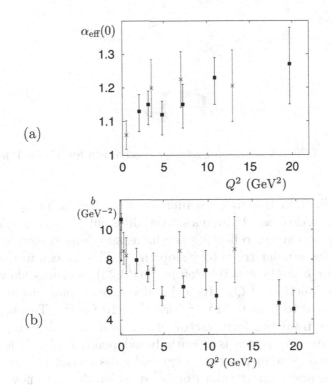

Figure 7.28. (a) Effective trajectory and (b) forward slope parameter for $\gamma^* p \to \rho p$. Data from H1 (black points)[354,355,348] and ZEUS (crosses)[356,353]

malisation of the soft-pomeron contribution can be determined only from the data. For the vertex function that couples the soft pomeron to the $\gamma \phi$ vertex we use a form that probably resembles the elastic form factor of the ϕ:

$$G_\phi = \frac{1}{1 - t/m_\phi^2}. \tag{7.69}$$

As for the J/ψ any contribution from f_2, a_2 exchange can be neglected. Fitting the soft-pomeron coupling to the data yields $A_{I\!P_1} = 1.49$ and the results are shown in figure 7.26.

As Q^2 increases the hard-pomeron component should become increasingly important relative to the soft-pomeron component and the total cross section should increase more rapidly with increasing energy. The cross sections for ρ^0 electroproduction[353,348] at various Q^2 are shown in figure 7.27. If the intercept α_{eff} is extracted from the rise of the cross section with energy,

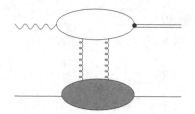

Figure 7.29.   Two-gluon exchange mechanism for $\gamma^* p \to V p$

there is an indication that it does increase with increasing Q^2, as shown in figure 7.28a. Also, as Q^2 increases the differential cross section $d\sigma/dt$ should become less steep, reflecting the increasing importance of the hard pomeron with its smaller trajectory slope $\alpha'_{I\!P_0}$. This is seen in the forward slope parameter $b = b(s, t = 0)$, defined in (3.23), which is shown in figure 7.28b as a function of Q^2. Its value decreases rapidly from about 10 GeV^{-2} at $Q^2 = 0$ to 5 to 6 GeV^{-2} for $Q^2 > 10$ GeV^2. The latter value is expected just from the form factor of the proton, which suggests that the $\gamma^* \to \rho^0$ transition vertex is essentially independent of t. It is apparent from the energy dependence of the integrated cross section that, even at the highest energies and highest values of Q^2 at which data are now available, the hard pomeron does not dominate the cross section at small $|t|$. Two-gluon-exchange models of ρ^0 electroproduction incorporating perturbative and nonperturbative effects[357,358] agree with this.

The two-gluon-exchange mechanism of figure 7.29 relates exclusive photoproduction and electroproduction of the J/ψ to the gluon structure function of the proton. The upper part of the diagram is calculated from perturbative QCD, while the lower part is related to the proton's gluon structure function and so is nonperturbative. The same mechanism is supposed to apply to electroproduction of the ρ and ϕ at sufficiently large Q^2. The photon dissociates into a $q\bar{q}$ pair which then scatters on the target. The probability that the $q\bar{q}$ pair recombines into a vector meson is determined by the projection on the meson wave function. The photon dissociation is described by the photon wave functions introduced in section 8.6, where it is shown that transverse photons will dominate at small Q^2, but for $Q^2 \geq m^2$, where m is the mass of the charm quark, longitudinal photons will dominate. The wave function of a vector meson is constructed in analogy to the photon wave function. In the nonrelativistic limit it can be assumed that the quark and antiquark are weakly bound in comparison with their mass and that their relative momentum is small. This means that the relative transverse momentum $\mathbf{k}_T$ is approximately zero and that the longitudinal momentum

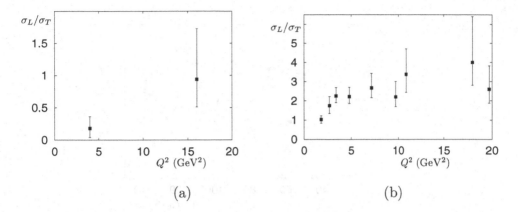

Figure 7.30.   Data[363,364] for σ_L/σ_T in (a) $\gamma p \to J/\psi\, p$ and (b) $\gamma p \to \rho\, p$

fraction z of the quark is approximately the same as that of the antiquark. This also means that the mass of the vector meson is approximately the sum of the masses of the constituent quark and antiquark, that is for the J/ψ we have $M_{J/\psi} \approx 2m_c$. In this simplified approach one can write the vertex function that couples the $c\bar{c}$ pair to the J/ψ as

$$\psi^V(z, \mathbf{k}_t) = C\delta(z - \tfrac{1}{2})\, \delta^2(\mathbf{k}_T) \tag{7.70}$$

and the constant C is obtained by using the same vertex function to calculate the leptonic decay width $\Gamma^V_{e^+e^-}$ of the vector meson[359,360]. The complete expression for the forward cross section for exclusive vector-meson production is then[361,362] approximately

$$\left.\frac{d\sigma^V}{dt}\right|_{t=t_{min}} = \frac{16\,\Gamma^V_{e^+e^-}\, M_V^3\, \pi^3}{3\alpha(Q^2 + M_V^2)^4}\left|(i+\rho)\,\alpha_s\, x_{I\!P}\, g(x_{I\!P}, \bar{Q}^2)\right|^2\left(1 + \frac{Q^2}{M_V^2}\right) \tag{7.71}$$

where $x_{I\!P} = (Q^2 + M_V^2)/(Q^2 + W^2)$, $\bar{Q}^2 = \tfrac{1}{4}(Q^2 + M_V^2)$, W is the $\gamma^* p$ centre-of-mass energy, $g(x, Q^2)$ is the gluon structure function of the proton and $\rho = \mathrm{Re}\, A/\mathrm{Im}\, A$ is the relative contribution of the real part of the amplitude. The approximation

$$\mathrm{Re}\, A = \frac{\pi}{2}\, x_{I\!P}\, \frac{\partial}{\partial x_{I\!P}}\, \mathrm{Im}\, A \tag{7.72}$$

is often used though if, as is the case, the gluon distribution rises steeply with $1/x$ at small x, it would be more accurate to calculate ρ using the signature factor ξ_α^+ of (2.18), with $(\alpha - 1)$ the W^2 effective power rise of the cross section.

The first term in the last bracket of (7.71) gives the transverse cross section σ_T and the term Q^2/M_V^2 gives the longitudinal cross section σ_L. So a

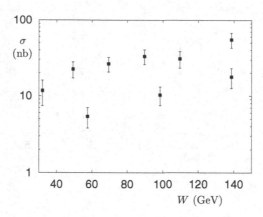

Figure 7.31.   Cross section[363] for $\gamma^* p \to J/\psi\, p$ at $Q^2 = 3.5$ GeV^2 (upper points) and 10.1 GeV^2 (lower points)

specific prediction of this simple model is that $\sigma_L/\sigma_T = Q^2/M_V^2$. The data for σ_L/σ_T for J/ψ electroproduction are shown in figure 7.30a and are not incompatible with a linear rise. Using more realistic wave functions slows the predicted rate of rise[365,357]. The ratio σ_L/σ_T for ρ electroproduction does not follow the linear rise, as can be seen in figure 7.30b. This ratio in fact is very sensitive to the choice of the vector-meson wave function, and provides the strongest constraint[357] of all data on ρ electroproduction.

The rise with energy of the total cross section is predicted by (7.71) to be related to the rise of $(g(x_{I\!P}, \bar{Q}^2))^2$ with $1/x_{I\!P}$. The data shown in figure 7.31 for J/ψ electroproduction are compatible with this.

Another feature of the simple formula (7.71) is that it is of the form $\Gamma_{e^+e^-}^V M_V$ times a universal function of $(Q^2 + M_V^2)$. Figure 7.32 shows that, if the cross sections for ρ, ω, ϕ and J/ψ electroproduction, and for J/ψ photoproduction, are plotted against $(Q^2 + M_V^2)$, divided by the numbers 9:1:2:8 associated with the quark charges and wave-function symmetry factors, they do seem to lie on a single curve[348]. However, this is not compatible with (7.71) and indeed it is somewhat mysterious, because when these quark-charge factors are removed experiment finds that $\Gamma_{e^+e^-}^V$ is rather constant: see figure 7.33. So (7.71) would predict that one should obtain a single curve only if one also divided the cross section by M_V.

We have explained that the derivation of (7.71) uses a very simple vector-meson wave function. More realistic calculations find that the cross section is very sensitive to the wave function[357,366,367]. This uncertainty affects the absolute normalisation, but has little effect on the energy dependence of the cross section. However, the latter is subject to a different source of

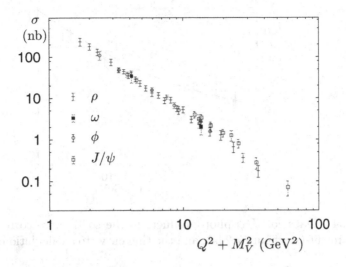

Figure 7.32. H1 and ZEUS data for $\sigma(\gamma^* p \to V p)$ with factors 9:1:2:8 removed, plotted against $Q^2 + M_V^2$

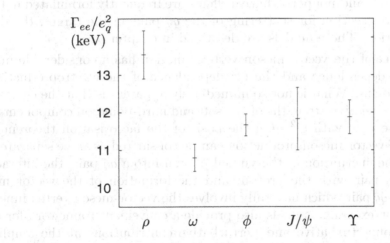

Figure 7.33. Data for $\Gamma^V_{e^+ e^-}$ with the factors 9:1:2:8:1 removed. Note the suppressed zero on the vertical scale.

uncertainty, as the scale which enters the structure function is in general not linked to the hard scale in a simple way. With a suitable choice of scale and structure function it is possible to obtain a satisfactory description of the forward differential cross section at high energy. A typical result[368] is shown in figure 7.34, which also includes the two-pomeron curve of figure 7.24a; note the different curvature. It can be seen that at low energy

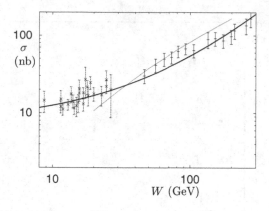

Figure 7.34.   Data for J/ψ photoproduction; the solid curve corresponds to the two-pomeron fit (see figure 7.24a), and the thin curve to a calculation[368] of (7.71)

the perturbative-QCD predictions under-estimate the cross section, and it is necessary to include nonperturbative effects[357,365]. Models combining perturbative and nonperturbative effects are frequently formulated in terms of the cross section for scattering of the $q\bar{q}$ pair on the target, the dipole cross section. These models are described in chapter 9.

The choice of the vector-meson vertex function has a considerable impact on the t dependence and the Q^2 dependence of the electroproduction of vector mesons. What is not so immediately apparent is that the choice also affects the relative strengths of the soft and hard-pomeron components and how these vary with Q^2. On the basis of the factorisation theorem[369], exclusive vector meson production can be considered as three separate processes: the fluctuation of the virtual photon into a $q\bar{q}$ pair; the interaction of the $q\bar{q}$ pair with the proton; and the formation of the vector meson from the $q\bar{q}$ pair which naturally involves the vector-meson vertex function. Two-gluon-exchange models also provide a convenient framework for combining nonperturbative and perturbative contributions at the amplitude level. This approach has been developed[370,357] in a simultaneous study of ρ, ϕ and J/ψ electroproduction and of J/ψ photoproduction. The perturbative gluons are described by the usual perturbative propagator and the gluon flux obtained from the unintegrated gluon density or from the model of [371]. A specific nonperturbative propagator[372] is used for the nonperturbative gluons and the gluon-proton interaction is treated by the nonperturbative model[7] described in section 8.1. The dynamics of the exchange is governed by two parameters, one for the nonperturbative term and one for the perturbative term. These parameters are the same for each of the ρ, ϕ and J/ψ. The only freedom in going from one vector meson to another is a parameter p_F which is related to the "size" of the meson in its rest frame. The wave function is assumed to be Gaussian:

$$\Phi(\mathbf{p}^2) = N \, \exp\left(-\frac{\mathbf{p}^2}{2p_F^2}\right). \tag{7.73}$$

The model gives a good global description of all the data, reproducing the combinations of the soft and hard pomerons determined phenomenologically. The role of wave functions is explored further in chapters 8 and 9.

An important aspect of ρ^0 electroproduction is the measurement[364,373] of the density-matrix elements from the ρ^0-decay angular distribution. The density-matrix elements provide tests of s-channel helicity conservation and of the assumed dominance of natural-parity exchange. (Natural parity is defined in section 2.5.) The ρ^0-decay angular distribution for unpolarised leptons is given in terms of the density-matrix elements ρ_{ik}^α in (E.3), and as bilinear combinations of the helicity amplitudes in (E.8). As the present data do not separate the longitudinal and transverse cross sections, the measured quantities are the r_{ik}^α of (E.6). The ρ^0-decay angular distribution then has the form

$$
\begin{aligned}
\frac{4\pi}{3} & W(\cos\theta, \phi, \psi) = \\
& \tfrac{1}{2}(1 - r_{00}^{04}) + \tfrac{1}{2}(3r_{00}^{04} - 1)\cos^2\theta - \sqrt{2}\,\mathrm{Re}\,r_{10}^{04}\sin 2\theta\cos\phi \\
& \quad - r_{1-1}^{04}\sin^2\theta\cos 2\phi \\
& \quad - \epsilon\cos 2\psi\,(r_{11}^1\sin^2\theta + r_{00}^1\cos^2\theta - \sqrt{2}\,\mathrm{Re}\,r_{10}^1\sin 2\theta\cos\phi \\
& \quad\quad - r_{1-1}^1\sin^2\theta\cos 2\phi) \\
& \quad - \epsilon\sin 2\psi\,(\sqrt{2}\,\mathrm{Im}\,r_{10}^2\sin 2\theta\sin\phi + \mathrm{Im}\,r_{1-1}^2\sin^2\theta\sin 2\psi) \\
& \quad + \sqrt{2\epsilon(1+\epsilon)}\,\cos\psi\,(r_{11}^5\sin^2\theta + r_{00}^5\cos^2\theta \\
& \quad\quad - \sqrt{2}\,\mathrm{Re}\,r_{10}^5\sin 2\theta\cos\phi - r_{1-1}^5\sin^2\theta\cos 2\phi) \\
& \quad + \sqrt{2\epsilon(1+\epsilon)}\,\sin\psi\,(\sqrt{2}\,\mathrm{Im}\,r_{10}^6\sin 2\theta\sin\phi \\
& \quad\quad + \mathrm{Im}\,r_{1-1}^6\sin^2\theta\sin 2\phi)
\end{aligned}
\tag{7.74}
$$

where ϵ is the polarisation of the virtual photon. For the HERA data $\langle\epsilon\rangle \approx 0.99$.

From (E.15) and (E.6) we see that the asymmetry P_σ between natural and unnatural-parity exchange is

$$P_\sigma = \frac{\sigma^N - \sigma^U}{\sigma^N + \sigma^U} = (1 + \epsilon R)(2r_{1-1}^1 - r_{00}^1) \tag{7.75}$$

	r^1_{1-1}	$-\operatorname{Im} r^2_{1-1}$	$\operatorname{Re} r^5_{10}$	$-\operatorname{Im} r^6_{10}$
ZEUS	0.334 ± 0.025	0.331 ± 0.023	0.142 ± 0.013	0.141 ± 0.006
ZEUS	0.098 ± 0.023	0.135 ± 0.039	0.142 ± 0.011	0.136 ± 0.011
H1	0.122 ± 0.019	0.119 ± 0.021	0.146 ± 0.010	0.140 ± 0.009

Table 7.1. s-channel helicity-matrix elements in ρ^0 electroproduction[364,373] showing that (7.77) is satisfied. The first ZEUS entry is from the low-Q^2 data, the second from the high-Q^2 data; see the text.

where R is the ratio of the longitudinal to transverse cross sections. The asymmetry is found to be compatible with 1. This implies that natural-parity exchange dominates in the data, at least for transverse photons. The measurement of the corresponding asymmetry for longitudinal photons requires two different values of ϵ.

For s-channel helicity conservation all matrix elements become zero, except five:

$$r^{04}_{00}, \quad r^1_{1-1}, \quad \operatorname{Im} r^2_{1-1}, \quad \operatorname{Re} r^5_{10}, \quad \operatorname{Im} r^6_{10} \qquad (7.76)$$

and they are not all independent. They satisfy

$$r^1_{1-1} = -\operatorname{Im} r^2_{1-1}, \quad \operatorname{Re} r^5_{10} = -\operatorname{Im} r^6_{10}. \qquad (7.77)$$

Of the remaining density-matrix elements only one, r^5_{00}, differs significantly from zero within the present measurement precision. The ZEUS data cover two ranges of Q^2, $0.25 \text{ GeV}^2 < Q^2 < 0.85 \text{ GeV}^2$ and $3 \text{ GeV}^2 < Q^2 < 30$ GeV^2. The corresponding energy ranges are $20 \text{ GeV} < W < 90 \text{ GeV}$ and $40 \text{ GeV} < W < 120 \text{ GeV}$, and both samples extend up to $|t| = 0.6 \text{ GeV}^2$. The H1 ranges are $1 \text{ GeV}^2 < Q^2 < 60 \text{ GeV}^2$, $30 \text{ GeV} < W < 140 \text{ GeV}$ and $|t| < 0.5 \text{ GeV}^2$. The values found for r^5_{00}, averaged over the complete kinematical ranges, are 0.051 ± 0.021, 0.095 ± 0.031 and 0.093 ± 0.029. The variation with Q^2, W and t is shown in figure 7.35. The r^5_{00} term is proportional to the interference between the helicity-non-flip amplitude T_{00} for longitudinal photons, and the helicity-single-flip amplitude T_{01} for the production of longitudinally polarised ρ^0 mesons from transverse photons. The s-channel-helicity-conserving relations (7.77) are well satisfied by the data which are given in table 7.1.

Relative to the dominant helicity-non-flip amplitudes T_{11} and T_{00}, T_{01} is small. From (E.6), (E.7) and (E.8),

$$r^5_{00} \approx \frac{\sqrt{R}}{1 + \epsilon R} \frac{|T_{01}|}{|T_{11}|} \qquad (7.78)$$

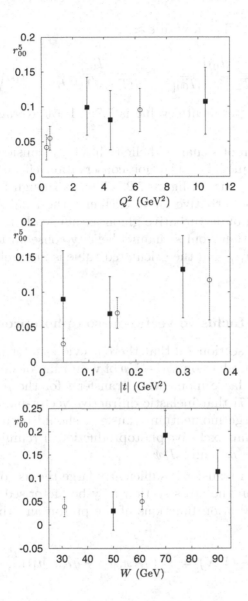

Figure 7.35. Data for the r_{00}^5 matrix element for $\gamma^* p \to \rho p$ from H1 (black points)[364] and ZEUS (open points)[373]

where the term $|T_{01}|^2$ has been neglected relative to $|T_{11}|^2$ in the denominator (see (E.8)) and it is assumed that the amplitudes T_{00} and T_{01} are both purely imaginary. Then, with $\epsilon \approx 1$,

$$\frac{|T_{01}|}{\sqrt{|T_{11}|^2 + |T_{00}|^2}} \approx \frac{|T_{01}|}{|T_{11}|\sqrt{1+R}} \approx r_{00}^5 \sqrt{\frac{1+R}{2R}} \qquad (7.79)$$

for which the experimental value is $7.7 \pm 1.6\%$ averaged over the present Q^2 domain.

Non-conservation of s-channel helicity has been considered[374,375] in two-gluon-exchange models. The non-conservation effects arise entirely from the $\gamma \rightarrow q\bar{q} \rightarrow V$ part of figure 7.29. Thus although the calculations were performed for perturbative gluon exchange the results are also applicable to models with nonperturbative gluon exchange. It is found that the only significant deviation from s-channel helicity conservation is in the density-matrix element r_{00}^5, and the calculated value is in excellent agreement with experiment[373].

7.9 Inclusive vector-meson photoproduction

We have seen in section 7.9 that there is evidence for the hard pomeron in exclusive photoproduction $\gamma p \rightarrow Vp$ of vector mesons, most notably for the J/ψ but also at large momentum transfer t for the ρ and ϕ. It has been suggested[376,377] that inelastic diffractive vector-meson photoproduction $\gamma p \rightarrow VX$ at large momentum transfer t should be more sensitive to the hard pomeron than exclusive photoproduction. Preliminary data have been presented for the ρ, ϕ and J/ψ.

If the momentum transfer is sufficiently large the use of perturbative QCD is justified. Then the cross section may be factorised into the product of the parton distribution functions of the proton and the parton-level cross sections:

$$\frac{d\sigma}{dt\,dx}(\gamma p \rightarrow VX) = \sum_f (q_f(x,|t|) + \bar{q}_f(x,|t|)) \frac{d\sigma}{dt}(\gamma q_f \rightarrow V q_f)$$

$$+ g(x,|t|)\frac{d\sigma}{dt}(\gamma g \rightarrow Vg) \qquad (7.80)$$

where $g(x,|t|)$ and $q_f(x,|t|)$ are the gluon and quark parton distribution functions, which are summed over flavour f and evaluated at $Q^2 = |t|$. It then remains to choose a model for the exchange and for the vector-meson wave functions.

A good simultaneous description of the preliminary ρ, ϕ and J/ψ data has been obtained[378] using the leading-log BFKL pomeron for the ex-

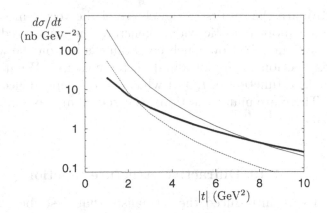

Figure 7.36. Theoretical curves[378] for large-t photoproduction of ρ (upper curve), ϕ (dashed curve) and J/ψ (thick curve)

change[376,377] and treating the vector-meson wave functions in the non-relativistic limit (7.70). This is an oversimplification as it allows production only of transversely polarised mesons. There is evidence in the data for production of longitudinally polarised mesons, although the contribution is small. The calculation involves only two parameters: the strong-coupling constant α_s, which was taken to be fixed, and the scale for the energy variable in the logarithms, which was taken to be $\beta M_V^2 + \gamma|t|$ with β and γ common parameters for all vector mesons. Acceptable fits are obtained for $\alpha_s \approx 0.2$ and $\beta \approx \gamma \approx 1.0$. Typical results are shown in figure 7.36.

In contrast to this succcess, modelling the exchange by two perturbative gluons provides only a qualitative description of the J/ψ data and a very poor one of the light mesons[378]. This failure results from a dip in the predictions at $t = -M_V^2$ where the two-gluon-exchange amplitude vanishes. It is only because this value of t is outside the range of the data for the J/ψ that the prediction of the two-gluon-exchange model looks reasonable. As the BFKL amplitude does not have this zero the dip must be filled by corrections to the two-gluon-exchange approximation, such as the resummation of the two-gluon amplitude to include virtual corrections and the addition of rungs.

It is certainly premature to conclude that these preliminary data support the BFKL pomeron, as the calculations are subject to many uncertainties. A complete description should use a running coupling α_s, but we do not yet know how to incorporate it into the BFKL approach. How much more complicated than two-gluon exchange must the colour-singlet exchange be to describe the data? Do we need the full complexity of the BFKL pomeron?

How sensitive are the results to the choice of the vector-meson wave functions? Does a more realistic choice generate the same predictions as the non-relativistic model? How much constraint is put on the wave functions by the cross section for longitudinal vector mesons? What is the energy dependence as a function of t, and what is the t dependence as a function of energy? There are many questions, theoretical and experimental, still to be answered.

7.10 Diffractive structure function

In section 3.3 we introduced the formalism that describes soft diffractive events, which are events in hadron-hadron scattering in which one of the initial hadrons appears in the final state having lost only a very small fraction ξ of its initial momentum. Ingelman and Schlein[8] suggested that it would similarly be useful to consider $\gamma^* p$ interactions in which the initial proton loses only very little momentum fraction ξ. That is, the hadron b in figure 3.16 is replaced with a γ^* which has been radiated from an electron. It is found that the diffraction dissociation contributes about 10% of the small-x deep-inelastic scattering events. Such events are examples of what are known as hard diffraction events.

The cross section for single diffraction dissociation in deep-inelastic scattering, $ep \to eXp$, defines a diffractive structure function $F_2^{D(4)}(\beta, Q^2, \xi, t)$:

$$\frac{d^4\sigma}{d\xi\, dt\, dx\, dy} = \frac{4\pi\alpha^2}{xyQ^2}\left(1 - y + \frac{y^2}{2(1 + R(x, Q^2, \xi, t))}\right)F_2^{D(4)}(\beta, Q^2, \xi, t) \quad (7.81)$$

where x and y are the usual variables of deep-inelastic scattering, given in (7.1), and R is the ratio of the cross sections for longitudinal and transversely polarised photons. The variables ξ and t in (7.81) are the ones customary in diffraction dissociation, see figure 3.16, and $\beta = x/\xi$ is the fraction of the exchanged object's momentum carried by the struck quark. ξ is often instead called $x_{I\!P}$. The centre-of-mass energy of the $\gamma^* p$ system is $W \approx \sqrt{Q^2(1/x - 1)}$ and the mass M of the system X is $M \approx \sqrt{\xi s}$. The formula is analogous to (7.5), when F_1 is eliminated through (7.7). The determination of $F_2^{D(4)}(\beta, Q^2, \xi, t)$ is not sensitive to the choice of R for small y, so it is customary to set $R = 0$ or to estimate an error on $F_2^{D(4)}$ by determining it at the two extremes $R = 0$ and $R = \infty$.

One may define[379], in analogy with (3.43),

$$F_2^{D(4)}(\beta, Q^2, \xi, t) = D^{I\!P/a}(t, \xi)F_2^{I\!P}(\beta, Q^2, t) \quad (7.82)$$

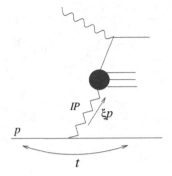

Figure 7.37. Diffractive electroproduction factorised as in (7.82)

with $D^{I\!P/a}(t,\xi)$ defined in (3.44), and regard $F_2^{I\!P}(\beta,Q^2,t)$ as the structure function of the pomeron. It is assumed, though it is difficult to prove theoretically, that $F_2^{I\!P}(\beta,Q^2,t)$ obeys perturbative evolution equations similar to those for a hadron structure function. If this is valid, although measurement of $F_2^{D(4)}(\beta,Q^2,\xi,t)$ directly corresponds only to the quark structure function of the pomeron, from its Q^2 evolution one may extract also the pomeron's gluon structure function. Notice that the interpretation of $F_2^{I\!P}(\beta,Q^2,t)$ as the structure function of the pomeron does not depend on the pomeron being a particle – indeed, it is not – but it makes sense because of the Regge factorisation. See figure 7.37. However, this factorisation is only plausible when ξ is rather small; otherwise there will be reggeon exchanges additional to the soft pomeron, and adding in their contributions spoils the factorisation. Further, there is the possibility, which seems to be supported by experiment, that there is also a significant contribution from hard-pomeron exchange, so factorisation is of doubtful validity for all values of ξ, even if it is valid for the hard and soft-pomeron exchanges separately.

To determine t unambiguously requires the detection of the recoil nucleon and measurement of its momentum[380,349]. However this is not always possible, so[381,382] the process being studied is in effect $ep \rightarrow eXY$. As t is not measured, results are quoted in terms of the diffractive structure function $F_2^{D(3)}(\beta,Q^2,\xi)$, the integral of $F_2^{D(4)}$ over t. This complicates comparison with theoretical models and, as the range of of mass of the unknown system Y is different in different experiments, for example $m_Y < 1.6$ GeV in [381] and $m_Y < 5.5$ GeV in [382], it is difficult to compare experiments directly.

We consider first the $ep \rightarrow eXY$ data and the structure function $F_2^{D(3)}$.

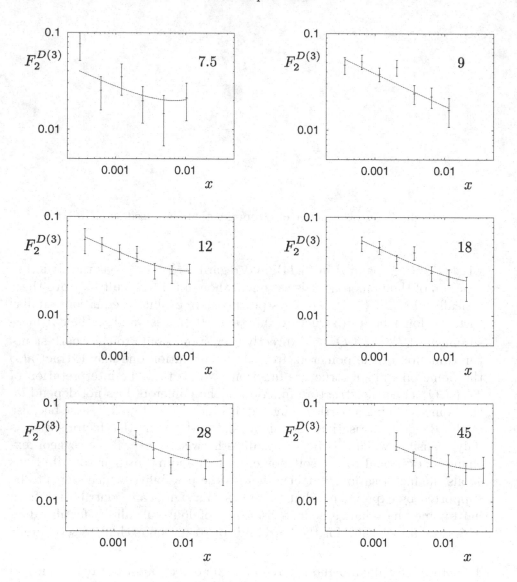

Figure 7.38.   H1 data[381] for the diffractive structure function $F_2^{D(3)}$ at $\beta = 0.65$ and Q^2 ranging from 7.5 to 45 GeV^2

Although there is no theoretical justification for it, it is usually assumed that the structure function of the pomeron is independent of t, in which case the factorisation (7.82) can be used at fixed β, Q^2 to find the effective pomeron intercept from $D^{I\!\!P/a}(t, \xi)$. The ZEUS data[382] on their own are compatible with a single exchange and, if this is assumed, the effective pomeron intercept is found to be $\alpha_{\text{eff}} = 1.127 \pm 0.009^{+0.039}_{-0.012}$. However, this result is dependent on the assumption that there is no contribution from reggeon exchange. The absence or presence of reggeon exchange cannot be inferred from these data as they are mainly at values of ξ which are too small to provide the necessary sensitivity. The H1 data[381] some of which are shown in figure 7.38, extend to larger values of ξ than do the ZEUS data and show clearly that a reggeon term is required. That is, one has to replace (7.82) with something like

$$F_2^{D(4)}(\beta, Q^2, \xi, t) = D^{\mathbb{P}/a}(t, \xi) F_2^{\mathbb{P}}(\beta, Q^2) + D^{\mathbb{R}/a}(t, \xi) F_2^{\mathbb{R}}(\beta, Q^2)$$

$$+ 2c_I D^I(t, \xi) \sqrt{F_2^{\mathbb{P}}(\beta, Q^2, t) F_2^{\mathbb{R}}(\beta, Q^2)} \qquad (7.83)$$

though the square-root term is extremely model-dependent. The interference flux $D^I(t, \xi)$ is obtained from the pomeron and reggeon fluxes with the signature factors as we explained after (3.47). The quantity c_I may be thought of as the degree of coherence. H1 made[381] two fits to their data for the two extremes of no interference, $c_I = 0$, and "full" interference, $c_I = 1$, with $\alpha_{\mathbb{P}}(0)$, $\alpha_R(0)$, $F_2^{\mathbb{P}}(\beta, Q^2)$, $F_2^{\mathbb{R}}(\beta, Q^2)$ as free parameters. It was assumed that the slopes of the pomeron and reggeon trajectories are the standard values for the soft pomeron and reggeon found in hadronic interactions and the flux factors $D^{\mathbb{P}/a}(t, \xi)$, $D^{\mathbb{R}/a}(t, \xi)$ were chosen to have the simplified form

$$D^{\mathbb{P}/a}(t, \xi) \propto e^{b_{\mathbb{P}} t} \xi^{1 - 2\alpha_{\mathbb{P}}(t)}$$

$$D^{\mathbb{R}/a}(t, \xi) \propto e^{b_{\mathbb{R}} t} \xi^{1 - 2\alpha_{\mathbb{R}}(t)} \qquad (7.84)$$

with $b_{\mathbb{P}} = 4.6$ GeV^{-2} and $b_{\mathbb{R}} = 2.0$ GeV^{-2} taken from hadron-hadron-scattering data. The fits were made for $y < 0.45$ to exclude any significant contribution from the longitudinal cross section. The result for the effective pomeron intercept is not affected by the choice of c_I, being $\alpha_{\text{eff}} = 1.200 \pm 0.017 \pm 0.011 \pm 0.032$ for $c_I = 0$ and $\alpha_{\text{eff}} = 1.206 \pm 0.022 \pm 0.013 \pm 0.033$ for $c_I = 1$. The errors are respectively statistical, systematic and an estimate of the model dependence. It is the inclusion of the reggeon term in the H1 analysis which results in α_{eff} for the pomeron term being larger than that obtained from the ZEUS data.

The effective intercept found in the fit to the H1 data is comparable with that found in deep-inelastic scattering over the same Q^2 range, 4.5 GeV$^2 \leq Q^2 \leq 75$ GeV^2, as can be seen from figure 7.5. This suggests a similar interpretation. That is, one should include both the soft and the hard pomeron in the fit.

The reggeon contribution is best studied by selecting values of ξ sufficiently large to suppress the pomeron contribution. Recall that for a single-reggeon

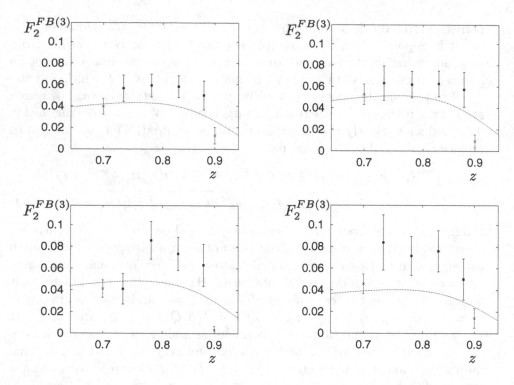

Figure 7.39.　H1 data[380] for forward proton production (black points) and neutron production (crosses) at $(x, Q^2) = (0.00010, 2.5), (0.00033, 4.4), (0.00104, 7.5)$ and $(0.00329, 13.3)$

exchange the differential cross section for forward baryon production as a function of ξ at fixed t is proportional to $\xi^{1-2\alpha(t)}$, where $\alpha(t)$ is the appropriate trajectory. Thus at small t the cross section behaves approximately as ξ^{-1} for soft-pomeron exchange, ξ^0 for the leading reggeon trajectories and ξ for pion exchange.

The region of large ξ has been studied[380] by H1, detecting the forward baryon at $p_T < 200$ MeV using a forward proton spectrometer and a forward neutron calorimeter. The ranges of ξ covered are $0.12 \leq \xi \leq 0.27$ for the proton data and $0.1 \leq \xi \leq 0.7$ for the neutron data. The data are all at very small β, $\beta \leq 0.01$. As these data are not diffractive it is convenient to use the notation $F_2^{FB(4)}$ rather than $F_2^{D(4)}$ of (7.81). For the neutron data with $0.1 < \xi < 0.3$ the shape of the cross section as a function of ξ, averaged over t, suggests that the average value of $\alpha(t)$ is consistent with zero. Although one might be uneasy about extending the theory to such large values of ξ, nevertheless this is naively the expectation from pion exchange. It is not surprising that pion exchange should dominate the neutron data, as we have seen in chapter 3 that the ρ and a_2 trajectories are coupled weakly to the proton. It has been estimated[383] that the contribution to forward neutron production from ρ and a_2 exchange is an order of magnitude smaller than that from pion exchange. Assuming factorisation analogous to (7.82), replacing the pomeron flux by the known pion flux and the pomeron structure function by the known pion structure function[384,385], gives[386] a good parameter-free description of the data. The data and the fit are shown in figure 7.39.

If pion exchange were the only contribution to $F_2^{FB(4)}$ for some range of ξ then we would expect $F_2^{FB(4)}$ measured with forward neutrons to be twice that obtained with forward protons. This is not found to be true. The forward-proton structure function is everywhere larger than the neutron one. The data are shown in figure 7.39. This figure plots $F_2^{FB(3)}$, which is $F_2^{FB(4)}$ integrated over t. The curves are the charged-pion-exchange contribution[380] to the forward-neutron structure function and clearly provide an acceptable description of these data. The leading-proton data do not depend strongly on ξ and are consistent with the dominance of a trajectory with $\alpha(0) \approx 0.5$, for which the f_2 is the natural choice. Recall that the pion-exchange contribution to the forward-proton data is half that to the forward neutron data. As β is so small in the $F_2^{FB(4)}$ data there is little overlap with the $F_2^{D(3)}$ measurements, so the structure function of f_2 is unknown and the data are insufficient to determine it. Assuming[380] that $F_2^{R} = F_2^{\pi}$, adding a pomeron contribution (which is significant only at large z) with[387] $F_2^{P} = (0.026/0.12)F_2^{R}$ and including the pion-exchange term provides an acceptable description of the data.

A global fit to the H1 and ZEUS $F_2^{D(3)}$ data has been performed[388], including contributions from pomeron exchange with the pomeron intercept as a free parameter, together with reggeon exchange and higher-twist terms[389–391], although the latter do not make a significant difference to the results. It was assumed that the reggeon structure function has the same Q^2 and β dependence as the pion structure function and interference between the pomeron and reggeon was not included. The pomeron's quark and gluon structure functions were parametrised at the scale $Q_0^2 = 3\ \text{GeV}^2$ and evolved to higher Q^2 using next-to-leading-order DGLAP evolution equations. The results confirm those of the earlier H1[381] and ZEUS[382] analyses. The pomeron structure functions extracted from the two data sets are very different, which is not surprising as there is a significant reggeon component in the fit to the H1 data and effectively none in the fit to the ZEUS data. The principal conclusions from the analyses[381,388] of the H1

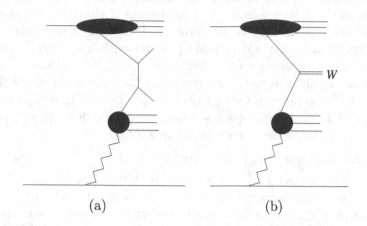

Figure 7.40. pp collisions with a very fast proton in the final state: production of (a) a high-P_T jet pair and (b) W boson.

data are that the gluon density of the pomeron is very much larger than the quark density, and that the gluon density is very much flatter, as a function of β, than is the gluon density of the proton as a function of x. An alternative solution in which the gluon density is strongly peaked near $\beta = 1$ cannot be excluded but is less favoured.

However it is not clear what significance one should attach to the pomeron structure function. If the effective pomeron in these fits is in fact a combination of the hard and soft pomeron, as its effective intercept would indicate, then the picture is much more complex.

The Ingelman-Schlein suggestion[8] mentioned at the beginning of this section was first tested[392] at the CERN $p\bar{p}$ collider, where events were found with a very fast proton or antiproton in which a pair of high-p_T jets is also produced. Similar experiments have since been performed at the Tevatron[393]. High-p_T jet production is calculated using the structure functions of the incident particles, so these special high-p_T jet production events involve a structure function of the proton for the emission of a quark or a gluon with also a very fast proton or antiproton in the final state. See figure 7.40a.

Similarly, W production has been observed at the Tevatron with a very fast proton or antiproton in the final state[394], which one might expect to calculate from the Drell-Yan mechanism, figure 6.1, using a similar structure function, as is shown in figure 7.40b.

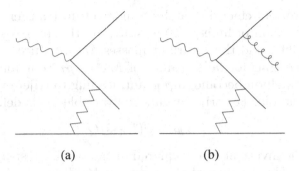

Figure 7.41. Diffractive dijet and trijet production

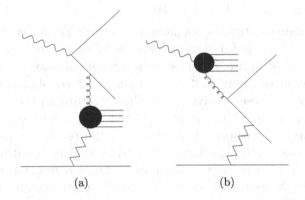

Figure 7.42. Diffractive jet production: (a) resolved-pomeron and (b) resolved-photon mechanisms

Comparison of the rates of occurrence of the various types of hard diffractive events does not support the notion of factorisation. We have already said that this is not surprising if a combination of the soft and hard pomerons is involved. But there is also another possibly-significant source of factorisation breakdown, which is soft initial and final-state interactions in the processes of figure 7.40. There is no reliable theory of these, though some authors believe that they give a very large contribution[395,396].

7.11 Diffractive jet production

In principle the pomeron-parton interaction can best be studied in the deep-inelastic diffraction-dissociation reaction $\gamma^* p \to X p$, when the proton emerges very fast and the system X is composed of two or three jets. Dijet events may be visualized as the dissociation of the photon into a quark-antiquark pair, one of which may then scatter elastically on the proton, so giving two jets emerging at high momentum transfer; see figure 7.41a. As the production of high-p_T final states by this processs is strongly suppressed[397,398] and the invariant masses M_X produced tend to be small, it is expected that for large values of M_X or p_T contributions due to the radiation of a gluon become important, leading to trijet events, as in figure 7.41b for example. For trijet events the variable $z_{I\!P}$ is defined by

$$z_{I\!P} = \beta(1 + \hat{s}/Q^2) \tag{7.85}$$

where $\hat{s}$ is the invariant-mass squared of the two highest-p_T partons. Analogously to β for the case of the lowest-order diagram, figure 7.41a, $z_{I\!P}$ is the longitudinal-momentum fraction of the pomeron which takes part in the hard interaction of figure 7.41b.

In practice the situation is not nearly so clear. Trijet events with one low-p_T jet cannot be separated from genuine dijet events. In addition to the direct reaction of figure 7.41a, significant contributions arise from the resolved pomeron of figure 7.42a, from the resolved photon of figure 7.42b, and from both. These involve either the structure function of the pomeron or photon or both. Thus it is expected that dijet events will typically exhibit a structure where, in addition to the reconstructed jets, the system X contains hadronic energy with transverse momentum below the jet scale. This is evident in the data[399]. Only a fraction of the available energy of the X system is contained in the dijet system or, equivalently, the squared dijet invariant mass M_{12}^2 is considerably smaller than M_X^2 on average. The differential dijet cross sections are shown in figure 7.43 as functions of Q^2, $p_{T\text{jet}}$, M_X, W, β and ξ. W is the $\gamma^* p$ centre-of-mass energy; the transverse momentum is measured relative to the $\gamma^* p$ axis and $p_{T\text{jet}} = \frac{1}{2}(p_{T\text{jet1}} + p_{T\text{jet2}})$ is the mean dijet transverse momentum. The histograms are the predictions[399] from a resolved pomeron, reggeon and photon model, using the parton distributions of the pomeron and reggeon obtained in the analysis[381] of $F_2^{D(3)}$ evolved to a scale $Q^2 + p_T^2$ and with the pomeron effective intercept and slope fixed at the values found in that analysis. These predictions use the "flat" gluon distribution which provides a better fit to the data than does the "hard" distribution. The result of varying the effective intercept of the pomeron is illustrated in the last of the figures 7.43, where the solid histogram is for intercept 1.08 and the other is for intercept 1.4. If the effective intercept is varied, a best fit is found with $\alpha_{\text{eff}}(0) = 1.17 \pm 0.03 \pm 0.06^{+0.03}_{-0.07}$ which is compatible both with the result found in [381] and with the standard soft-pomeron intercept. The errors are respectively statistical, systematic and an estimate of the model dependence.

The dijet cross sections are sensitive to differences between phenomenolog-

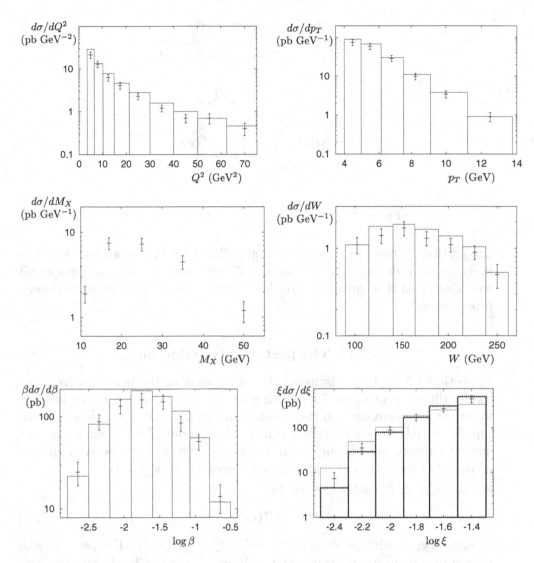

Figure 7.43. Diffractive dijet production[399]: Q^2, p_T, M_X, W, β and ξ distributions. The histograms are described in the text.

ical models which all give reasonable descriptions of $F_2^{D(3)}$. In addition to favouring the flat gluon distribution of the pomeron over the hard one, the data also pose difficulties for other models. The soft-colour model[400–402] is not able to reproduce simultaneously the overall dijet rate and the shapes of the differential cross sections. The saturation model[403,404], which takes into account only k_T-ordered configurations, describes the shapes of

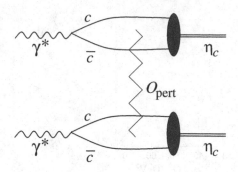

Figure 7.44. $C = -1$ exchange in $\gamma^*\gamma^* \to \eta_c + \eta_c$

the jet distributions but under-estimates the overall cross section. However if strong-k_T ordering is not imposed[405,406] then the data are reasonably well described if a cutoff of 1.5 GeV is imposed for the gluon transverse momentum.

7.12 The perturbative odderon

In section 7.3 we have discussed the summation of the perturbative-gluon ladder diagrams of figure 7.8, which give $C = +1$ exchange. In the leading-logarithmic approximation the result of this summation is obtained by solving the BFKL equation. It is natural to ask about $C = -1$ exchange in a similar context, that is the perturbative odderon, a perturbative analogue of the phenomenological odderon introduced in section 3.10.

As an example consider the reaction

$$\gamma^*(Q^2) + \gamma^*(Q^2) \to \eta_c + \eta_c \tag{7.86}$$

at high energy, with photons of virtuality $Q^2 \gg 1$ GeV^2. The pseudoscalar $c\bar{c}$ bound state η_c has mass $m_{\eta_c} \approx 3$ GeV and $C = +1$. Clearly we must have $C = -1$ exchange in the t-channel in this reaction, as indicated in figure 7.44. Since Q^2 and m_{η_c} are large, a perturbative analysis of (7.86) should be applicable. In leading order in α_s the exchange in figure 7.44 consists of three perturbative gluons in a colour-singlet $C = -1$ state. To see how this arises, first consider the colour-neutral states of two and three gluons. From two gluon-potential matrices (6.4) we can construct one colour-singlet operator

$$\mathcal{P}_{\lambda\mu}(x,y) = \text{Tr}\left(\mathbf{A}_\lambda(x)\mathbf{A}_\mu(y)\right)$$
$$= \tfrac{1}{2}A_\lambda^a(x)\,A_\mu^b(y)\,\delta_{ab}. \tag{7.87}$$

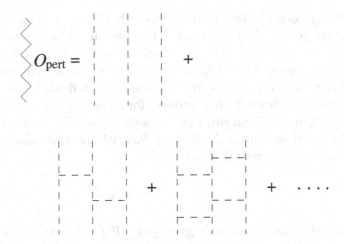

Figure 7.45. Ladder diagrams contributing to the perturbative odderon

Under charge conjugation

$$\mathbf{A}_\lambda(x) \rightarrow -\mathbf{A}_\lambda^T(x) \tag{7.88}$$

which leaves the operator $\mathcal{P}_{\lambda\mu}(x,y)$ invariant. The exchange of two perturbative gluons in a colour-neutral state is described by the free-field correlation function

$$\langle 0| \text{T}\, \mathcal{P}_{\lambda'\mu'}(x',y')\, \mathcal{P}_{\lambda\mu}(x,y)|0\rangle. \tag{7.89}$$

The C-parity of the exchange equals the C-parity of $\mathcal{P}_{\lambda\mu}$, that is $C = +1$.

In a similar way we can analyse the colour-singlet exchange of three perturbative gluons. With three gluon-potential matrices we can construct two colour-singlet operators

$$\mathcal{P}_{\lambda\mu\rho}(x,y,z) = -i\, \text{Tr}\, ([\mathbf{A}_\lambda(x), \mathbf{A}_\mu(y)]\mathbf{A}_\rho(z))$$
$$= \tfrac{1}{2} f_{abc} A_\lambda^a(x) A_\mu^b(y) A_\rho^c(z) \tag{7.90}$$

and

$$\mathcal{O}_{\lambda\mu\rho}(x,y,z) = \text{Tr}\, (\{\mathbf{A}_\lambda(x), \mathbf{A}_\mu(y)\}\mathbf{A}_\rho(z))$$
$$= \tfrac{1}{2} d_{abc} A_\lambda^a(x) A_\mu^b(y) A_\rho^c(z). \tag{7.91}$$

The constants f_{abc} and d_{abc} are defined in appendix B. Using (7.88) we see that $\mathcal{P}_{\lambda\mu\rho}$ has $C = +1$ and $\mathcal{O}_{\lambda\mu\rho}$ has $C = -1$. The correlation function

$$\langle 0| \text{T}\, \mathcal{O}_{\lambda'\mu'\rho'}(x',y',z')\, \mathcal{O}_{\lambda\mu\rho}(x,y,z)|0\rangle$$

evaluated in free-field theory gives perturbative three-gluon exchange in the colour-singlet $C = -1$ state, that is odderon exchange, to lowest order in α_s. In reaction (7.86) this exchange leads to a constant cross section for large s. In Regge terminology it corresponds to a pole which, at least until higher-order corrections are included, is fixed at $l = 1$. Clearly we expect corrections from higher orders, for example from ladder diagrams as shown in figure 7.45, to produce a moving pole. The equation summing up these ladder diagrams in the leading-logarithmic approximation[407,408] has been solved[409] with the result

$$\alpha_{O_{\text{pert}}}(0) = 1 - 0.2472\,\frac{3\alpha_s}{\pi}. \tag{7.92}$$

Taking $\alpha_s = 0.20$ as an example gives $\alpha_{O_{\text{pert}}}(0) \approx 1 - 0.05$, that is an intercept slightly less than 1. In contrast, the intercept of the BFKL pomeron was found in leading order to be $\alpha_{I\!P_{\text{pert}}}(0) \approx 1.5$ using the same value of α_s as in section 7.3. Another solution of the equation has been found[410] which gives a perturbative-odderon intercept exactly equal to 1. It was also found that the solution corresponding to (7.92) decouples in the reaction (7.86). Nevertheless one can deduce from these calculations that to leading order in α_s there is one odderon pole with intercept at $l = 1$ and at least one, maybe even a whole family[410], of odderon poles with intercepts slightly below $l = 1$. Of course, as for the pomeron case, nonleading terms may be large and change the perturbative-odderon intercepts substantially.

A number of high-energy reactions have been proposed[411,412] for detecting the perturbative odderon, for example $\gamma^{(*)} + p \rightarrow \eta_c + p$ and $\gamma + \gamma \rightarrow M + M$ at $|t| > 3$ GeV^2, where M stands for a pseudoscalar or tensor meson. In all cases three-gluon exchange is the leading perturbative contribution. Since it is clear that the exchange of three gluons occurs, the real issue is that of how to observe the special effects associated with the gluon-gluon interaction in the ladder diagrams of figure 7.45. This will not be easy, as we can see from the example of diffractive η_c photoproduction, $\gamma + p \rightarrow \eta_c + p$. In two lowest-order calculations the total cross section was estimated to be[148] $\sigma \approx 11$ pb and[413] $\sigma \approx 47$ pb. Summation of the leading logarithmic terms leads[414] to the estimate $\sigma \approx 50$ pb, which is essentially the same as the larger lowest-order estimate.

The theoretical and experimental status of the perturbative odderon are reviewed in [411] and [412].

8

Soft diffraction and vacuum structure

In this chapter we discuss an approach whereby soft diffractive phenomena are treated from a microscopic point of view starting from the scattering of the hadrons constituents, that is quarks and gluons, and we relate scattering phenomena to properties of the QCD vacuum. We have argued in chapter 6 that in QCD total cross sections are essentially nonperturbative quantities. Thus it is quite natural to think about a possible connection between the nontrivial vacuum structure of QCD, which is a typical nonperturbative phenomenon, and soft high-energy reactions.

8.1 The Landshoff-Nachtmann model

The Landshoff-Nachtmann model[7] seeks to understand some features of diffractive phenomena in hadron-hadron scattering in terms of the exchange of two nonperturbative gluons between quarks. It was shown that this model is capable of reproducing the additive-quark rule for total cross sections[89,415–417], which we introduced in chapter 3. If one calculates two-gluon exchange in QCD perturbation theory, one does not obtain such a result[418,298]. By making detailed assumptions about the nature of the wave functions of mesons and baryons, it is possible to obtain the additive-quark rule for total cross sections from perturbative two-gluon exchange [419,420]. However, this perturbative exchange of two gluons gives the elastic hadron-hadron scattering amplitudes a singularity at $t = 0$ and does not reproduce the t dependence found in experiment. The observed t dependence is rather related to the elastic form factor and is obtained naturally when one makes the pomeron couple to single quarks like an even-signature isoscalar photon.

Thus in this section we will study a model containing one quark field $\psi(x)$ and one Abelian gluon field whose potential and field-strength tensor will be denoted by $A_\mu(x)$ and $F_{\mu\nu}(x) = \partial_\mu A_\nu(x) - \partial_\nu A_\mu(x)$. Let the Lagrangian be

$$\mathcal{L} = \mathcal{L}_0 + \mathcal{L}' \qquad (8.1)$$

where

$$\mathcal{L}_0 = \mathcal{L}_G + \tfrac{1}{2}i(\bar{\psi}\gamma^\mu\partial_\mu\psi - (\partial_\mu\bar{\psi})\gamma^\mu\psi) - m\bar{\psi}\psi \qquad (8.2)$$

and

$$\mathcal{L}' = -gA_\mu\bar{\psi}\gamma^\mu\psi. \qquad (8.3)$$

The term $\mathcal{L}_G$ is the pure "gluon" part of the Lagrangian which will not need to be specified explicitly, except that we will assume it to contain interaction terms producing a gluon condensate in the vacuum, in analogy to what apparently happens in QCD. The interaction term $\mathcal{L}'$ contains a coupling constant g which, contrary to the QCD case, can be varied independently of the pure-gluon interaction in this Abelian model.

We consider the quantum theory of the quark and gluon fields in the Dirac picture with only $\mathcal{L}'$ as perturbing Lagrangian. To zeroth order in $\mathcal{L}'$ or g we have the $\mathcal{L}_0$ theory describing a free quark field and a self-interacting gluon field. We suppose the $\mathcal{L}_0$ theory to be quantised in Feynman gauge, for example by the method of Gupta and Bleuler. Our basic assumptions are now as follows.

Assumption 1. It makes sense to split the complete gluon propagator of the $\mathcal{L}_0$ theory into a perturbative part Δ_p and a nonperturbative part Δ_{np} summarising the vacuum-condensate effects:

$$\langle 0|T(A_\mu(x)A_\nu(0))|0\rangle = ig_{\mu\nu}\Delta_p(x^2) + ig_{\mu\nu}\Delta_{np}(x^2). \qquad (8.4)$$

Here we define the perturbative part of the gluon propagator as the expectation value of the T-product of two gluon potentials in the perturbative vacuum state:

$$ig_{\mu\nu}\Delta_p(x^2) = \langle 0|T(A_\mu(x)A_\nu(0))|0\rangle_{\text{pert}}. \qquad (8.5)$$

The nonperturbative part in (8.4) can be identified with the expectation value of a nonlocal condensate (see section 6.5):

$$ig_{\mu\nu}\Delta_{np}(x^2) = \langle 0| : A_\mu(x)A_\nu(0) : |0\rangle. \qquad (8.6)$$

We assume that for small distances $(x \to 0)$ the perturbative part dominates over the nonperturbative one in (8.4) and that for large distances the opposite is true.

Assumption 2. We assume factorisation of higher-point functions of the gluon field, for instance

$$\langle 0|\mathrm{T}(A_{\mu_1}(x_1)A_{\mu_2}(x_2)A_{\mu_3}(x_3)A_{\mu_4}(x_4))|0\rangle$$
$$= \langle 0|\mathrm{T}(A_{\mu_1}(x_1)A_{\mu_2}(x_2))|0\rangle\langle 0|\mathrm{T}(A_{\mu_3}(x_3)A_{\mu_4}(x_4))|0\rangle$$
$$+ \text{ all other pairings.} \tag{8.7}$$

This assumption again is made for simplicity: it will allow us to develop a model which reproduces the experimental facts. With these hypotheses the Feynman diagrams of the full theory described by $\mathcal{L}$ can be constructed from the following elements:

- free quark propagators and external lines

- quark-gluon vertices from $\mathcal{L}'$ of (8.3), that is corresponding to $-ig\gamma^\mu$

- gluon lines corresponding to the sum of the perturbative and nonperturbative propagators of the gluon field.

Renormalisation of this theory is straightforward and from now on we assume that it has been carried out.

We study now the nonperturbative gluon propagator and its Fourier transform:

$$\langle 0| : A_\mu(x)A_\nu(y) : |0\rangle = ig_{\mu\nu}\Delta_{\mathrm{np}}((x-y)^2)$$
$$= -ig_{\mu\nu}\int \frac{d^4k}{(2\pi)^4}e^{-ik(x-y)}D_{\mathrm{np}}(k^2) \tag{8.8}$$

which leads to the correlation function

$$\langle 0| : F_{\mu\nu}(x)F_{\lambda\rho}(y) : |0\rangle = -i\int \frac{d^4k}{(2\pi)^4}e^{-ik(x-y)}D_{\mathrm{np}}(k^2)$$
$$\times (k_\mu k_\lambda g_{\nu\rho} - k_\nu k_\lambda g_{\mu\rho} - k_\mu k_\rho g_{\nu\lambda} + k_\nu k_\rho g_{\mu\lambda}) \tag{8.9}$$

for the field strengths. Note that the analyticity structure of $D_{\mathrm{np}}(k^2)$ and the $i\epsilon$ prescription are as usual, since the starting point was a time-ordered product (8.4). Setting $x = y$ in (8.9) we make contact with the analogue of the gluon condensate (6.38) for this Abelian theory:

$$M_c^4 = \langle 0| : g^2 F_{\mu\nu}(x)F^{\mu\nu}(x) : |0\rangle = -ig^2\int \frac{d^4k}{(2\pi)^4}6k^2 D_{\mathrm{np}}(k^2)$$
$$= -g^2\int \frac{d^4K}{(2\pi)^4}6K^2 D_{\mathrm{np}}(-K^2) \tag{8.10}$$

where the $i\epsilon$ prescription has allowed us to make a Wick rotation so that the four-vector K in the last line is Euclidean, $k^2 \rightarrow -K^2$. We see that a finite gluon condensate requires $D_{np}(-K^2)$ to vanish faster than $(K^2)^{-3}$ for large K^2, which we will assume to hold in the following.

Gluons in QCD are supposed to be confined and we assume the same to hold in this Abelian model. So on-shell gluons do not propagate, which implies that the gluon propagator must be less singular[421] than $1/k^2$ for $k^2 \rightarrow 0$. Since in this case the gluon propagator for small k^2 is, by assumption, dominated by $D_{np}(k^2)$, we find from (8.9)

$$\int d^4x \, \langle 0| : F_{\mu\nu}(x)F_{\lambda\rho}(y) : |0\rangle$$
$$= -iD_{np}(k^2) \left(k_\mu k_\lambda g_{\nu\rho} - k_\nu k_\lambda g_{\mu\rho} - k_\mu k_\rho g_{\nu\lambda} + k_\nu k_\rho g_{\mu\lambda}\right)\Big|_{k=0}$$
$$= 0 \tag{8.11}$$

which in the domain picture would be interpreted as zero correlation of the nonperturbative fluctuations of the gluon field in different domains. We shall find that experiment indeed requires $D_{np}(k^2)$ to be finite at $k^2 = 0$ and we will formulate this here as our third basic assumption:

Assumption 3. $D_{np}(k^2)$ is finite for $k^2 = 0$.

The gluon correlation function (8.8) may be regarded as the nonperturbative propagator and will play a central role. We find it convenient to write it as

$$g^2 D_{np}(k^2) = -\frac{M_c^4}{6} a^6 F(k^2 a^2) \tag{8.12}$$

where M_c is defined in (8.10), F is a dimensionless function and a is a parameter of dimension length. Inserting (8.12) into (8.10), we see that F must satisfy

$$\frac{1}{16\pi^2} \int_0^\infty dz \, z^2 F(-z) = 1. \tag{8.13}$$

To define a, we require also

$$\frac{1}{8\pi} \int_0^\infty dz \, (F(-z))^2 = 1. \tag{8.14}$$

Inserting (8.12) into (8.9), we get

$$\langle 0| : g^2 F_{\mu\nu}(x)F^{\mu\nu}(y) : |0\rangle = M_c^4 \, f\left(\frac{(x-y)^2}{a^2}\right) \tag{8.15}$$

where

$$f\left(\frac{z^2}{a^2}\right) = i \int \frac{d^4k}{(2\pi)^4} e^{-ikz} \, k^2 a^6 \, F(k^2 a^2) \tag{8.16}$$

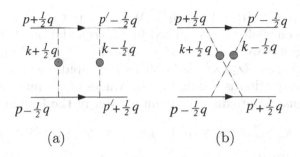

Figure 8.1. Exchange of two nonperturbative gluons between two quarks; (a) direct exchange and (b) crossed diagram.

is again a dimensionless function with $f(0) = 1$. Compare (8.10) and (8.14). From (8.15) we see that the parameter a is a measure of the correlation length in the vacuum.

We now explore the consequences of our assumptions first for quark-quark and then for hadron-hadron scattering, assuming that hadrons exist in this model as bound states of quarks and antiquarks as in QCD. Since diffractive exchange corresponds to the exchange of vacuum quantum numbers, in particular $C = +1$ and colour singlet, it corresponds to at least two gluons, so we will concentrate on two-gluon exchange in the following.

Consider then the diagrams of figure 8.1. Since we are concerned with small momentum transfer $t = q^2$, say $|t| \leq 1$ GeV^2, the long-distance nonperturbative gluon propagators are involved. There will also be a contribution from the perturbative propagators, but by assumption this is relatively small. Because it is simpler, we calculate the imaginary part of the quark-quark scattering amplitude, to which only the diagram of figure 8.1a contributes. The other diagram, figure 8.1b, in which the two gluons cross one another, contributes only to the real part. Because the exchange is $C = +1$, or even-signature, we may readily calculate the complete amplitude from its imaginary part; see (3.25). In fact, the diagram of figure 8.1b just cancels the real part of the diagram of figure 8.1a, so that in this approximation the amplitude is purely imaginary.

In this section we sketch the calculations of the diagram of figure 8.1a. More details of this calculation are given in [7]. With the labelling of figure 8.1 the momentum transfer q is transverse to both p and p', so that $t = q^2 = -\mathbf{q}_T^2$. We use the standard technique[33] of writing the loop momentum k as

$$k = \frac{xp}{2\nu} + \frac{yp'}{2\nu} + k_T \tag{8.17}$$

where $\nu = p.p' = \frac{1}{2}(s + \frac{1}{2}t - 2m^2)$. When we put the two intermediate-state quark lines on shell, as is required for the calculation of the imaginary part of the diagram, we find that $x = -y \sim \mathbf{k}_T^2 - \frac{1}{4}\mathbf{q}_T^2$. Because, according to our assumptions, $D_{\mathrm{np}}(k^2)$ falls off rather rapidly when k^2 becomes large, the loop integration is dominated by values of $\mathbf{k}_T^2$ much less than order ν. Hence we may approximate the numerator of the Feynman integral:

$$(\gamma^\mu \, \gamma.(p - k) \, \gamma^\nu) \otimes (\gamma_\mu \, \gamma.(p' + k) \, \gamma_\nu) \approx [\gamma^\mu \, \gamma.p \, \gamma^\nu) \otimes (\gamma_\mu \, \gamma.p' \, \gamma_\nu). \quad (8.18)$$

We may verify that if we perform the traces necessary to calculate either elastic scattering or the total cross section, then for $\nu \gg |t|$ the result is the same as if we were to replace (8.18) with

$$4\nu \, \gamma_\mu \otimes \gamma^\mu. \quad (8.19)$$

This is just the type of coupling of quarks to the pomeron which seems to be supported by experiment, as we explained in section 3.2. In the Abelian-gluon model the pomeron couples like a $C = +1$ isoscalar photon. The two δ-functions that put the intermediate quarks on shell make the integrations over the longitudinal components of k trivial. The result is that figure 8.1 yields the imaginary part

$$\gamma_\mu \otimes \gamma^\mu \, G^4 a^2 \, I(a^2 t) \quad (8.20)$$

where (see (8.12)) the dimensionless constant G^2 is

$$G^2 = \tfrac{1}{6} M_c^4 a^4 \quad (8.21)$$

and the dimensionless function I is

$$I(a^2 t) = \tfrac{1}{2} a^2 \int \frac{d^2 k_T}{(2\pi)^2} F(-a^2 (\mathbf{k}_T + \tfrac{1}{2}\mathbf{q}_T)^2) \, F(-a^2 (\mathbf{k}_T - \tfrac{1}{2}\mathbf{q}_T)^2). \quad (8.22)$$

It is easy to see that I is indeed a function of $a^2 t$, with $I(0) = 1$ according to the second condition of (8.14). The constant G^2 may be regarded as the expansion parameter: contributions to the quark-quark scattering amplitude that involve additional gluons will have a structure that differs from (8.20) in that they have higher powers of G^2, together with some large logarithms. Note that the integral in (8.22) converges at $\mathbf{k}_T = \pm\frac{1}{2}\mathbf{q}_T$ since $F(0)$ is finite according to assumption (3), whereas a free-gluon propagator, $F(k^2) \propto 1/k^2$ for $k^2 \to 0$, would lead to an infrared divergence. Also we have seen that a finite value for the gluon condensate requires the function $F(-z)$ to fall off faster than z^{-3} for $z \to \infty$, so that the integral (8.22) converges rapidly for $|\mathbf{k}_T| \to \infty$. Hence the approximation (8.18) is justified and the ensuing properties (8.19) and (8.20) of the pomeron-exchange amplitude find their physical origin in this finite value of the gluon condensate, that is in a fundamental property of the vacuum in this Abelian-gluon model.

Consider now the two-gluon-exchange term in hadron-hadron scattering, figure 8.2. If, as in figure 8.2a, the two gluons couple to the same quark in each hadron, the imaginary part of the amplitude is again given by (8.20) with an extra factor $(F_1(t))^2$ to take account of the hadron wave functions (see section 3.2), where $F_1(t)$ is the $C = +1$ isoscalar elastic form factor of the hadron. However, if in either or both of the hadrons the gluons couple to two different quarks as in figure 8.2b, then the loop momentum k has to pass through at least one hadron wave function. So the integral (8.22) becomes modified. At $t = 0$ it is now given by

$$\tfrac{1}{2}a^2 \int \frac{d^2 k_T}{(2\pi)^2} (F(-a^2 \mathbf{k}_T^2))^2 \, V(-R^2 \mathbf{k}_T^2) \qquad (8.23)$$

where V is a hadron wave function or a product of two of them. Presumably the scale R contained in V is the hadron radius. In section 3.2 we have argued that the pomeron seems to couple to single quarks, indicating that this contribution to the amplitude is relatively unimportant for light hadrons. The way to achieve this is to impose the condition

$$a^2 \ll R^2 \qquad (8.24)$$

so that the integral (8.23) is approximately

$$\tfrac{1}{2}a^2 \big(F(0)\big)^2 \int \frac{d^2 k_T}{(2\pi)^2} V(-R^2 \mathbf{k}_T^2) = \frac{a^2}{8\pi R^2} \big(F(0)\big)^2 \int_0^\infty dz \, V(-z). \qquad (8.25)$$

It is reasonable to expect that the last integral converges. Also, $F(0)$ should be finite by assumption (3), for otherwise the two-gluon-exchange terms would give the hadron-hadron scattering amplitude a singularity at $t = 0$. This is not allowed by the usual arguments of analyticity and unitarity. Indeed, we have seen in chapter 1 that the scattering amplitude is an analytic function of t for $t < m^2$, where m is the mass of the lightest hadron or hadronic system which can be exchanged in the t-channel. Usually this lightest hadron is the pion. We conclude that, because of (8.24) and the factor a^2/R^2 in (8.25), it is preferred that in each hadron the two gluons couple to a single quark.

Put in more physical terms, what we have found is that in this very simple model two quarks can exchange a nonperturbative gluon only if they pass within distance a of each other, where a is the correlation length associated with the vacuum. If a is much less than the hadron radius R, it is unlikely

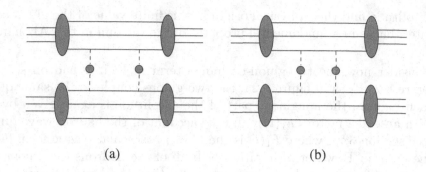

Figure 8.2. Two-gluon exchange in nucleon-nucleon scattering. Examples of diagrams with exchange of two gluons (a) between the same quarks and (b) where in one nucleon the gluons couple to separate quarks.

that two quarks in a hadron will satisfy this condition, so that it is most likely that only one quark will participate in the scattering.

We summarise the main results of this section so far. We have investigated high-energy diffractive quark-quark and hadron-hadron scattering in a model where quarks are coupled to Abelian gluons which we assumed to form a gluon condensate as in real QCD. We argued that for small-t elastic scattering the exchange of nonperturbative gluons should give the dominant contribution. We calculated quark-quark scattering with exchange of two nonperturbative gluons and found for the invariant T-matrix element for large ν

$$\langle q(p_3)q(p_4)|T|q(p_1)q(p_2)\rangle \sim i(\bar{u}(p_3)\gamma^\mu u(p_1))\,(\bar{u}(p_4)\gamma_\mu u(p_2))\,G^4 a^2\, I(a^2 t).$$

$$(8.26)$$

From this the optical theorem gives the spin-averaged total quark-quark cross section:

$$\sigma^{\mathrm{Tot}}(qq) \sim 2G^4 a^2.$$

$$(8.27)$$

These results are consistent with a dimensional analysis of the diagrams of figure 8.1. Each gluon propagator together with the coupling to the quarks gives a factor $G^2 a^2$; see (8.12) and (8.21). One factor a^2 is cancelled by the loop integral which is dominated by small k_T for these nonperturbative gluons, leading in total to the factor $G^4 a^2$ in (8.26) and (8.27).

For nucleon-nucleon scattering we considered the diagrams of figure 8.2. The gluon lines, together with the vertices for both types of diagram, figures 8.2a and 8.2b, give a factor $G^4 a^4$ in the amplitude. The loop integration, however, gives a factor $1/a^2$ for the diagrams of type figure 8.2a and a factor $1/R^2$ for diagrams of type figure 8.2b. This leads to a relative suppression factor a^2/R^2 for diagrams where the gluons do not couple to the same quark in a hadron. Here the assumption that the correlation length a of the vacuum is much smaller than the hadron radius R is crucial.

Thus we have shown in this section how in this Abelian model some properties of the pomeron arise which are similar to those observed in real hadron-hadron scattering: the vector-like coupling to quarks of (8.20) or (8.26), the additive-quark rule and the constant cross section (8.27) to lowest order in G^2. The interesting feature of the Landshoff-Nachtmann model is that it provides a quantitative connection between high-energy diffractive reactions and properties of the nonperturbative vacuum. The model has been applied phenomenologically to various reactions, for example to dijet production in diffractive virtual-photon-proton collisions[372].

There have been two interesting theoretical developments concerning the model. The first assumes that in real QCD the simple formulae discussed above also hold and all the nonperturbative aspects of quark-quark and hadron-hadron scattering are determined by the gluon propagator, which can then be calculated in lattice gauge theory in a given gauge. The results obtained in this way compare quite well with phenomenology[422]. The second one tries to extend the model to higher orders in the coupling using the BFKL techniques described in section 7.3, summing gluon ladders but inserting now the nonperturbative propagator. This leads to an energy dependence of total cross sections which is closer to experiment than the one obtained from gluon ladders with perturbative propagators, but is still far from being satisfactory[6]. Of course in the real world quarks are coupled to non-Abelian gluons. In the following sections we will describe attempts to deal with this essential complication. We will see that in an approximation scheme for QCD suitable for high-energy scattering, some of the results are similar to the ones obtained in this section for the Abelian model, but some are different. This is explained in section 8.5.

8.2 Functional-integral approach

In this section we discuss high-energy hadron-hadron scattering using a functional-integral representation[423,244] of the parton-parton scattering amplitudes. Consider as an example the elastic scattering of two hadrons

$$h_1 + h_2 \rightarrow h_1 + h_2 \qquad (8.28)$$

at high energy and small momentum transfer. We look at reaction (8.28) imagining that we have a microscope with resolution much better than

1 fm for observing what happens during the collision. Of course, we should choose an appropriate resolution for our microscope. If we choose the resolution to be much too good, we see too many details of the internal structure of the hadrons which are irrelevant for the reaction considered and we miss the essential features. If the resolution is too poor, we see too few details and the same is true.

A series of simple arguments based on the uncertainty relation has been used[423] to estimate the appropriate resolution. Let $\tau = 0$ be the time in the centre-of-mass system at which the hadrons h_1 and h_2 have maximal spatial overlap. Let $\frac{1}{2}\tau_0$ be the time when, in an inelastic collision, the first hadrons produced appear. It may be estimated from the Lund model of particle production[424,158] that $\tau_0 \approx 2$ fm. Then the appropriate resolution, that is the cutoff in transverse parton momenta k_T of the hadronic wave functions to be chosen for describing reaction (8.28) in an economical way, is estimated to be

$$k_T^2 \leq \tfrac{1}{2}\sqrt{s}/\tau_0 \qquad\qquad (8.29)$$

where $\sqrt{s}$ is the centre-of-mass energy. Modes with higher k_T can be assumed to be integrated out. Then it is argued[423] that over the time interval

$$-\tfrac{1}{2}\tau_0 \leq \tau \leq \tfrac{1}{2}\tau_0 \qquad\qquad (8.30)$$

the following can be assumed:

- the parton state of the hadrons does not change qualitatively, that is parton annihilation and production processes can be neglected

- the partons undergo only soft elastic scattering and so travel on almost-straight lightlike world lines.

The strategy is first to study soft parton-parton scattering. The relevant interaction will turn out to be mediated by the gluonic vacuum fluctuations. We have argued in section 6.2 that these have a nonperturbative character. In this way the nonperturbative QCD vacuum structure will govern the high-energy soft hadronic reactions. Once we have solved the problem of parton-parton scattering we have to fold the partonic S-matrix with the hadronic wave functions of the appropriate resolution (8.29) to get the hadronic S-matrix elements.

We now give an outline of the various steps in this programme, in which we use standard methods of quantum field theory like the reduction formula and the path-integral formalism. The reader who is interested only in the final result may go to section 8.4.

Consider first quark-quark scattering:

$$q(p_1) + q(p_2) \to q(p_3) + q(p_4).$$ (8.31)

Of course, free quarks do not exist, but let us close our eyes to this at the moment. We should calculate the scattering of the quarks over the finite time interval (8.30) of length $\tau_0 \approx 2$ fm. We assume that 2 fm is long enough to use the standard reduction formula of Lehmann, Symanzik and Zimmermann to relate the S-matrix element for (8.31) to an integral over the four-point function of the quark fields. With this we get

$$\langle 3, 4|S|1, 2\rangle = \langle q(p_3, s_3, a_3)\, q(p_4, s_4, a_4)|S|q(p_1, s_1, a_1)\, q(p_2, s_2, a_2)\rangle$$

$$= \langle 3, 4|1, 2\rangle + Z_\psi^{-2} \int d^4x_1\, d^4x_2\, d^4x_3\, d^4x_4$$

$$e^{ip_4 . x_4}\bar{u}(p_4)(i\gamma.\overrightarrow{\partial}_{x_4} - m_q^R) \otimes e^{ip_3 . x_3}\bar{u}(p_3)(i\gamma.\overrightarrow{\partial}_{x_3} - m_q^R)$$
$$\times \langle 0|Tq(x_4)q(x_3)\bar{q}(x_1)\bar{q}(x_2)|0\rangle$$
$$\times (i\gamma.\overleftarrow{\partial}_{x_1} + m_q^R)u(p_1)e^{-ip_1 . x_1} \otimes (i\gamma.\overleftarrow{\partial}_{x_2} + m_q^R)u(p_2)e^{-ip_2 . x_2}.$$ (8.32)

Here $q(x)$ is the unrenormalised quark field, Z_ψ is the quark wave-function renormalisation constant and m_q^R the renormalised quark mass. In (8.32) spin and colour indices are not written out explicitly.

We can represent the four-point function of the quark fields as a functional integral; see section 6.1. With the shorthand notation $q(x_j) = q(j)$, we get

$$\langle 0|Tq(4)q(3)\bar{q}(1)\bar{q}(2)|0\rangle$$
$$= Z^{-1} \int \mathcal{D}A\, \mathcal{D}q\, \mathcal{D}\bar{q}\ q(4)q(3)\bar{q}(1)\bar{q}(2) \exp\left(i \int d^4x\, \mathcal{L}_{\text{QCD}}(x)\right)$$ (8.33)

where

$$Z = \int \mathcal{D}A\, \mathcal{D}q\, \mathcal{D}\bar{q}\ \exp\left(i \int d^4x\, \mathcal{L}_{\text{QCD}}(x)\right).$$ (8.34)

The QCD Lagrangian (6.1) is bilinear in the quark and antiquark fields. Thus the functional integration over q and $\bar{q}$ can be carried out immediately. After some standard manipulations we arrive at

$$\langle 0|Tq(4)q(3)\bar{q}(1)\bar{q}(2)|0\rangle = \frac{1}{Z} \int \mathcal{D}A\ \exp\left(-i \int d^4x\, \tfrac{1}{2}\, \text{tr}\,(\mathbf{F}_{\lambda\rho}(x)\mathbf{F}^{\lambda\rho}(x))\right)$$
$$\times \prod_q \det[-i(i\gamma^\lambda D_\lambda - m_q + i\epsilon)]\left(\frac{1}{i}S_F(4, 2; A)\frac{1}{i}S_F(3, 1; A) - (3 \leftrightarrow 4)\right).$$ (8.35)

Here $S_F(j, k; A) = S_F(x_j, x_k; A)$ is the unrenormalised quark propagator in the given gluon potential $A_\lambda(x)$. This propagator satisfies

$$(i\gamma.D_x - m_q)S_F(x, y; A) = -\delta^4(x - y). \tag{8.36}$$

Functional integrals as in (8.35) will occur frequently further on. Let $F(A)$ be some functional of the gluon potentials. We will denote the functional integral of $F(A)$ by

$$\langle F(A) \rangle_A = \frac{1}{Z} \int \mathcal{D}A \; F(A) \exp\left(-i \int d^4x \, \tfrac{1}{2} \, \mathrm{tr}\left(\mathbf{F}_{\lambda\rho} \mathbf{F}^{\lambda\rho}\right)\right)$$
$$\times \prod_q \det[-i(i\gamma^\lambda D_\lambda - m_q + i\epsilon)]. \tag{8.37}$$

Here the product runs over all active quark flavours. Gauge-fixing and Faddeev-Popov-determinant terms are included in the measure $\mathcal{D}A$.

We insert (8.35) into (8.32) and get

$$\langle 3, 4|S|1, 2\rangle = \langle 3, 4|1, 2\rangle - Z_\psi^{-2}\langle \mathcal{M}_{31}^F(A)\mathcal{M}_{42}^F(A) - (3 \leftrightarrow 4)\rangle_A \tag{8.38}$$

where, with $k = 3, 4$ and $j = 1, 2$,

$$\mathcal{M}_{kj}^F(A) = \int d^4x \, d^4y \, e^{ip_k.x} \bar{u}(p_k)(i\gamma.\overrightarrow{\partial}_x - m_q^R)$$
$$S_F(x, y; A)(i\gamma.\overleftarrow{\partial}_y + m_q^R)u(p_j)e^{-ip_j.y}. \tag{8.39}$$

The term $\mathcal{M}_{31}^F \mathcal{M}_{42}^F$ on the right-hand side of (8.38) corresponds to the t-channel-exchange diagrams of figure 8.3a. The second term, in which the roles of the quarks 3 and 4 are interchanged, corresponds to the u-channel-exchange diagrams of figure 8.3b and should be unimportant for high-energy, small-t scattering; thus we neglect it. For the scattering of different quark flavours it is absent anyway. So

$$\langle 3, 4|S|1, 2\rangle \cong \langle 3, 4|1, 2\rangle - Z_\psi^{-2}\langle \mathcal{M}_{31}^F(A)\mathcal{M}_{42}^F(A)\rangle_A. \tag{8.40}$$

We can interpret $\mathcal{M}_{kj}^F(A)$ as the scattering amplitude for quark j going to k in the fixed gluon potential $A_\lambda(x)$. To see this, define the wave function

$$\psi_{p_j}^F(x) = \int d^4y \, S_F(x, y; A) \, (i\gamma.\overleftarrow{\partial}_y + m_q^R) \, u(p_j)e^{-ip_j.y} \tag{8.41}$$

which satisfies the Dirac equation with the gluon potential $A_\lambda(x)$:

$$(i\gamma.D - m_q)\psi_{p_j}^F(x) = 0. \tag{8.42}$$

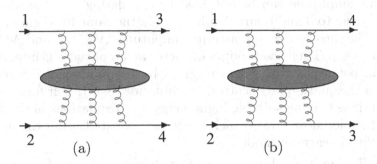

Figure 8.3. (a) t-channel and (b) u-channel exchanges in quark-quark scattering

We use the Lippmann-Schwinger equation for S_F:

$$S_F(x, y; A) = S_F^{(0)}(x - y) - \int d^4z \, S_F^{(0)}(x - z)(g\gamma.\mathbf{A}(z) - \delta m_q)S_F(z, y; A) \tag{8.43}$$

where $S_F^{(0)}$ is the free-quark propagator for mass m_q^R and $\delta m_q = m_q^R - m_q$ is the quark mass shift. Inserting (8.43) and (8.41) into (8.39) gives, after some simple algebra,

$$\mathcal{M}_{kj}^F(A) = \int d^4x \, e^{ip_k.x}\bar{u}(p_k)(g\gamma.\mathbf{A}(x) - \delta m_q)\psi_{p_j}^F(x). \tag{8.44}$$

This represents $\mathcal{M}_{kj}^F$ in the form a scattering amplitude should have: a complete incoming wave is folded with the potential and the free outgoing wave. However, there is a small problem. The wave function $\psi_{p_j}^F(x)$ defined in (8.41) does not satisfy the boundary condition which we should have for using it in the scattering amplitude, that is it does not go to a free incoming wave for time $x^0 \to -\infty$. The wave function $\psi_{p_j}^{\text{ret}}(x)$ with this boundary condition is obtained by replacing the Feynman propagator S^F in (8.41) with the retarded one S^{ret}. The function $\psi_{p_j}^{\text{ret}}(x)$ satisfies (8.42) and

$$\psi_{p_j}^{\text{ret}}(x) \sim u(p_j)e^{-ip_j.x} \quad \text{for} \quad x^0 \to -\infty. \tag{8.45}$$

It has been shown[423] that in the high-energy limit the replacement of $\psi_{p_j}^F(x)$ with $\psi_{p_j}^{\text{ret}}(x)$ in (8.44) can indeed be justified if the gluon potential $A_\lambda(x)$ contains only a limited range of frequencies. This leads to

$$\mathcal{M}_{kj}^F(A) \approx \mathcal{M}_{kj}^{\text{ret}}(A) = \int d^4x \, e^{ip_k.x}\bar{u}(p_k)(g\gamma.\mathbf{A}(x) - \delta m_q)\psi_{p_j}^{\text{ret}}(x). \tag{8.46}$$

We summarise the results so far. At high energy and small t the quark-quark scattering amplitude can be obtained by calculating first the scattering of quark 1 going to 3 and quark 2 going to 4 in the same fixed gluon potential $A_\lambda(x)$. The corresponding scattering amplitudes $\mathcal{M}_{31}^{\text{ret}}(A)$ and $\mathcal{M}_{42}^{\text{ret}}(A)$ are given in (8.46). Then the product of these two amplitudes is integrated over all gluon potentials with measure given by the functional integral (8.37); this gives the quark-quark scattering amplitude (8.40). Our further strategy is to continue making suitable high-energy approximations in the integrand of this functional integral, which finally is evaluated using the methods of the stochastic-vacuum model.

In the following it will be convenient to choose a coordinate system for the description of reaction (8.31) where the quarks 1, 3 move with high velocity essentially in the positive x^3 direction, the quarks 2, 4 in the negative x^3 direction. We define the light-cone coordinates $x_\pm = x^0 \pm x^3$ and in a similar way the $\pm$ components of any four-vector. Then we have for the $\pm$ components of the four-momenta of our quarks $p_{j+} \to \infty$, $p_{j-} \to 0$ for $j = 1, 3$ and $p_{k-} \to \infty$, $p_{k+} \to 0$ for $k = 2, 4$.

8.3 Quark-quark scattering amplitudes

We must now solve the Dirac equation (8.42) for an arbitrary external gluon potential $A_\lambda(x)$. Of course, we cannot do this exactly. But we are interested only in the high-energy, small-t limit. This suggests using a semiclassical or WKB-type approach, which is well known in quantum mechanics; see for instance [425,426]. This does indeed work, but it is not as straightforward as one would at first think, since the Dirac equation is of first order in the derivatives, whereas the semiclassical expansion is easy to make for a second-order differential equation. What was done in [423] was to make an ansatz for the Dirac field $\psi_{p_j}^r(x)$ in terms of a "potential" $\phi_j(x)$:

$$\psi_{p_j}^{\text{ret}}(x) = (i\gamma^\lambda D_\lambda + m_q)\phi_j(x). \tag{8.47}$$

Thus the problem for the Dirac field is essentially reduced to a scalar-field type of problem which is handled more easily.

Now it is fairly straightforward to obtain the semiclassical approximations for $\phi_j(x)$ and $\psi_{p_j}^{\text{ret}}(x)$. Take $j = 1$ as an example, that is the quark coming from the left with large p_{1+}. The final formula for $\psi_{p_1}^{\text{ret}}(x)$ reads

$$\psi_{p_1}^{\text{ret}}(x) = \mathbf{V}_-(x_+, x_-, \mathbf{x}_T)\left(1 + O\left(1/p_{1+}\right)\right)e^{-ip_1 \cdot x}u(p_1) \tag{8.48}$$

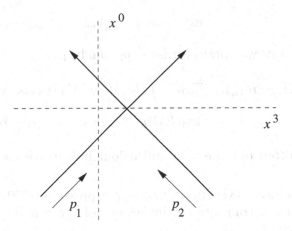

Figure 8.4. Projection on to the (x^0, x^3) plane of the world lines of the quarks 1 and 2 moving at high velocity in positive and negative x^3 directions in Minkowski space

where

$$\mathbf{V}_-(x_+, x_-, \mathbf{x}_T) = \mathrm{P}\exp\left(-\tfrac{1}{2}ig\int_{-\infty}^{x_+} dx'_+ \, \mathbf{A}_-(x'_+, x_-, \mathbf{x}_T)\right). \qquad (8.49)$$

As usual P means path ordering. When the quark comes in, it picks up a non-Abelian phase factor, just the ordered integral of the gluon potential matrix A_- along the path. $\mathbf{V}_-$ is a connector, as introduced in section 6.5, for a straight lightlike line $x_-, \mathbf{x}_T = $ constant, running from $-\infty$ to x; see figure 8.4. In a similar way we obtain, for the quark coming in from the right with large p_{2-},

$$\psi_{p_2}^{\mathrm{ret}}(x) = \mathbf{V}_+(x_+, x_-, \mathbf{x}_T)(1 + O(1/p_{2-}))e^{-ip_2.x}u(p_2) \qquad (8.50)$$

where

$$\mathbf{V}_+(x_+, x_-, \mathbf{x}_T) = \mathrm{P}\exp\left(-\tfrac{1}{2}ig\int_{-\infty}^{x_-} dx'_- \, \mathbf{A}_+(x_+, x'_-, \mathbf{x}_T)\right). \qquad (8.51)$$

Here the integral runs along the line $x_+, \mathbf{x}_T = $ constant. A solution for $\psi_{p_1}^{\mathrm{ret}}$ and $\psi_{p_2}^{\mathrm{ret}}$ as series expansions in powers of $1/p_{1+}$ and $1/p_{2-}$, respectively, has been obtained to all orders[427].

We can now insert our high-energy approximations (8.48) and (8.50) for $\psi_{p_{1,2}}^{\mathrm{ret}}(x)$ into (8.46). The resulting integrals are easily done and we get for large p_{1+} and p_{3+}

$$\mathcal{M}_{31}^{\mathrm{ret}}(A) \sim i\delta_{s_3,s_1}\sqrt{p_{3+}p_{1+}}\int dx_- \, d^2x_T \left(\mathbf{V}_-(\infty, x_-, \mathbf{x}_T) - 1\right)_{a_3,a_1}$$

$$\times \exp \left(\tfrac{1}{2} i (p_3 - p_1)_+ x_- - i(\mathbf{p}_3 - \mathbf{p}_1)_T . \mathbf{x}_T \right). \quad (8.52)$$

In a similar way we obtain for large p_{2-} and p_{4-}

$$\mathcal{M}_{42}^{\text{ret}}(A) \sim i \delta_{s_4, s_2} \sqrt{p_{4-} p_{2-}} \int dy_+ \, d^2 y_T \left(\mathbf{V}_+(y_+, \infty, \mathbf{y}_T) - 1 \right)_{a_4, a_2}$$

$$\times \exp \left(\tfrac{1}{2} i (p_4 - p_2)_- y_+ - i(\mathbf{p}_4 - \mathbf{p}_2)_T . \mathbf{y}_T \right). \quad (8.53)$$

We have written out the spin and colour indices with $s_i = \pm \tfrac{1}{2}$ and $a_i = 1, 2, 3$.

Now we can insert everything into our expression (8.40) for the S-matrix element and use translational invariance of the functional integral. This finally gives

$$\langle 3, 4 | S | 1, 2 \rangle = \langle 3, 4 | 1, 2 \rangle + i(2\pi)^4 \delta(p_3 + p_4 - p_1 - p_2) \langle 3, 4 | T | 1, 2 \rangle$$

$$\langle 3, 4 | T | 1, 2 \rangle = -2i \sqrt{p_{3+} p_{1+} p_{4-} p_{2-}} \, \delta_{s_3, s_1} \delta_{s_4, s_2} Z_\psi^{-2} \int d^2 z_T \, e^{i \mathbf{q}_T . \mathbf{z}_T}$$

$$\times \left\langle \left(\mathbf{V}_-(\infty, 0, \mathbf{z}_T) - 1 \right)_{a_3, a_1} \left(\mathbf{V}_+(0, \infty, 0) - 1 \right)_{a_4, a_2} \right\rangle_A. \quad (8.54)$$

Here $q = p_1 - p_3 = p_4 - p_2$ is the momentum transfer, which in the high-energy limit is purely transverse, $q^2 \to -\mathbf{q}_T^2$. The formula (8.54) has also been obtained[428] using different techniques.

We still have to calculate the wave-function renormalisation constant Z_ψ in (8.54). This can be done by considering a suitable matrix element of the baryon-number current $\tfrac{1}{3} \sum_q \bar{q}(x) \gamma^\mu q(x)$, which is conserved and, therefore, needs no renormalisation. The result is[423]

$$Z_\psi = \tfrac{1}{3} \langle \text{tr} \, \mathbf{V}_-(\infty, 0, 0) \rangle_A. \quad (8.55)$$

We summarise the results obtained so far. The quark-quark scattering amplitude (8.54) is diagonal in the spin indices. Thus we get helicity conservation in high-energy quark-quark scattering. For large p_{1+}, p_{3+}, p_{2-} and p_{4-} we can write the spin factor in (8.54) as

$$2 \sqrt{p_{3+} p_{1+} p_{4-} p_{2-}} \, \delta_{s_3, s_1} \delta_{s_4, s_2} \sim (\bar{u}_{s_3}(p_3) \gamma^\mu u_{s_1}(p_1))(\bar{u}_{s_4}(p_4) \gamma_\mu u_{s_2}(p_2)). \quad (8.56)$$

This $\gamma^\mu \otimes \gamma_\mu$ structure was postulated for high-energy quark-quark scattering in the phenomenological-pomeron model; see section 3.2. Thus, at first sight, the results of the present section seem to pave a road from QCD to the phenomenology of the soft pomeron described in chapter 3. However,

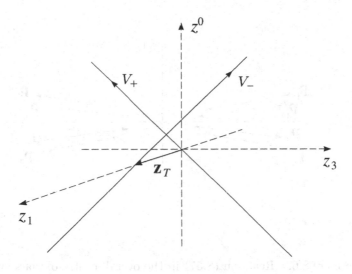

Figure 8.5. Two lightlike lines on which the string operators $\mathbf{V}_\pm$ in (8.54) are evaluated. Their correlation function governs quark-quark scattering at high energies.

as we will see below, things do not turn out to be so simple when we go from quarks to hadrons.

We now study the result (8.54) in more detail. As already emphasised in section 3.2, one should not be misled and interpret the $\gamma^\mu \otimes \gamma_\mu$ coupling in (8.56) as indicating a Lorentz-vector that is purely-spin-1 exchange between the quarks. Indeed, a study[423] of quark-antiquark scattering shows that the amplitude (8.54) has both C-even and C-odd-exchange contributions. The C-even part corresponds to the pomeron and this is not a spin-1 exchange, but the coherent sum of spin 2, 4, 6, ... exchanges. The C-odd part corresponds to the odderon and this is indeed a purely-spin-1 exchange.

The quark-quark scattering amplitude (8.54) is governed by the correlation function of two connectors or string operators $\mathbf{V}_\pm$ associated with two lightlike Wilson lines, figure 8.5. The first numerical evaluations of (8.54) using the methods of the stochastic-vacuum model were performed in [429,430]. However, it turned out that quark-quark scattering was calculable from (8.54) for Abelian gluons only. Indeed, the results of the Landshoff-Nachtmann model described in section 8.1 are easily reproduced in this way[423]. For the non-Abelian gluons difficulties arise, associated with our neglect of quark confinement. A solution[430,58] is to consider directly hadron-hadron scattering, representing the hadrons as $q\bar{q}$ or qqq

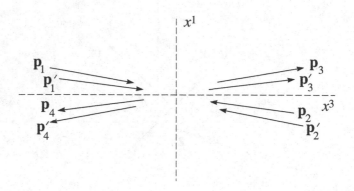

Figure 8.6.   Reaction (8.57) in the overall centre-of-mass system

wave packets. We will see below how this is done.

8.4 Scattering of systems of quarks, antiquarks and gluons

Consider now the scattering of systems of partons. As an example we study
the scattering of two $q\bar{q}$ pairs on each other:

$$(q(1) + \bar{q}(1')) + (q(2) + \bar{q}(2')) \rightarrow (q(3) + \bar{q}(3')) + (q(4) + \bar{q}(4')) \qquad (8.57)$$

where we write $q(i) = q(p_i, s_i, a_i)$, $\bar{q}(i') = \bar{q}(p_i', s_i', a_i')$, with p_i, s_i, a_i and
p_i', s_i', a_i' the momentum, spin, and colour labels for quarks and antiquarks;
see figure 8.6. We suppose that the momentum components p_{i+}, p_{i+}' are
large for i odd and p_{i-}, p_{i-}' are large for i even. The transverse momenta are
supposed to stay limited. Of course, the reduction formula can be applied
to the reaction (8.57). We have to be careful to keep disconnected pieces
because finally such pieces become connected through the wave functions.
The further strategy is completely analogous to the one used in section 8.2
for deriving (8.32) to (8.40). Just as we dropped the u-channel-exchange diagrams there, we drop now all terms which are estimated to give a very small
contribution to high-energy small-momentum-transfer scattering. These
terms are characterised by large momenta of order of the centre-of-mass
energy $\sqrt{s}$ flowing through gluon lines, leading to suppression factors of
order $1/s$. Keeping only the t-channel-exchange terms and performing all
the steps as done in sections 8.2 and 8.3 for quark-quark scattering leads
finally in the high-energy limit to a very simple answer for the S-matrix

element corresponding to reaction (8.57):

$$\langle 3, 3', 4, 4' | S | 1, 1', 2, 2' \rangle \sim$$

$$\langle ((\langle 3|1\rangle - iZ_\psi^{-1}\mathcal{M}_{31}^{\mathrm{ret}}(A))\,(\langle 3'|1'\rangle - iZ_\psi^{-1}\mathcal{M}'^{\mathrm{ret}}_{3'1'}(A))$$

$$\times\;((\langle 4|2\rangle - iZ_\psi^{-1}\mathcal{M}_{42}^{\mathrm{ret}}(A)])\,(\langle 4'|2'\rangle - iZ_\psi^{-1}\mathcal{M}'^{\mathrm{ret}}_{4'2'}(A))\rangle_A\,. \qquad (8.58)$$

Here $\mathcal{M}_{31}^{\mathrm{ret}}(A)$ and $\mathcal{M}_{42}^{\mathrm{ret}}(A)$ are as in (8.46) and $\mathcal{M}'^{\mathrm{ret}}_{3'1'}(A)$ and $\mathcal{M}'^{\mathrm{ret}}_{4'2'}(A)$ are the corresponding amplitudes for the scattering of antiquarks by the gluon potential $A_\lambda(x)$. We have, with S^{ret} the retarded Green's function for quarks in the gluon potential $A_\lambda(x)$,

$$\mathcal{M}'^{\mathrm{ret}}_{k'j'}(A) \sim -\int d^4x\,d^4y\,e^{-ip_{j'}\cdot x}\bar{v}(p_j)(i\gamma.\overrightarrow{\partial}_x - m_q^R)S^{\mathrm{ret}}(x, y; A)$$

$$\times\,(i\gamma.\overleftarrow{\partial}_y + m_q^R)v(p_{k'})e^{ip_{k'}\cdot y} \qquad (8.59)$$

where $(k', j') = (3', 1')$ or $(4', 2')$.

In the high-energy limit the scattering amplitudes (8.59) can again be obtained in the semiclassical approximation. Indeed, we can just use the C-invariance of QCD to get

$$\mathcal{M}'^{\mathrm{ret}}_{3'1'}(A) \sim i\delta_{s_3', s_1'}\sqrt{p_{3+}'p_{1+}'}\int dx_-\,d^2x_T\left(\mathbf{V}_-^*(\infty, x_-, \mathbf{x}_T) - 1\right)_{a_3', a_1'}$$

$$\times\,\exp\left(\tfrac{1}{2}i(p_3' - p_1')_+x_- - i(\mathbf{p}_3' - \mathbf{p}_1')_T.\mathbf{x}_T\right) \qquad (8.60)$$

and a similar expression for $\mathcal{M}'^{\mathrm{ret}}_{4'2'}(A)$. In section 8.2 we have argued that over the time interval (8.30) we can neglect parton production and annihilation processes. For the scattering over such a time interval we should have an effective wave-function renormalisation constant $Z_\psi = 1$, since the deviation of Z_ψ from 1 is just a measure of the strength of the quark splitting processes, $q \to q+G$ etc; see section 7.2. In the calculation of Z_ψ in the framework of the stochastic-vacuum model, one does indeed find $Z_\psi = 1$, showing the consistency of this approach with the simple physical picture of section 8.2. With this result we see from (8.58), (8.53) and (8.60) that in the S-matrix element (8.58) the $\langle k|j\rangle$ and $\langle k'|j'\rangle$ terms cancel with the 1 terms in $\mathcal{M}_{kj}^{\mathrm{ret}}(A)$ and $\mathcal{M}'^{\mathrm{ret}}_{k'j'}(A)$.

This leads us to the following simple rules for obtaining the S-matrix element in the high energy limit. For the right-moving quark $(1 \to 3)$ we have to insert the factor

$$S_{q+}(3, 1) = \delta_{s_3, s_1}\sqrt{p_{3+}p_{1+}}\int dx_-\,d^2x_T\,V_-(\infty, x_-, \mathbf{x}_T)_{a_3, a_1}$$

$$\times\,\exp\left(\tfrac{1}{2}i(p_3 - p_1)_+x_- - i(\mathbf{p}_3 - \mathbf{p}_1)_T.\mathbf{x}_T\right). \qquad (8.61)$$

For the right-moving antiquark $(1' \to 3')$ we have to insert

$$S_{\bar{q}+}(3',1') = \delta_{s'_3,s'_1} \sqrt{p'_{3+}p'_{1+}} \int dx_- \, d^2x_T \, V_-^*(\infty, x_-, \mathbf{x}_T)_{a'_3,a'_1}$$

$$\times \exp\left(\tfrac{1}{2}i(p'_3 - p'_1)_+ x_- - i(\mathbf{p}'_3 - \mathbf{p}'_1)_T \cdot \mathbf{x}_T\right). \quad (8.62)$$

For the left-moving quark $(2 \to 4)$ and antiquark $(2' \to 4')$ we have to exchange the $+$ and $-$ labels everywhere in (8.61) and (8.62). Finally we have to multiply together the factors $S_{q\pm}, S_{\bar{q}\pm}$ and integrate over all gluon potentials with the functional-integral measure (8.37) to get

$$\langle 3, 3', 4, 4'|S|1, 1', 2, 2'\rangle = \langle S_{q+}(3,1)S_{\bar{q}+}(3',1')S_{q-}(4,2)S_{\bar{q}-}(4',2')\rangle_A \; . \quad (8.63)$$

Going from quarks to antiquarks corresponds just to changing from the fundamental representation (3) of the colour group $SU(3)$ to the complex-conjugate representation (3^*), as we see by comparing (8.61) with (8.62).

These rules can be generalised in an obvious way to the scattering of arbitrary systems of quarks and antiquarks on each other. Here we always suppose that we have one distinct collision axis and that one group of partons moves to the right with large momenta, the other group to the left. The transverse momenta are supposed to stay limited.

It has been shown[244] that these rules can also be extended to any gluons that participate in the scattering. One simply has to change the colour representation in (8.61) from the fundamental to the adjoint one. In detail one finds that for a right-moving gluon making the transition $1 \to 3$, the factor

$$S_{G+}(3,1) = \delta_{j_3,j_1} \sqrt{p_{3+}p_{1+}} \int dx_- \, d^2x_T \, \mathcal{V}_-(x_-, \mathbf{x}_T)_{a_3,a_1}$$

$$\times \exp\left(\tfrac{1}{2}i(p_3 - p_1)_+ x_- - i(\mathbf{p}_3 - \mathbf{p}_1)_T \cdot \mathbf{x}_T\right) \quad (8.64)$$

has to be inserted into the S-matrix element. Here j_1, j_3 are the spin indices, which are purely transverse, $j_1, j_3 = 1, 2$. a_1 and a_3 are colour indices with $a_1, a_3 = 1, 2, \ldots, 8$, and $\mathcal{V}_-(x_-, \mathbf{x}_T)$ is the connector for the adjoint representation of $SU(3)$:

$$\mathcal{V}_-(x_-, \mathbf{x}_T) = \mathrm{P} \exp\left(-\tfrac{1}{2}ig \int_{-\infty}^{\infty} dx_+ \, G_-^a(x_+, x_-, \mathbf{x}_T)T_a\right)$$

$$(T_a)_{bc} = -if_{abc}. \quad (8.65)$$

For a left-moving gluon we have again to exchange $+$ and $-$ labels in (8.64).

In sections 8.5 and 8.6 we will go from parton-parton to hadron-hadron scattering. Our strategy will be to represent hadrons by wave packets of partons, where we make simple ansätze for the wave functions. Then the partonic S-matrix element will be constructed by the rules derived in this section, leaving us with functional integrals of the type (8.63). These will be evaluated using the stochastic-vacuum model, suitably continued to Minkowski space-time.

8.5 Evaluation of the dipole-dipole scattering amplitude

As has been stressed in section 6.2, purely-perturbative methods cannot explain the essential parameters of soft high-energy scattering. We must at present have recourse to models. In section 8.1 we have presented such a model which lacks, however, an essential feature of QCD, namely the non-Abelian structure. We now apply the stochastic-vacuum model to high-energy reactions. This model, introduced in section 6.7, describes the vacuum structure in the quenched approximation, in which internal quark loops are neglected. In applying the model to high-energy scattering we also use the quenched approximation to set the fermion determinant in the functional integral (8.37) to 1.

Our starting point is (8.63), where the scattering matrix is expressed as a functional integral over line integrals along quark paths. Using (8.61), (8.62) and (8.56) we can write the part describing the $(3, 1)$-quark system and the $(3', 1')$-antiquark system as

$$
S_{q+}(3, 1)\, S_{\bar{q}+}(3', 1') \approx \int dx_-\, dx'_-\, d^2x_T\, d^2x'_T
$$

$$
\bar{u}_{s_3}(p_3)e^{i(\frac{1}{2}p_{3+}x_- - \mathbf{p}_{3T}\cdot\mathbf{x}_T)}\, \gamma_\mu\, u_{s_1}(p_1)e^{-i(\frac{1}{2}p_{1+}x_- - \mathbf{p}_{1T}\cdot\mathbf{x}_T)}
$$

$$
\times\, \bar{v}_{s'_1}(p'_1)e^{-i(\frac{1}{2}p'_{1+}x'_- - \mathbf{p}'_{1T}\cdot\mathbf{x}'_T)}\, \gamma^\mu\, v_{s'_3}(p'_3)e^{i(\frac{1}{2}p'_{3+}x'_- - \mathbf{p}'_{3T}\cdot\mathbf{x}'_T)}
$$

$$
\times \lim_{T\to\infty}\Big\langle \Big(\mathrm{P}\exp\Big(-\tfrac{1}{2}ig\int_{-2T}^{2T} dx_+\, \mathbf{A}_-(x_+, 0, \mathbf{x}_T)\Big)\Big)_{a_3, a_1}
$$

$$
\times \Big(\mathrm{P}\exp\Big(-\frac{i}{2}g\int_{2T}^{-2T} dx_+\, \mathbf{A}_-(x_+, 0, \mathbf{x}'_T)\Big)\Big)_{a'_1, a'_3} \Big\rangle_A \tag{8.66}
$$

where we have used the hermiticity of $\mathbf{A}_\mu$ and the translational invariance of the integrals.

To ensure that the quark and antiquark which travel along parallel lightlike lines asymptotically form a colour singlet, we have to connect them with a Wilson line at $x_+ = 2T$ in the transverse direction,

$$
\mathrm{P}\exp\Big(-\tfrac{1}{2}ig\int_0^1 d\lambda\, \mathbf{A}_T(2T, 0, \mathbf{x}_T + \lambda(\mathbf{x}'_T - \mathbf{x}_T)\cdot(\mathbf{x}'_T - \mathbf{x}_T)\Big) \tag{8.67}
$$

and an analogous expression at $-2T$, and finally the trace has to be performed. The Wilson lines of the quark scattering amplitude $S_{q+}(3,1)$ and the antiquark amplitude $S_{\bar{q}+}(3',1')$ in (8.66), together with the lines (8.67), form a Wilson loop $W[C_+]$. Analogously, the loop $W[C_-]$ is constructed from $S_{q-}(4,2)$ and $S_{\bar{q}-}(2',4')$. The four corners of C_+ have the light-cone coordinates

$$(-2T,0,\mathbf{x}_T),\ (2T,0,\mathbf{x}_T),\ (2T,0,\mathbf{x}'_T),\ (-2T,0,\mathbf{x}'_T) \qquad (8.68)$$

where $\mathbf{x}_T$ and $\mathbf{x}'_T$ are the transverse coordinates of the quark $(3,1)$ and antiquark $(3',1')$, and T goes to infinity. Correspondingly we have for C_-

$$(-2T,0,\mathbf{y}_T),\ (2T,0,\mathbf{y}_T),\ (2T,0,\mathbf{y}'_T),\ (-2T,0,\mathbf{y}'_T) \qquad (8.69)$$

where $\mathbf{y}_T$ and $\mathbf{y}'_T$ are the transverse coordinates of the quark $(4,2)$ and antiquark $(4',2')$.

We introduce the relative and centre coordinates:

$$\begin{aligned}
\mathbf{R}_1 &= \mathbf{x}_T - \mathbf{x}'_T & \mathbf{X}_1 &= \mathbf{x}'_T + z_1\mathbf{R}_1, & 0 \le z_1 \le 1 \\
\mathbf{R}_2 &= \mathbf{y}_T - \mathbf{y}'_T & \mathbf{X}_2 &= \mathbf{y}'_T + z_2\mathbf{R}_2, & 0 \le z_2 \le 1.
\end{aligned}$$
$$(8.70)$$

The quantity z_i will later be identified with the longitudinal momentum fraction of the quark. The impact vector $\mathbf{b}$ is defined through

$$\mathbf{b} = \mathbf{X}_1 - \mathbf{X}_2. \qquad (8.71)$$

With this definition the t-dependent scattering amplitude can be obtained as a two-dimensional Fourier transform with respect to the impact parameter.

The basic element of the scattering matrix for colour-singlet quark-antiquark dipoles is the expectation value of two loops,

$$S(\mathbf{b},1,2) = \frac{\frac{1}{9}\langle W[C_+]W[C_-]\rangle}{\frac{1}{3}\langle W[C_+]\rangle\frac{1}{3}\langle W[C_-]\rangle} \qquad (8.72)$$

where 1 stands for $\mathbf{R}_1, z_1$ and 2 for $\mathbf{R}_2, z_2$. The denominator is the loop renormalisation constant that replaces the quark-field renormalisation (8.55). We will discuss later how we come from the dipole amplitudes to hadronic scattering or electroproduction amplitudes by integrating over light-cone wave functions.

Our next step is to calculate the expectation value by applying the stochastic vacuum model. We again use the non-Abelian Stokes theorem to express

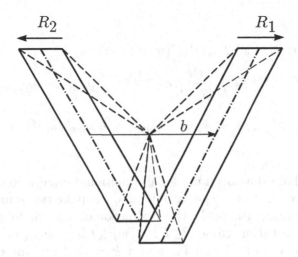

Figure 8.7. Wilson loops formed by the paths of quarks and antiquarks inside two mesons

the line integrals over the colour potential in the loop by a surface integral over the field strength. The reference point w must be common to both surfaces bordered by the loops C_+ and C_-. We choose it at $\mathbf{X}_2 + \frac{1}{2}\mathbf{b}$. As the surface $\mathcal{S}_+$ emerging from the loop C_+ we choose the sliding sides of a pyramid with the loop C_+ as basis and the point w as apex; see figure 8.7. The same holds analogously for $\mathcal{S}_-$.

Before we apply it to high-energy scattering the stochastic-vacuum model must be translated from Euclidean space-time, in which it is naturally formulated and can be compared with lattice calculations, into the Minkowski continuum. Fortunately it turns out that the details of the continuation are irrelevant and that in high-energy scattering we need the invariant functions D and D_1 of the correlator only in the Euclidean region[268]. The continuation of the correlators is made in the following way. Introduce the Euclidean Fourier transform of the Euclidean-space-time correlator

$$i\tilde{D}(k^2) = \int d^4z\, D(z^2)\, e^{ik.z}. \tag{8.73}$$

The correlator functions $\tilde{D}_M(k^2)$ in Minkowski momentum space coincide with $\tilde{D}(-k^2)$ for space-like k and are analytically continued to time-like vectors k. The full correlator in Euclidean space is then obtained by using the correlator functions $\tilde{D}_M(k^2)$ and replacing $\delta_{\mu\nu}$ with the Minkowskian metric tensor $g_{\mu\nu}$:

$$\langle: g^2 F^a_{\mu\nu}(x; C(w, x)) \, F^b_{\rho\sigma}(x'; C(w, x')) :\rangle =$$

$$\frac{1}{96} \delta^{ab} \langle g^2 FF \rangle_A \int \frac{d^4 k}{(2\pi)^4} \, e^{-ik.(x-x')} \Big((g_{\mu\rho} g_{\nu\sigma} - g_{\mu\sigma} g_{\nu\rho}) \, \kappa \, i\tilde{D}_M(k^2)$$

$$+ (-g_{\nu\sigma} k_\mu k_\rho + g_{\nu\rho} k_\mu k_\sigma - g_{\mu\rho} k_\nu k_\sigma + g_{\mu\sigma} k_\nu k_\rho)(1 - \kappa) \, i\frac{d\tilde{D}_{1M}(k^2)}{dk^2} \Big).$$

$$(8.74)$$

After this choice has been made, all functional integrations can in principle be performed. Before we go into details, we make two remarks which facilitate further calculations. In the evaluation of a single loop, say $\langle W[C_+]\rangle_A$, only the expectation value $\langle e^\mu_+ F_{\mu i}(x, w) e^\rho_+ F_{\rho k}(y, w)\rangle_A$ occurs, where e^μ_+ is the lightlike vector $(1, 0, 0, 1)$, and $i, k = 1, 2$ are indices of the transverse plane. These correlators are zero by virtue of the tensor structures given in (8.74) since $g_{++} = g_{--} = g_{\pm i} = 0$. Note that $x.y = \frac{1}{2}x_+ y_- + \frac{1}{2}x_- y_+ - \mathbf{x}_T.\mathbf{y}_T$. Therefore in $W[C_\pm]$ only the unit term contributes, leading to

$$\langle \text{tr } W[C_+]\rangle_A = \langle \text{tr } W[C_-]\rangle_A = 3. \qquad (8.75)$$

This does not contradict the area law in Euclidean space-time, since the area of a loop with lightlike sides becomes zero in Euclidean metric. The denominator in (8.72) equals 1, as anticipated in the discussion following (8.55). In the same way only terms from different pyramids contribute to the expectation value in the numerator of (8.72).

Secondly we show that in the end we need the fundamental correlator only for space-like k, that is in the region accessible directly in Euclidean metric. Applying Stokes theorem as discussed above we obtain for the dipole-dipole amplitude

$$S(\mathbf{b}, 1, 2) = \frac{1}{9} \Big\langle \text{tr } \exp\Big(-ig \int_{\mathcal{S}_+} d\sigma^{\mu\nu} \, \mathbf{F}_{\mu\nu}\Big)$$

$$\times \text{tr } \exp\Big(-ig \int_{\mathcal{S}_-} d\sigma^{\mu\nu} \mathbf{F}_{\mu\nu}\Big)\Big\rangle_A. \qquad (8.76)$$

Using Gauss's theorem one can transform the surface integrals over the four sliding surfaces $\mathcal{S}_i$ of each pyramid into a volume integral over the interior of the pyramid minus a surface integral over its basis. If this is done it is easily seen that in the surface integral $\mathcal{S}_+$ in (8.76), based on the loop with lightlike sides $x^0 = x^3$, we have a distribution $\delta(x^0 - x^3)$, and in the $\mathcal{S}_-$ integral we have $\delta(x'^0 + x'^3)$. Together with the exponential from the

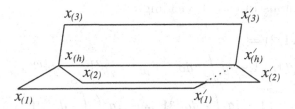

Figure 8.8. Wilson lines describing a baryon path in the three-body picture. The line from $x_{(i)}$ to $x'_{(i)}$ represents the path for the quark labelled i. The line from x_h to x'_h is the path for the central point of the baryon.

Fourier transform in (8.74) we thus have in (8.76) the integration

$$\int_{-\infty}^{\infty} dx^0 \, dx^3 \, dx'^0 \, dx'^3 \, \delta(x^3 - x^0)\delta(x'^3 + x'^0)e^{i\left(k^0(x^0 - x'^0) - k^3(x^3 - x'^3)\right)}. \quad (8.77)$$

This integration yields a factor $(2\pi)^2\delta(k^0 - k^3)\,\delta(k^0 + k^3)$ and hence the correlator $\tilde{D}_M(k^2)$ is evaluated only for space-like vectors with $k^0 = k^3 = 0$, that is in a region where the Euclidean and Minkowski definitions of the correlator coincide.

For the treatment of baryons in QCD two pictures are adopted, a genuine three-body configuration and a diquark picture. In the latter the baryon is described as a "meson" where the diquark replaces the antiquark. In the three-body picture the baryon is described as shown in figure 8.8. There are three quark paths leading from $x_{(i)}$ to $x'_{(i)}$, $i = 1, 2, 3$. The coordinates x_h and x'_h refer to the central point of the baryon. The paths from x_h to $x_{(i)}$ and x'_h to $x'_{(i)}$ must ensure that the baryon is a colour singlet under local gauge transformations. This is done by parallel-transporting the colour from the quark positions $x_{(i)}$ to x_h, and coupling the colours antisymmetrically in the form

$$\frac{1}{\sqrt{6}} \, \epsilon_{abc} \, \phi_{aa'}(x_h, x_{(1)}) \, \phi_{bb'}(x_h, x_{(2)}) \, \phi_{cc'}(x_h, x_{(3)}) \quad (8.78)$$

where $\phi_{a,a'}$ are the components of the Wilson lines defined in (6.43). An analogous factor occurs at the end. It turns out that the diquark picture is not only much simpler to treat both algebraically and numerically, but also that it leads generally to somewhat more consistent results.

The formalism set up in the previous sections allows us to calculate the scattering matrix in terms of the correlator (8.74) using the assumptions of the stochastic-vacuum model. The most straightforward way is to expand the exponentials in (8.76), yielding

$$S(\mathbf{b}, 1, 2) =$$

$$\frac{1}{9}\left\langle \mathrm{tr}\left(1 - ig\int_{\mathcal{S}_+} d\sigma^{\mu\nu}\, \mathbf{F}_{\mu\nu} - \frac{1}{2}g^2\Big(\int_{\mathcal{S}_+} d\sigma^{\mu\nu}\, \mathbf{F}_{\mu\nu}\Big)^2 + \cdots\right)\right.$$

$$\left. \times \mathrm{tr}\left(1 - ig\int_{\mathcal{S}_-} d\sigma^{\kappa\lambda}\mathbf{F}_{\mu\nu} - \frac{1}{2}g^2\Big(\int_{\mathcal{S}_-} d\sigma^{\kappa\lambda}\mathbf{F}_{\kappa\lambda}\Big)^2 + \cdots\right)\right\rangle_A \quad (8.79)$$

where $1 = \mathbf{R}_1, z_1$ and $2 = \mathbf{R}_2, z_2$. We use factorisation in the colour components of the field-strength tensor (6.59) and, recalling (8.75) and the discussion below it, we notice that the first nontrivial nonvanishing term is the one where two field-strength tensors come from each loop and hence the correlators have arguments from different loops:

$$S(\mathbf{b}, 1, 2) = \frac{1}{144}\,\mathrm{tr}\,(\lambda^a\lambda^b)\mathrm{tr}(\lambda^c\lambda^d)$$

$$\times\left(\Big\langle g^4\int_{\mathcal{S}_+} d\sigma^{\mu\nu} F^a_{\mu\nu}\int_{\mathcal{S}_-} d\sigma^{\kappa\lambda} F^c_{\kappa\lambda}\int_{\mathcal{S}_+} d\sigma^{\mu\nu} F^b_{\mu\nu}\int_{\mathcal{S}_-} d\sigma^{\kappa\lambda} F^d_{\kappa\lambda}\Big\rangle_A\right)$$

$$+ \cdots \quad (8.80)$$

where the dots represent products of more than four field-strength tensors. Inserting the correlator and parametrising the surface integrals finally leads to

$$S(\mathbf{b}, 1, 2) = 1 - \tfrac{1}{9}\chi^2(\mathbf{b}, 1, 2) \quad (8.81)$$

with

$$\chi(\mathbf{b}, 1, 2) = \frac{1}{96}\,\langle g^2 FF\rangle\,\Big(I(\mathbf{x}_T, \mathbf{y}_T) + I(\mathbf{x}'_T, \mathbf{y}'_T) - I(\mathbf{x}_T, \mathbf{y}'_T) - I(\mathbf{x}'_T, \mathbf{y}_T)\Big). \quad (8.82)$$

Using the special form of the correlators (6.50) we obtain

$$I(\mathbf{x}_T, \mathbf{y}_T) = \tfrac{1}{2}\pi\kappa\int_0^1 dv\Big(|v\mathbf{y}_T - \mathbf{x}_T|^2\, K_2(\lambda^{-1}|v\mathbf{r}_2 - \mathbf{x}_T|)$$

$$+ \, |\mathbf{y}_T - v\mathbf{x}_T|^2\, K_2(\lambda^{-1}|\mathbf{y}_T - v\mathbf{r}_1|)\Big)$$

$$+ (1 - \kappa)\pi\lambda^2|\mathbf{y}_T - \mathbf{x}_T|^2\, K_3(\lambda^{-1}|\mathbf{y}_T - \mathbf{x}_T|) \quad (8.83)$$

with $\lambda = 3\pi/(8a)$. If the exponentials in (8.79) are expanded further the next nonvanishing term will contain the expectation value of six gluon fields. It contains a C-even and a C-odd contribution. The C-odd contribution is symmetric in colour and therefore path ordering of the colour matrices is no restriction and the expectation value can be evaluated in a way similar to before. The result for the leading C-odd contribution is

$$S_{C=-1}(\mathbf{b}, 1, 2) = -\frac{5i}{324} \chi^3 . \tag{8.84}$$

The C-odd contribution corresponds to odderon exchange; see section 3.9. A more refined method has been developed[431], in which the main idea is to interpret the product of the two traces of 3×3 matrices (generators of $SU(3)$ in the fundamental representation) as one trace tr_2 in the product space of the two fundamental representations of $SU(3)$; that is tr_2 acts in $SU(3) \otimes SU(3)$. So

$$S(\mathbf{b}, 1, 2) = \tfrac{1}{9} \, \mathrm{tr}_2 \Big\langle \exp\Big(-ig \int_{\mathcal{S}_+} d\sigma^{\mu\nu} F^a_{\mu\nu} (\tfrac{1}{2}\lambda^a \otimes 1) \Big)$$
$$\times \exp\Big(-ig \int_{\mathcal{S}_-} d\sigma^{\mu\nu} F^c_{\mu\nu} (1 \otimes \tfrac{1}{2}\lambda^c) \Big) \Big\rangle_A . \tag{8.85}$$

The two exponentials commute in the product space and we can write the right-hand side of this equation as a single exponential:

$$S(\mathbf{b}, 1, 2) = \tfrac{1}{9} \, \mathrm{tr}_2 \Big\langle \exp\Big(-ig \int_{\mathcal{S}} d\sigma^{\mu\nu} \mathbf{F}_{\mu\nu} \Big) \Big\rangle_A \tag{8.86}$$

where the surface integral extends over the surfaces $\mathcal{S}_+$ and $\mathcal{S}_-$, and $\mathbf{F}_{\mu\nu}$ takes its values in the product algebra of $SU(3) \otimes SU(3)$. We can now make a cluster expansion with stochastic variables from the product algebra analogous to the one introduced below (6.63) and, neglecting the higher cumulants in that expansion, obtain again

$$S(\mathbf{b}, 1, 2) = \tfrac{1}{9} \, \mathrm{tr}_2 \, \exp(C_2(\mathbf{b}, \mathbf{R}_1, \mathbf{R}_2))$$
$$C_2(\mathbf{b}, 1, 2) = -\tfrac{1}{8} g^2 \int_{\mathcal{S}} d\sigma^{\mu\nu} d\sigma^{\kappa\lambda} \Big\langle \mathbf{F}_{\mu\nu} \mathbf{F}_{\kappa\lambda} \Big\rangle_A . \tag{8.87}$$

In order to evaluate this expectation value, we use the correlator (8.74) and obtain finally

$$S(\mathbf{b}, 1, 2) = \tfrac{2}{3} e^{-\frac{1}{3}i\chi} + \tfrac{1}{3} e^{\frac{2}{3}i\chi} \tag{8.88}$$

where $\chi = \chi(\mathbf{b}, 1, 2)$ is given in (8.82).

In the following we refer to the first method leading to (8.81) and (8.84) as the expansion method and to the second method leading to (8.88) as the matrix-cumulant method. An expansion of (8.88) in powers of χ yields

$$S(\mathbf{b}, 1, 2) = 1 - \tfrac{1}{9}\chi^2 - \tfrac{1}{81}i\chi^3 + \cdots . \tag{8.89}$$

Comparison of this equation with (8.81) and (8.84) shows that the terms proportional to χ^2 agree, whereas in the leading C-odd contribution there is a discrepancy by a factor $\frac{5}{4}$. This is a consequence of the noncommutativity of the stochastic variables[270]; the higher cumulants with variables from the algebra of $SU(3)$ are different from those of the algebra of $SU(3) \otimes SU(3)$.

The matrix element for dipole-dipole scattering with momentum transfer $\mathbf{q}$ is then given by

$$T_{fi}(s, t, \mathbf{R}_1, z_1, \mathbf{R}_2, z_2) = -2is \int d^2b \, e^{i\mathbf{q} \cdot \mathbf{b}} \Big(S(\mathbf{b}, 1, 2) - 1 \Big) \qquad (8.90)$$

with $t = -\mathbf{q}^2$.

The treatment of three quark loops in a colour-singlet state is analogous, but technically more involved. We refer to the literature[259,431] for the corresponding results.

If the parameters of the stochastic-vacuum model are taken to be independent of the energy, the resulting cross sections turn out to be independent of the scattering energy too. This result is not unexpected since also in the stochastic-vacuum model the fundamental interaction is the vector interaction of the gluon field with the quarks. We shall address later the question of how to incorporate an energy dependence into the model. At present we concentrate on a fixed energy scale $\sqrt{s} \approx 20$ GeV, which is appropriate to the formalism developed in section 8.2.

A striking feature of the model is the dependence of the amplitude on the size of the hadrons. The Abelian part of the correlator yields a cross section independent of the size of the hadrons if this size is very large compared with the correlation length a of the correlator (6.50). The non-Abelian part, however, which is responsible for string formation and hence confinement, yields a hadron-hadron total cross section which increases linearly for large hadron sizes. In figure 8.9 we show the confining non-Abelian contribution ($\kappa = 1$) and the Abelian contribution ($\kappa = 0$) to the total cross section as functions of the transverse radius R of one of the quark-antiquark pairs, with the radius of the other pair kept fixed at 0.7 fm. The saturation of the cross section for the Abelian part is in agreement with the Landshoff-Nachtmann model, which indeed turns out to be contained in the treatment presented here for the case $\kappa = 0$. The continuing rise of the cross section coming from the non-Abelian part can be traced back to an interaction of the gluonic strings between the quark and antiquark; see figure 6.10.

We are not taking into account internal fermion loops. The continuing rise of the cross section with the size of the hadrons is due to this approximation; it is related to the unlimited rise of the interquark potential in quenched QCD; see figure 6.9. We expect that for large dipoles, say larger than 2 fm,

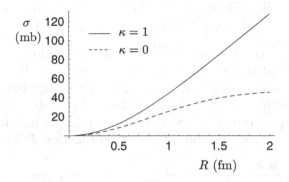

Figure 8.9. Contribution of the non-Abelian correlator ($\kappa = 1$) and the Abelian correlator ($\kappa = 0$) to the total cross section of two quark-antiquark pairs, one with fixed radius 0.7 fm and one with radius R.

the increase of the cross section becomes unrealistic.

8.6 Wave functions of photons and hadrons

To come from the dipole-dipole scattering amplitude (8.90) to scattering amplitudes which are accessible experimentally, one has to represent the hadrons and photons as superpositions of dipoles of different transverse sizes $\mathbf{R}_i$. A meson or a photon is described by a wave function $\psi_n(\mathbf{R}_i, z_i)$ of a quark and an antiquark with relative transverse coordinates $\mathbf{R}_i$ and longitudinal momentum fraction z_i of the quark. The meson or photon scattering amplitude is then obtained by integrating the dipole-dipole scattering amplitude $T_{fi}(s, t, \mathbf{R}_1, z_1, \mathbf{R}_2, z_2)$ of (8.90) over all radii, with the overlap densities of the wave functions as weight:

$$T_{1,2 \to 3,4}(s,t) = \int d^2 R_1 d^2 R_2 \int_0^1 dz_1 dz_2 \, \psi_3^*(\mathbf{R}_1, z_1) \, \psi_1(\mathbf{R}_1, z_1)$$
$$\times \psi_4^*(\mathbf{R}_2, z_2) \, \psi_2(\mathbf{R}_2, z_2) \, T_{fi}(s, t, \mathbf{R}_1, z_1, \mathbf{R}_2, z_2). \quad (8.91)$$

It is a great challenge to construct hadronic wave functions from first principles. In this section we summarise some practical attempts to arrive at more-or-less plausible expressions. We start with the most model-independent case, photon wave functions. The photon wave function can be derived directly in light-cone perturbation theory. We indicate here the derivation in covariant perturbation theory. Consider a photon with momentum q and helicity λ. If it dissociates into a quark and an antiquark with momenta k and $q - k$, each of mass m_f, the term

$$\frac{i(\gamma.k + m_f)\,(-i\gamma.\epsilon_\lambda)\,i(\gamma.k - \gamma.q + m_f)}{(k^2 - m_f^2 + i\epsilon)\,((k - q)^2 - m_f^2 + i\epsilon)} \tag{8.92}$$

occurs in all perturbative expressions. The integration over the internal momentum k is best performed in light-cone coordinates $k_\pm = k_0 \pm k_3$, $\mathbf{k}_T = (k_1, k_2)$. The wave function is the vertex function integrated over one of the variables[432], in this case k_-. One sees that the k_- integration contributes in the high-energy limit $q_+ \to \infty$ only in an infinitesimal interval around $k_- = 0$. This yields for any suitably-behaved function $F(k)$

$$\int \frac{d^4 k}{(2\pi)^4} \frac{1}{(k^2 - m_f^2 + i\epsilon)((k - q)^2 - m_f^2 + i\epsilon)}\, F(k)$$

$$\sim \int \frac{d^2 k_T\, dk_-\, dz}{(2\pi)^4} \frac{i\pi\delta(k_-)}{\mathbf{k}_T{}^2 + m_f^2 - z(1 - z)q^2}\, F(k) \tag{8.93}$$

where z is the longitudinal-momentum fraction of the quark, $k_+ = zq_+$. The numerators in (8.92) can be expressed through sums over products of spinors:

$$\gamma.k + m_f = \sum_h u(\mathbf{k}, h)\bar{u}(\mathbf{k}, h)$$

$$\gamma.k - \gamma.q + m_f = -\sum_{\bar{h}} v(\mathbf{q} - \mathbf{k}, \bar{h})\bar{v}(\mathbf{q} - \mathbf{k}, \bar{h}) \tag{8.94}$$

where $u(\mathbf{k}, h)$ and $v(\mathbf{q} - \mathbf{k}, \bar{h})$ are spinors for mass m_f of a quark with helicity h and an antiquark with helicity $\bar{h}$. We insert (8.94) into (8.92) and absorb the part $\bar{u}(\mathbf{k}, h)(-i\gamma.\epsilon_\lambda)v(\mathbf{q} - \mathbf{k}, \bar{h})$ in the wave function. The remaining spinors contribute to the quark and antiquark scattering amplitudes; see (8.56).

After Fourier transformation with respect to $\mathbf{k}_T$, we arrive at the following expressions for the quark-antiquark wave functions of photons with polarisation λ and virtuality $Q^2 = -q^2$:

$$\psi_\gamma^{\lambda=0, h, \bar{h}}(Q^2, \mathbf{R}, z) = \frac{\sqrt{3\alpha}}{2\pi}\hat{e}_f(-2z(1 - z)\delta_{h, -\bar{h}})Q\, K_0(\epsilon R)$$

$$\psi_\gamma^{\lambda=\pm 1, h, \bar{h}}(Q^2, \mathbf{R}, z) = \pm\frac{\sqrt{6\alpha}}{2\pi}\hat{e}_f\Big(i\, e^{\pm i\phi_T}(z\delta_{h, \frac{1}{2}}\delta_{\bar{h}, -\frac{1}{2}}$$

$$-(1 - z)\delta_{h, -\frac{1}{2}}\delta_{\bar{h}, \frac{1}{2}})\,\epsilon K_1(\epsilon R) + m_f\,\delta_{h, \pm\frac{1}{2}}\delta_{\bar{h}, \pm\frac{1}{2}}K_0(\epsilon R)\Big) \tag{8.95}$$

with

$$\epsilon = \sqrt{z(1 - z)Q^2 + m_f^2}\,. \tag{8.96}$$

Here R and ϕ_T are the plane-polar coordinates of the transverse separation $\mathbf{R}$ of the quark-antiquark pair, $\hat{e}_f$ is the charge of the quark in units of the elementary charge, m_f its mass, and h and $\bar{h}$ the helicities of the quark and the antiquark; $\lambda = 0$ indicates a longitudinal photon, $\lambda = \pm 1$ a transverse photon. The functions K_i are the modified Bessel functions.

Consider the scattering amplitude (8.91) when particles 1 and 3 are both photons. The amplitude then involves one of the hadronic densities

$$\rho_\gamma(\lambda = 0, Q^2, R, z) = \sum_{h+\bar{h}=0} |\psi_\gamma^{\lambda=0,h,\bar{h}}(Q^2, \mathbf{R}, z)|^2$$

$$= \frac{6\alpha}{4\pi^2} \hat{e}_f^2 \, 4Q^2 z^2 (1-z)^2 \left(K_0(\epsilon R)\right)^2$$

$$\rho_\gamma(\lambda = \pm 1, Q^2, R, z) = \sum_{h+\bar{h}=\pm 1} |\psi_\gamma^{\lambda=\pm 1,h,\bar{h}}(Q^2, \mathbf{R}, z)|^2$$

$$= \frac{6\alpha}{4\pi^2} \hat{e}_f^2 \left((z^2 + (1-z)^2)\epsilon^2 \left(K_1(\epsilon R)\right)^2 + m_f^2 \left(K_0(\epsilon R)\right)^2 \right). \quad (8.97)$$

We collect some mathematical properties of the densities which will be important for later applications. The asymptotic behaviour of the Bessel functions for large x, $K_n(x) \sim \sqrt{\pi/(2x)} \exp(-x)$, leads to an exponential decay $\exp(-R\sqrt{4z(1-z)Q^2 + 4m_f^2})$ of the density for large R. For heavy quarks the transverse extension is small enough for confinement effects to be neglected and to apply perturbation theory even for low values of Q^2. For massless quarks the scale is set by $1/(4z(1-z)Q^2)$, and the contribution from small values of z or $(1-z)$ leads to a power-like behaviour for large R. For transverse photons we obtain for large R and large Q^2

$$\int_0^1 dz\, \rho_\gamma(\lambda = \pm 1, Q^2, R, z) \sim$$

$$\frac{6\alpha}{4\pi^2} \hat{e}_f^2 \frac{4}{Q^2 R^4} \int_0^\infty du\, u^3 \left(K_1(u)\right)^2 = \frac{6\alpha}{4\pi^2} \hat{e}_f^2 \frac{8}{3} \frac{1}{Q^2 R^4}. \quad (8.98)$$

In the case of longitudinal photons the factor $z^2(1-z)^2$ in the density suppresses small values of z and $(1-z)$ and therefore the power-like term is of higher twist and falls off faster with R:

$$\int_0^1 dz\, \rho_\gamma(\lambda = 0, Q^2, R, z) \sim$$

$$\frac{6\alpha}{4\pi^2} \hat{e}_f^2 \frac{16}{Q^4 R^6} \int_0^\infty du\, u^5 \left(K_0(u)\right)^2 = \frac{6\alpha}{4\pi^2} \hat{e}_f^2 \frac{256}{15} \frac{1}{Q^4 R^6}. \quad (8.99)$$

The singularities of the Bessel functions at the origin

$$K_0(x) = -\log(x) + O(1) \qquad K_1(x) = \frac{1}{x} + O(x) \quad (8.100)$$

lead to singularities of the hadronic densities (8.95) which make the integral

$$\int d^2R \int_0^1 dz\, \rho_\gamma(\lambda = \pm 1, Q^2, R, z)$$

divergent; the transverse photon wave function is not normalisable. This does not cause problems, since for the evaluation of observable amplitudes the density is multiplied by the dipole scattering amplitude, which vanishes for $R \to 0$.

Whereas the photon wave function at high virtualities Q^2 can be calculated by perturbation theory, this is not the case for low virtualities. Often vector-meson dominance is applied in order to take confinement effects into account. Another possibility is to introduce a constituent-quark mass as infrared regulator in the photon wave function. This approach[433] will be justified in the following.

Consider first a model with scalar photons and quarks. In such a model the "photon" wave function at high virtualities is given in momentum and position spaces by:

$$\tilde{\psi}_\gamma(\mathbf{k}_T) = \frac{1}{k_T^2 + \epsilon^2}$$

$$\psi_\gamma(\mathbf{R}) = \frac{1}{2\pi} K_0(\epsilon\, R) \tag{8.101}$$

where ϵ is defined in (8.96). For low values of Q^2 we expect confinement to modify these perturbative expressions considerably. The structure of (8.101) is the same as that of a nonrelativistic quantum-mechanical Green's function for the relative motion of a free two-body system with reduced mass m, continued to negative energies $E = -M$:

$$G_0(\mathbf{R}, 0, M) = \int \frac{d^2k_T}{(2\pi)^2} \frac{\exp(i\mathbf{k}_T.\mathbf{R})}{k^2/(2\,m) + M} = \frac{m}{\pi} K_0(\sqrt{2mM}\, R). \tag{8.102}$$

The quantity ϵ is replaced with $\sqrt{2mM}$. In order to impose confinement we go from the free-particle state to a system bound by a harmonic-oscillator potential. The harmonic oscillator is a very useful model[250] for QCD, as it shows both confinement and asymptotic freedom. We therefore investigate the effects of confinement by comparing the free Green's function (8.102) with that of the full harmonic oscillator in two dimensions,

$$G_H(\mathbf{R}, 0, M) = \sum_{n_1, n_2} \frac{\psi_{\vec{n}}(\mathbf{R})\psi_{\vec{n}}(\mathbf{R})}{(n_1 + n_2 + 1)\omega + M} \tag{8.103}$$

which can be calculated easily. It turns out that for non-negative values of M a free Green's function in which M is replaced with a suitable function of M yields an excellent fit to the exact Green's function (8.103) over the relevant range of R. We transfer this to QCD by replacing the quark mass m_f with a Q^2-dependent effective mass $m_{\text{eff}}(Q^2)$. The function $m_{\text{eff}}(Q^2)$ can be fixed by a fit to the phenomenologically-known vector-current two-point function and thus no new parameter is introduced. The linear parametrisations

$$m_{\text{eff}}(Q^2) = \begin{cases} m_f + m_{0q}\,(1 - Q^2/Q_0^2) & Q^2 \leq Q_0^2 \\ m_f & Q^2 \geq Q_0^2 \end{cases} \qquad (8.104)$$

can be used[433] with

$$m_{0q} = 0.21 \pm 0.02 \text{ GeV} \quad m_f = 0.007 \text{ GeV} \quad Q_0^2 = 1.05 \text{ GeV}^2 \quad (8.105)$$

for the up and down quarks, and

$$m_{0q} = 0.31 \pm 0.02 \text{ GeV} \quad m_f = 0.15 \text{ GeV} \quad Q_0^2 = 1.6 \text{ GeV}^2 \quad (8.106)$$

for the strange quark. These effective masses replace the quark masses m_f in the wave functions (8.95) and the densities (8.97).

The perturbative expressions for the wave functions with a Q^2-dependent quark mass are very economical descriptions of the photon wave functions. The part where the separation of the quark and antiquark is large describes what is often called the hadronic part of the photon. The small-distance part with its singularity at the origin is due to the pointlike coupling of the electromagnetic current to the quarks and is referred to as the hard part of the photon. The planar hadronic density of the real photon

$$\tilde{\rho}_\gamma(\pm 1, 0, R) = 2\pi R \int_0^1 dz\, \rho_\gamma(\lambda = \pm 1, 0, R, z) \qquad (8.107)$$

is plotted in figure 8.10. As can be seen from (8.97) and (8.100) the behaviour of $\tilde{\rho}_\gamma(R)$ near the origin is independent of the effective quark mass. Radial densities of virtual photons are displayed in figure 9.4.

For hadronic wave functions we have no perturbative expression as a starting point, therefore they are even more model-dependent than the photon wave functions are at low virtualities. There are ways to calculate the distribution of the longitudinal momentum (see for instance [434] and the literature quoted there) but for light quarks the dependence on k_T or R is purely a question of models. Even for heavy quarks the scale for the extension of the wave function is set by $1/(m_f \alpha_s)$, which for charmed quarks is of the order of the scale parameter Λ of QCD and confinement effects are

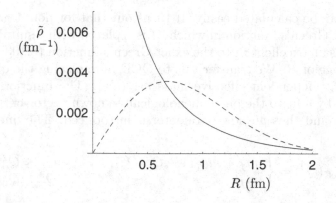

Figure 8.10. The planar density $\tilde{\rho}_\gamma(\pm 1, 0, R)$ defined in (8.107) of a real photon (solid line), and the planar density of a meson (dashed line) corresponding to (8.108) with $R_h = 0.68$ fm multiplied by the factor 0.004

by no means negligible. This contrasts with the charm content of a real or virtual-photon wave function, where one is far away from the mass of the bound state and the scale is set by the heavy-quark mass directly.

In principle for hadronic wave functions one could start from similar considerations to those for the photon and replace the elementary photon-quark-antiquark vertex by a Bethe-Salpeter hadron-quark-antiquark amplitude. With a simple model for this amplitude it has been shown[432] that such an approach is feasible and leads to factorisation for hard processes. In practice however, and especially for soft processes, one starts rather from (8.91) and makes a plausible ansatz for the wave functions $\psi_i(\mathbf{R}, z)$.

The following strategy has been adopted[435] to construct expressions for hadronic wave functions. The spin structure and polynomial dependence on z were taken to be those of the simplest corresponding interpolating field composed of a quark and an antiquark; an exponential z dependence $h(z)$ suppressing the values near 0 and 1 was introduced following the prescription of [436]. The matrix element T_{fi} of (8.91) depends only weakly on z, as can be seen from figure 9.2, and therefore hadronic cross sections depend very little on the exact choice of $h(z)$. The dependence on the transverse extension of the ground states was approximated by a Gaussian wave function

$$\psi_h(\mathbf{R}, \mathbf{z}) = \exp\left(-\frac{R^2}{4R_h^2}\right) h(z). \tag{8.108}$$

In the approach described in section 8.5 the hadronic wave function has to be normalised:

$$\int_0^1 dz \int d^2 R \, |\psi_h(\mathbf{R}, z)|^2 = 1. \tag{8.109}$$

The S-matrix element for a single particle in this approach is given by

$$S = \int_0^1 \int d^2 R \, |\psi_h(\mathbf{R}, z)|^2 \, \tfrac{1}{3} \langle W[C] \rangle \tag{8.110}$$

analogously to (8.76) and (8.91), where C is a Wilson loop with lightlike sides and width $\mathbf{R}$. The S-matrix element for a single particle is 1 and since according to (8.75) $\langle W[C] \rangle = 3$, the normalisation condition (8.109) is a consequence of (8.110). The transverse radius R_h in (8.108) was fitted[259,435,437] to such hadronic properties as the electromagnetic radius or the electromagnetic decay width. As discussed above, for purely-hadronic reactions the z dependence is not important and the value of z can be fixed to $\frac{1}{2}$. In figure 8.10 we display the planar density $2\pi R |\psi_h(\mathbf{R})|^2$ for a hadron with transverse radius 0.68 fm (dashed line). We note the difference between the hadronic density of the photon and of a typical hadron. It will be discussed in detail in section 9.2. For baryons the wave function depends on the position of the three valence quarks in transverse space; see figure 8.8. For practical purposes it is convenient to work in the diquark picture. In that case the wave function of the baryon is like that of a meson, the diquark replacing the antiquark.

In another approach[438] one starts from an ansatz for the wave function $\tilde{\psi}_R(\mathbf{k})$ in the rest frame of the hadron, and substitutes the dependence of the relative momentum $|\mathbf{k}|$ in the rest frame with light-cone coordinates according to

$$\mathbf{k}_T^2 \to \frac{\mathbf{k}_T^2 + m_f^2}{4z(1-z)} - m_f^2. \tag{8.111}$$

This approach makes sense only for heavy quarks, or for light quarks if a constituent mass is introduced. This is because, even if the quarks in the hadron are confined, the transverse radius of the light-cone wave function $\psi(\mathbf{k}_T, z)$ obtained with the substitution (8.111) diverges as z tends to 0 or 1. Only for a finite quark mass is there a suppression of the wave function near to $z = 0$ or 1, such that this divergence does not create problems. We refer to [357] and [439] for a discussion and comparison of different approaches for the construction of wave functions. A more sophisticated treatment of the wave functions[434] would certainly be desirable but would lead to more free parameters.

8.7 Applications to high-energy hadron-hadron scattering

In this section we apply the functional approach to hadron-hadron scattering and photoproduction of vector mesons, with the functional integrals evaluated approximately by the stochastic-vacuum model. In principle the lattice results for the correlator together with low-energy data are sufficient to calculate the high-energy dipole-dipole scattering amplitude (8.90), and it has been shown[440] that this approach leads to qualitatively-satisfactory results. The facts that the extraction of the gluon condensate $\langle g^2 FF \rangle$ from lattice data depends crucially on an extrapolation, that the data have finite errors, and that the model neglects all higher cumulants, make it advisable to include some high-energy data in the determination of the effective parameters. Most applications[435,433,441–443] are based on the diquark picture of the proton and the expansion method, using a parameter set which has as input

• lattice data for the forms of the correlators (6.47) and (6.50), including the value for κ
• the relation (6.65) between the gluon condensate and the string tension
• the pomeron part of the total cross section and the logarithmic slope of the elastic cross section for proton-proton scattering at a centre-of-mass energy $\sqrt{s} = 20$ GeV.

In this way the following set of values for the gluon condensate $\langle g^2 FF \rangle$, the correlation length a, the parameter κ of the gluon correlator and the transverse radius of the proton R_p were obtained:

$$\langle g^2 FF \rangle = (1.26 \text{ GeV})^4 \quad a = 0.346 \text{ fm} \quad R_p = 0.75 \text{ fm} \quad \kappa = 0.74 \quad (8.112)$$

The amplitude T_{fi} of (8.90) at $t = 0$ averaged over all orientations of the dipoles $\mathbf{R}_1$, $\mathbf{R}_2$ factorises approximately into the product of two functions $\sigma(R_1)\,\sigma(R_2)$. The total cross sections can be evaluated easily once some ansatz for the wave functions has been made.

In figure 8.11 we show the hadron-proton cross section from the model with parameter set (8.112) as a function of the ratio R_h/R_p, where R_p is the transverse radius of the proton and R_h that of the other hadron. The data points are the phenomenological values for the pomeron contribution to pp, πp, and Kp scattering, given in figures 3.1 and 3.2. The value for the $J/\psi p$ total cross section at $\sqrt{s} \approx 20$ GeV is estimated from perturbation theory[444] using a phenomenological gluon distribution, and from J/ψ photoproduction[445,446] near that energy. As we can see from figure 8.11 the ratio $\sigma_{\pi p} : \sigma_{pp} \approx 2 : 3$ is a consequence of the model. Also the ratio $\sigma_{Kp} : \sigma_{pp} = 0.55$ coincides with the phenomenological value 0.55 from the pomeron contribution given in figure 3.2. The ratio for the pomeron part

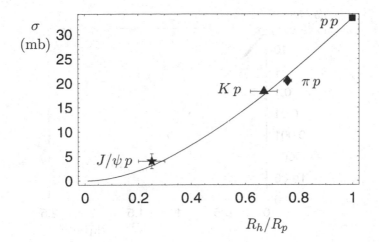

Figure 8.11. Pomeron contribution at $\sqrt{s} = 20$ GeV to the hadron-proton total cross section as a function of the hadron size. The curve is a prediction of the model with the parameter set (8.112).

of the $\pi\,\pi$ and $p\,p$ total cross section comes out to $\sigma_{\pi\pi} : \sigma_{pp} = 0.43$ which is close to the value obtained from $\sigma_{\pi p}$ and σ_{pp} with the assumption of Regge factorisation as in (2.32):

$$\frac{\sigma_{\pi\pi}}{\sigma_{pp}} \approx \left(\frac{\sigma_{\pi p}}{\sigma_{pp}}\right)^2 \approx 0.44. \qquad (8.113)$$

The approach does not lead to quark additivity since the cross section depends crucially on the sizes of the hadrons. This radius dependence is strong for the non-Abelian term in the ($\kappa = 1$) correlator shown in figure 8.9. But from the same figure we see that even the Abelian part of the ($\kappa = 0$) correlator, which corresponds to the Landshoff-Nachtmann model, leads to a distinct dependence on the size of the dipole for typical hadron sizes in the range 0.5 fm $\leq R_h \leq 1$ fm. This signals that, in this region, contributions for which the gluons couple to separate quarks in the same hadron, figure 8.2b, still play an important role. Interference effects become negligible only if the hadron size is really very large compared with the correlation length a. This is due to the strong enhancement of small exchanged momenta typical for this nonperturbative model. This may also explain why quark additivity and the non-Abelian approach yield very similar results. The exchanged momentum is typically so small that it cannot distinguish between pointlike quarks with a string and extended constituent quarks.

The matrix-cumulant method introduced in section 8.5 allows an evaluation of the scattering amplitude for a larger range of t values. The differential

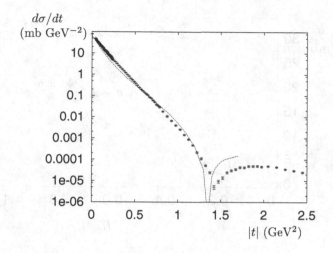

Figure 8.12. pp scattering at $\sqrt{s} = 23$ GeV; the solid curve is obtained with parameter set (8.114). Experimental points are from [62].

cross section $d\sigma/dt$ for pp scattering in the energy range $\sqrt{s} = 23$ to 1800 GeV has been compared[431] with the results obtained with this method. The calculated values of $d\sigma/dt$ depend again on the vacuum parameters $\langle g^2 FF \rangle$, a and κ, and on the s-dependent proton-radius parameter $R_p(s)$. The latter was fixed by requiring $d\sigma/dt|_{t=0}$ at each s to be equal to the value obtained from the phenomenological pomeron discussed in chapter 3. Then, the vacuum parameters were varied and a reasonable description of $d\sigma/dt$ for the whole energy interval investigated was obtained with

$$\langle g^2 FF \rangle = (1.33 \text{ GeV})^4 \quad a = 0.32 \text{ fm} \quad R_p = 0.87 \text{ fm} \quad \kappa = 0.74 \quad (8.114)$$

Here the value $\kappa = 0.74$ is a result of the fit to the elastic pp cross section whereas in (8.112) it is taken from lattice calculations. In figure 8.12 the differential cross section for elastic pp scattering at $\sqrt{s} = 23$ GeV, calculated with the matrix-cumulant method and the parameter set (8.114), is compared with experimental data.

The quality of the fit to the data in figure 8.12 is not as good as that obtained with the phenomenological-pomeron approach of [95]. It must however be kept in mind that in [431] only the pomeron contribution to $d\sigma/dt$ is calculated whereas in [95] the reggeon, Regge cut and perturbative contributions are included which improve the fit. In [431] the emphasis was not on obtaining a very good fit, but on trying to give a description of $d\sigma/dt$, and hence of the pomeron, in a calculation starting from the QCD Lagrangian (6.1). Taking the proton radius parameter $R_p(s)$ as phe-

parameter	lattice	static potential	scattering	
			expansion method	matrix cumulant
$\sqrt{\sigma_q}$ (MeV)	420	415	425	435
$\langle g^2 FF \rangle^{1/4}$ (GeV)	1.22	1.22	1.26	1.33
κ	0.89	0.89	0.74	0.74
a (fm)	0.33	0.33	0.35	0.32

Table 8.1. Values for the string tension σ_q and the vacuum parameters $\langle g^2 FF \rangle, \kappa, a$ obtained with four methods. Underlined numbers are input. Errors on the other numbers are estimated to be about 10%.

nomenological input, it is possible to describe $d\sigma/dt$ up to $|t| \approx 1$ GeV^2 in terms of the properties of the QCD vacuum. In table 8.1 we compare the values for the string tension σ_q and the vacuum parameters $\langle g^2 FF \rangle$, κ and a determined by various methods. The second column lists the values from the lattice calculation[447] in the quenched approximation, using the phenomenological value for the string tension σ_q as input. The third column gives the result of the calculation of the static quark-antiquark potential in the stochastic-vacuum model; see section 6.6. Here the vacuum parameters are taken as input and the string tension is calculated from (6.65) in good agreement with the input value of the lattice calculation. This gives confidence that the stochastic-vacuum model describes the QCD vacuum properties quite well. The fourth column lists the values (8.112) obtained from high-energy scattering using the expansion method. As stated above, here κ is taken as input. Furthermore (6.65) is used to calculate the string tension σ_q. In the fifth column we list the values of the vacuum parameters obtained from high-energy pp scattering using the matrix-cumulant method, with the corresponding value of $\sqrt{\sigma_q}$ obtained from (6.65). The agreement with the values from the lattice is quite reasonable and gives support to the view that high energy diffractive scattering is governed by the QCD vacuum properties, as is the confinement phenomenon. In particular, in the matrix-cumulant approach a non-zero κ and thus a non-zero string tension are essential[431] to obtain a reasonable description of $d\sigma/dt$. The resulting numerical value for the string tension is within errors compatible with the one determined[271] from the linear confinement potential between a heavy quark-antiquark pair.

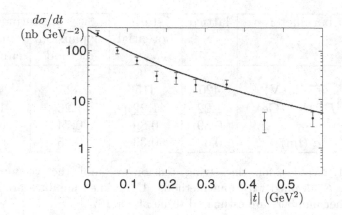

Figure 8.13. Differential cross section[435] for $\gamma^* + p \to \rho + p$ at $Q^2 = 6$ GeV^2 and $\sqrt{s} \approx 20$ GeV with data[448]

8.8 Application to photoproduction of vector mesons

The model has also been applied to photon-induced reactions. In principle no new free parameters need to be introduced, since the wave functions of the photon and the vector mesons can be determined from other constraints, as described in the preceding section. There is however a serious caveat: the photon-hadron overlap integrals occurring in these reactions are much more sensitive to details of the wave functions than the quark-antiquark densities of the photon or hadrons. Therefore the results depend strongly on the assumptions made for constructing the wave functions.

As an example of electroproduction of vector mesons, we show in figure 8.13 the differential cross section for the reaction $\gamma^* p \to \rho p$ obtained with the set (8.112) and the expansion method[435]; it can be seen that the parameter-free result of the model agrees well with the data for $|t| \leq 0.6$ GeV^2. Photoproduction of excited light vector mesons in principle offers an excellent test for the approach discussed here, and a way to test the cross section of a quark-antiquark pair with large separation since the excited mesons are large objects. Furthermore it is supposed to show drastic effects in the transition from the soft to the hard regime.

In figure 8.14 we display data for the reactions $e^+ e^- \to \pi^+ \pi^-$ and $\gamma p \to p\pi^+\pi^-$ as a function of the invariant mass of the two pions. In the annihilation cross section one notes a dip in the region 1.4 GeV $\leq M_{\pi\pi} \leq$ 1.8 GeV. This can be explained in the following way[452]. The branching ratio into two π-mesons of the excited states $\rho(1450)$ and $\rho(1700)$ is small, therefore these channels are mainly visible through their interference with the tail

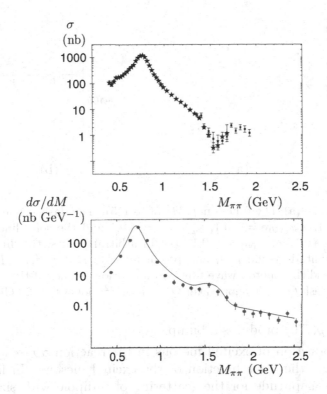

Figure 8.14.   Data[449,450] for $e^+e^- \to \pi^+\pi^-$ and[451] $\gamma p \to \pi^+\pi^- p$ as a function of the invariant mass of the two pions. The curve is from [435].

of the $\rho(770)$ meson. If the coupling of the $\rho(1450)$ to the electromagnetic current has the opposite sign from that of the $\rho(770)$ and if the $\rho(1700)$ coupling has the same sign as the $\rho(770)$ coupling, the excited mesons interfere constructively among each other in the region between 1450 and 1700 GeV, but their combined interference with the $\rho(770)$ is destructive.

In figure 8.15a we show the function

$$f(M) = \left| \sum_n \frac{c_n}{M^2 - M_n^2 + iM_n\Gamma_n} \right|^2 \tag{8.115}$$

where n runs over the $\rho(770)$, $\rho(1450)$, and $\rho(1700)$ with experimental values for the masses and the total widths and the two sign patterns discussed above. The dashed line corresponds to $c_{\rho(770)} = 1$, $c_{\rho(1450)} = -0.1$ and $c_{\rho(1700)} = 0.1$. For the solid line the signs of the excited-meson couplings are reversed. We see that the dashed line of figure 8.15a reproduces the pattern of the annihilation cross section in figure 8.14. If the signs of the couplings of the excited mesons are reversed, the constructive interference

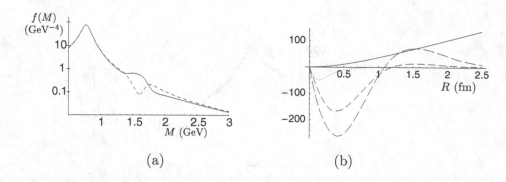

Figure 8.15. (a) The function $f(M)$ of (8.115). The dashed line is for the choice $c_{\rho(770)} = 1$, $c_{\rho(1450)} = -0.1$, $c_{\rho(1700)} = +0.1$ and the solid line for $c_{\rho(770)} = 1$, $c_{\rho(1450)} = +0.1$, $c_{\rho(1700)} = -0.1$. (b) In arbitrary units, the dipole-proton scattering amplitude (solid line) and products $R\psi_{2S}^*(R)\psi_\gamma(Q^2, 0.5, R)$ of a $2S$ wave function and the photon wave function for various values of the photon virtuality Q^2: long dash $Q^2 = 0$; short dash $Q^2 = 1$ GeV^2; dots $Q^2 = 20$ GeV^2.

with the $\rho(770)$ produces a bump.

Photoproduction of excited mesons in the reaction $\gamma p \to \pi^+\pi^- p$ depends strongly on the wave function of the excited mesons. In figure 8.15b we show the amplitude for the scattering of a dipole with size R on a proton as calculated in section 8.5. We also display $R\psi_{2S}^*(R)\psi_\gamma(R, 0.5, Q^2)$ for various values of Q^2, where $\psi_{2S}(R)$ is the wave function of a meson in a $2S$ state and $\psi_\gamma(R, z, Q^2)$ is the photon wave function (8.95). If the two excited mesons have a sizable component which can be described as a quark-antiquark $2S$ state we observe two general features. Firstly, photoproduction of two pions in the excited-resonance region is even more suppressed than in the annihilation process since there is a sizable cancellation between contributions from small and large distances in the integral over R in (8.91). Secondly, due to the increase of the dipole scattering amplitude at large R the contributions of large distances are dominant. In the annihilation process $e^+e^- \to \pi^+\pi^-$ the pion pair is produced by a highly-virtual photon. It tests the wave functions at the origin, which for the excited mesons have the signs opposite from those of the long-distance part. Therefore we expect for annihilation just the opposite interference pattern with the $\rho(770)$ from that for photoproduction. This is exactly what is seen for the photoproduction, as shown in figure 8.14, and also in a high-statistics experiment[453]: a bump in the region between 1400 and 1800 GeV. In the case of electroproduction the inner part of the wave function acquires increasing weight with rising virtuality Q^2 of the photon, as shown in figure 8.15b. Therefore the model predicts that for high virtualities of the photon the interference pattern for electroproduction is the same as in annihilation.

8.9 Photoproduction of pseudoscalar and tensor mesons

The odderon, a C-odd trajectory with an intercept near 1, was introduced in section 3.10. There it was pointed out that a soft odderon appears very naturally in QCD, but there is no evidence for its presence in pp and $\bar{p}p$ scattering in the near-forward direction. This is a puzzle and it is important to look for the odderon in other reactions.

Exclusive photoproduction or electroproduction of mesons with even C-parity is particularly suited to detecting odderon exchange since, because of the odd C-parity of the photon, only trajectories with negative C-parity can contribute[145]. Reggeon exchanges are strongly suppressed at HERA energies because their intercepts are well below 1. However, photon exchange is important at small momentum transfer. It can be calculated in a fairly model-independent way[454]. It has been argued[455] that the odderon should not contribute to elastic baryon or antibaryon scattering if the baryons have a spatially linear structure, consisting of a quark and a diquark. However this suppression mechanism does not work if in an inelastic diffractive process the baryon or antibaryon breaks up into a state of negative relative parity, such as the $N^*(1520)$ or $N^*(1535)$. It was furthermore estimated that a diquark of size less than 0.3 fm is small enough to explain the bound given through (3.87). This result was obtained in the model for high-energy scattering described in section 8.5, with a baryon structure as depicted in figure 8.8. The cross sections for[454] $\gamma p \to \pi^0 N^*$ and[456] $\gamma p \to f_2(1270) N^*$ have been calculated in the same model. The results for the integrated cross sections at centre-of-mass energy $W = 20$ GeV are

$$\sigma_{\gamma p \to \pi^0 N^*} = 290 \text{ nb} \qquad \sigma_{\gamma p \to f_2 N^*} = 21 \text{ nb}. \qquad (8.116)$$

The difference between the cross sections for the two reactions can be understood easily. For an isoscalar exchange, the isovector pion couples to the isovector photon and the isoscalar f_2 couples to the isoscalar photon. This already makes pion production a factor 9 larger than f_2 production. We also consider $\gamma p \to a_2(1320) N^*$, which similarly should be a factor 9 larger than f_2 production because a_2 is an isovector. The uncertainties are large, an estimated $\pm 50\%$, since the parameters (8.112), (8.114) occur in powers which are larger by a factor 3 than in total cross sections. Furthermore, higher cumulants will most probably play a more important role in C-odd exchange, as has been discussed below (8.89).

Preliminary data[457] at $W \approx 200$ GeV give an upper bound for $\sigma_{\gamma p \to \pi^0 N^*}$ which is a factor 4 below the value given in (8.116). So the puzzle of why we have not yet seen the odderon persists and we have a clear contradiction to the model prediction[454]. We summarise the essential points on which the model calculation is based.

• Nonperturbative pomeron and odderon exchange are purely gluonic effects. Soft gluons can be described by a Gaussian process.

• The γ-π^0-odderon coupling has been calculated with a wave function for the pion of the type described in section 8.6.

• An intercept of the odderon near 1 has been assumed in order to compare the model predictions at $W = 20$ GeV with the experimental data at $W \approx 200$ GeV.

The model as it stands yields intercept 1 both for the pomeron and for the odderon trajectory. Pomeron and odderon trajectories have been estimated[458] in the time-like region from predictions for glueball masses. In this approach the odderon intercept comes out as approximately -1.5, that of the pomeron as approximately 0.5. However in this model quark-gluon mixing increases the value of the pomeron intercept to approximately 1.1, but there is no such mixing for the odderon. This model[458] predicts that the contribution of the C-odd exchange at high energies is much too small to be detectable.

8.10 The pomeron trajectory and nonperturbative QCD

To understand the pomeron and reggeon trajectories is one of the most challenging goals of QCD. In this section we describe some attempts in this direction. One connection between nonperturbative QCD and Regge theory is a relation between the slope of the linear confining potential and the slope of the almost-degenerate (f_2, a_2, ω, ρ) trajectories shown in figure 2.7. In an intuitive picture of confinement the quark and the antiquark in a meson are connected by a string with a constant tension σ_q which is the slope of the linear confining potential. This simple picture is supported by lattice calculations and is also a result of models for nonperturbative QCD such as the one described in section 6.7. The angular-momentum excitations of this open string lie on a linear trajectory with slope

$$\alpha' = \frac{1}{2\pi\sigma_q}. \tag{8.117}$$

See for instance [459]. The string tension σ_q calculated from this relation, using the observed reggeon slope $\alpha' \approx 0.9$ GeV^{-2}, agrees remarkably well

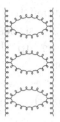

Figure 8.16. A next-to-leading-order contribution to the gluon ladder, the non-perturbative part of which gives rise to a soft-pomeron intercept larger than 1

with the values obtained from quarkonium spectroscopy[271].

This can be extended to the pomeron trajectory. If we interpret the soft pomeron as the trajectory of hadrons consisting of two gluons, as suggested in section 3.8, the string tension σ_q has to be replaced by σ_g, the string tension between two gluons. Lattice calculations[273] find that σ_g is a factor $\frac{9}{4}$ larger[274] than σ_q. From (8.117) we then obtain a slope for the soft-pomeron trajectory which is a factor $\frac{4}{9}$ smaller than the slope of the meson trajectory. This very simple consideration gives a result which goes in the right direction, but the resulting numerical value $\alpha'_{I\!P} \approx 0.4$ GeV^{-2} is still considerably larger than the phenomenological value 0.25 GeV^{-2}.

In this approach to the soft pomeron the glueballs appear naturally on the pomeron trajectory as composites of two gluons; compare figure 3.28. Therefore one can extend to the pomeron the duality considerations applied in section 4.5 to the reggeons. One has to consider glueball-glueball scattering amplitudes and will encounter daughter trajectories of the pomeron; see section 2.7.

Understanding the soft-pomeron intercept is even more challenging. Simple two-gluon exchange and the nonperturbative approaches discussed previously in this chapter lead to constant cross sections and hence to a pomeron intercept $\alpha_{I\!P}(0) = 1$. In section 7.3 we have explained that in perturbation theory summation of leading ($\alpha_s \log s$) terms from generalised ladder diagrams can lead to a cross section which increases with a power of s. Recently there have been interesting first attempts to calculate ϵ_1, the deviation from 1 of the soft-pomeron intercept, by nonperturbative dynamics related to the properties of the QCD vacuum. Though rather strong assumptions are involved the approach looks promising and the resulting numerical values are encouraging.

The starting point of [460] is similar to the treatment of the BFKL pomeron, namely a gluon ladder. In contrast to that approach the rungs of the ladder are not gluons but colour-singlet states. A contribution to this is shown in figure 8.16. Formally summing up the ladder contributions leads to an energy-dependent cross section

$$\sigma(s) = \sigma_0 \left(\frac{s}{s_0}\right)^\epsilon \qquad (8.118)$$

where ϵ can be expressed through an integral over the spectral density ρ^θ of the correlator of the divergence of the energy-momentum tensor (6.39). In perturbation theory this integral is infrared and ultraviolet divergent. The principal idea in [460] was to split this integral into a divergent perturbative and a finite nonperturbative piece. The perturbative piece is part of the next-to-leading-order contribution to the BFKL pomeron, but the nonperturbative part is supposed to give rise to the energy dependence of the soft-pomeron contribution. The intercept $1 + \epsilon_1$ is an integral over the nonperturbative part ρ^θ_{np} of the spectral density ρ^θ:

$$\epsilon_1 = \frac{144\pi^2}{\beta_0^2(N_C^2 - 1)} \int \frac{dM^2}{M^6} \rho^\theta_{np}(M^2) \qquad (8.119)$$

where β_0 is the first coefficient of the β function of QCD; see (6.19). The quantity ρ^θ_{np} is related by a low-energy theorem[461] to the trace of the energy-momentum tensor:

$$\int \frac{dM^2}{M^2} \rho^\theta_{np}(M^2) = -4\langle: \theta^\mu_\mu :\rangle = -2\frac{\beta(g)}{g^3} \langle g^2 FF \rangle \qquad (8.120)$$

where we have used (6.39). If we stick to the quenched approximation discussed so far in this section the only hadronic states in the gluon channel are glueballs. In the spirit of QCD sum rules[250] the nonperturbative contribution is approximated by the lowest-lying resonance, a scalar glueball with mass M_G, whereas the higher resonances are related to the perturbative contribution by quark-hadron duality. This leads to

$$\epsilon_1 = \frac{18\langle g^2 FF \rangle}{\beta_0(N_C^2 - 1)M_G^4} \qquad (8.121)$$

where we have approximated $\beta(g)/g^3$ by $-\beta_0/(16\pi^2)$. Using $M_G = 1.611$ GeV for the lightest 0^{++} glueball mass, as obtained from quenched lattice calculations[27], and the large value of the gluon condensate (6.52) of quenched QCD, $\langle g^2 FF \rangle = (1.22 \text{ GeV})^4$, one obtains $\epsilon_1 = 0.067$, which is close to the phenomenological value of about 0.08. In a more realistic treatment with light quarks the dominant contribution to the rungs of the ladder will come from pions. Using the effective chiral Lagrangian one can estimate[460] $\epsilon_1 \approx 0.1$.

More detailed investigations have been performed using the instanton model for the nonperturbative QCD vacuum. Instanton-induced amplitudes have been calculated in electroweak theory[462,463] in which they could lead to baryon-number violation and also[464] in QCD in which they should lead to interesting but less spectacular effects. Elastic and quasi-elastic scattering of quarks and quark-antiquark pairs in the instanton field have been investigated[465]. Since the instanton is a classical solution of Euclidean QCD, the scattering amplitude has to be calculated in Euclidean metric and the expressions in Minkowski metric require analytic continuation as discussed in section 8.11. The structure of the resulting amplitudes is similar to the one obtained in the stochastic-vacuum model and in particular the cross sections come out as independent of energy. In a further paper[466], instanton-induced inelastic collisions were considered and were shown to produce a rising total cross section. The estimated value $\epsilon_1 \approx 0.03$ is smaller than the phenomenological value.

The nonperturbative couplings of the colour-singlet rungs of the ladder of figure 8.16 have been evaluated[467] in the instanton field and inserted into a BFKL-type equation. The structure of the resulting trajectory shows interesting effects. A decoupling of the soft pomeron at high values of Q^2 is found, but numerical estimates are still very rough, since they depend crucially on several poorly-known parameters such as the maximum instanton size. Nevertheless, values of ϵ_1 around 0.1 are obtained with plausible assumptions and the ratio of the slope of the pomeron trajectory to ϵ_1, $\alpha'_{I\!P}/\epsilon_1 \approx 2$ GeV^{-2}, is not too far from the phenomenological value of about 3 GeV^{-2}.

We have already stated in section 4.4 that Regge theory was the starting point of string theory[468], and now starting from super-string theory attempts have been made to understand the pomeron trajectory[469]. However there appeared to be serious obstacles to applying string theory to QCD. Massless spin-two and spin-one hadrons, not present in QCD, appear unavoidably and quantum anomalies require the number of space-time dimensions to be 10 or 26. Now there is hope that these obstacles can be overcome. The microscopic theory is a four-dimensional gauge theory, a supersymmetric QCD. It lives on the boundary of a five dimensional anti-de Sitter space. For a short review see [470]. The massless 2^{++} graviton in a five-dimensional world can appear in four dimensions as a massive glueball on the pomeron trajectory. The imaginary part of the scattering amplitude is determined by small impact parameters and in confining theories is governed by minimal surfaces between the loops. A linear trajectory of a

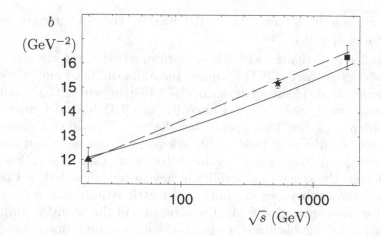

Figure 8.17.   Forward-slope parameter $b = b(s, 0)$ for $\bar{p}p$ elastic scattering. The solid curve is from the model of section 8.5, using the parameter set (8.112), and the dashed curve is from Regge theory with a linear pomeron trajectory (3.19). The experimental points are triangle[471], star[472], and box[52].

Regge double pole is found[469] with intercept 1. The slope is determined by the horizon, that is the penetration into the five-dimensional anti-de Sitter space, and is inversely proportional to the coupling in the microscopic gauge theory.

We end with a much more down-to-earth method to incorporate the s dependence of soft high-energy scattering into the model discussed in section 8.7. In it the transverse radius of the hadrons depends on the scattering energy. For values of the transverse radius of the proton between 0.5 and 1.5 fm, the radius dependence of the total pp cross section can be well described by the power behaviour $\sigma^{\text{Tot}}(R_h) \approx \sigma^{\text{Tot}}(R_0)(R_h/R_0)^3$, and hence an energy dependence of the radius

$$R_h(s) = R_0 \left(\frac{s}{s_0} \right)^{\epsilon_1/3} \tag{8.122}$$

leads to the observed energy dependence $\sigma^{\text{Tot}} \sim \sigma(s_0)(s/s_0)^{\epsilon_1}$ of the cross section. The values of R_0 and s_0 should be chosen in conformity with the conditions set up in section 8.7. In the same model the logarithmic slope of the elastic cross section, the forward slope parameter $b(s, 0)$ of (3.23), can be expressed by the simple relation[259]

$$b(s, 0) \approx Aa^2 + B\Big(R_h(s)\Big)^2 \tag{8.123}$$

where a is the correlation length of the gauge-invariant gluon correlator introduced in section 6.4. In figure 8.17 we display $b(s, 0)$ in the range $20 \text{ GeV} \leq \sqrt{s} \leq 2000 \text{ GeV}$ for both sets of parameters given in table 8.1 and compare it with data and with Regge theory.

8.11 Scattering amplitudes in Euclidean space

In the approach described in sections 8.2 to 8.4 the parton-parton scattering amplitudes were represented as expectation values of Wilson-line operators along lightlike paths as in (8.54). We said that in order to achieve this, the scattering of the partons was to be considered only over a finite time interval τ_0 (see (8.29) and (8.30)) and thus all the time integrals should run from $-\tau_0/2$ to $+\tau_0/2$. What happens if we take the integrals from $-\infty$ to $+\infty$? This leads[428] to infinities and one way of regularising is to take Wilson lines which are not exactly lightlike. Consider quark-quark scattering (8.31) in the centre-of-mass system. Let m be the quark mass and $\sqrt{s}$ the centre-of-mass energy. Then non-interacting classical quarks q of momentum p_1 and p_2 travel on world lines

$$x_i(\tau) = x_i(0) + \frac{p_i}{m}\tau, \quad i = 1, 2; \quad -\infty < \tau < \infty. \tag{8.124}$$

The momenta p_1, p_2 can be parametrised by a hyperbolic angle χ:

$$\frac{p_{1,2}}{m} = \begin{pmatrix} 1/\sqrt{1-v^2} \\ 0 \\ 0 \\ \pm v/\sqrt{1-v^2} \end{pmatrix} = \begin{pmatrix} \cosh\frac{1}{2}\chi \\ 0 \\ 0 \\ \pm\sinh\frac{1}{2}\chi \end{pmatrix}$$

$$v = (1 - 4m^2/s)^{1/2} = \tanh\frac{1}{2}\chi \tag{8.125}$$

where the upper sign goes with p_1 and the lower sign with p_2. For large s, $\chi \sim \log(s/m^2)$.

It has been suggested[428] that one should take the Wilson lines along the paths (8.124) and consider the limit $s \to \infty$, which corresponds to the limit of the Wilson lines becoming lightlike. It was shown that inserting these lines into (8.54) and evaluating the functional integral $\langle \cdots \rangle_A$ in perturbation theory leads to the standard results[5] for quark-quark scattering.

The Wilson lines at a hyperbolic angle χ were also[473,474] the starting point to show that at fixed t the scattering amplitude can be continued analytically in the angle χ to imaginary values

$$\chi \to i\theta. \tag{8.126}$$

With this the scattering amplitude

$$A(s, t) = g(\chi, t) \tag{8.127}$$

written as a function of χ and t using (8.125) is related to a correlation function $E(\theta, t) = g(i\theta, t)$ of connectors along a pair of Euclidean-space Wilson lines forming an angle θ. It was proposed to evaluate $E(\theta, t)$ directly in lattice QCD, then to make the reverse analytic continuation $\theta \to -i\chi$ and finally to take the limit $\chi \to \infty$ corresponding to $s \to \infty$. One can also write down a partial-wave expansion for $E(\theta, t)$ for fixed t and perform a Sommerfeld-Watson transformation. This is completely analogous to the t-channel partial-wave expansion for $A(s, t)$ discussed in sections 2.1 and 2.2. Again, if we assume the presence of poles in the complex l-plane, we find power behaviour in s for the analytically-continued amplitude $A(s, t)$ of (8.127). Thus, this approach offers a nice bridge between the functional-integral and the Regge approaches. Of course, we really should like to see from a lattice calculation that poles in the complex l-plane do indeed arise for $E(\theta, t)$.

The analytic continuation is also essential for estimating the contribution of instanton background fields[465] in the functional integral (8.54) and also for the anti-de-Sitter-space approach[475] discussed in section 8.10. Both these calculations must begin in Euclidean space.

Another type of analytic continuation has been considered[476] for the structure functions of deep-inelastic scattering. The forward virtual Compton amplitude was considered in a model with scalar fields. Instead of studying the limit of large ν on the real axis, the amplitude for large imaginary ν was investigated. This is equivalent, because of the Phragmén-Lindelöf theorem[477]. Then the amplitude was related to the correlation function of two currents in an effective field theory. Finally it was argued that there should be a critical point of this effective theory where the correlation length in one direction becomes very large, and that from this critical behaviour one should be able to get information on the pomeron properties. It was also shown how to continue the effective field theory from Minkowski to Euclidean space, where one has the possibility of doing calculations on a lattice. Thus, this approach offers the promise of doing nonperturbative calculations for the deep-inelastic structure functions at small x from first principles.

9

The dipole approach

In the dipole model photons and hadrons are treated as superpositions of colourless quark-antiquark pairs of different sizes. This approach allows one to relate deep-inelastic scattering to other reactions involving photons and to purely-hadronic scattering.

9.1 Deep-inelastic scattering

The dipole approach to deep-inelastic scattering is very close to the treatment of hadron-hadron scattering of section 8.5. The photon dissociates into a quark-antiquark pair in a colour-singlet state which interacts strongly with the target proton. The approach can be used if the dissociation time of the photon is large compared with the interaction time of the hadronic systems. The dissociation time τ_{dis} of a photon into a quark and an antiquark, each of mass m_f, can easily be estimated from the energy uncertainty principle. Let $q = (q_0, 0, 0, q_3)$ be the four-momentum of the photon, and

$$k_1 = \left(E_1, \mathbf{k}_T, z q_3 \right) \qquad k_2 = \left(E_2, -\mathbf{k}_T, (1-z) q_3 \right) \qquad 0 \le z \le 1 \qquad (9.1)$$

the four-momenta of the quark and antiquark. Then

$$\tau_{\text{dis}} = \frac{1}{|q_0 - E_1 - E_2|} \sim \frac{2 q_0 z(1-z)}{Q^2 z(1-z) + m_f^2 + \mathbf{k}_T^2} \qquad (9.2)$$

for large q_3, where $Q^2 = -q^2$, the virtuality of the photon. In the rest frame of the target proton we expect the interaction time of the quark-antiquark pair with the proton to be of the order of magnitude of the typical confinement radius of light quarks, which is approximately the inverse QCD

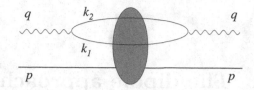

Figure 9.1.   Interaction of a proton with a highly-energetic photon

scale $1/\Lambda$. Hence in this frame the condition that the dissociation time is large compared with the interaction time is

$$W^2 = m_p^2 - Q^2 + 2m_p q_0 \gg \frac{m_p}{\Lambda}\left(Q^2 + \frac{m_f^2 + \Lambda^2}{z(1-z)}\right) \qquad (9.3)$$

where m_p is the target mass and we have put $|\mathbf{k}_T|$ approximately equal to the inverse confinement radius $1/\Lambda$. Unless z is close to 0 or 1 this implies that $W^2/Q^2 \gg 1$ and so according to (7.1) the Bjorken variable x is small. We thus come to a very intuitive picture of deep-inelastic scattering at small x which is visualised in figure 9.1.

We have seen from (8.97) that for high values of Q^2 the transverse size of the hadronic state into which the photon dissociates is small and the quark-antiquark pair acts as a colour dipole. Therefore this approach is generally called the dipole model although no dipole approximation is involved, that is one does not restrict considerations to very small quark-antiquark separations. For deep-inelastic scattering[478] the dipole model was introduced mainly to explain the nuclear shadowing observed by the EMC collaboration[479].

The interpretation of figure 9.1 is that the forward amplitude $T_{\gamma p}(\lambda, W^2, Q^2)$ for scattering a photon with helicity λ and virtuality Q^2 on a proton is written as an integral of a product of the quark-antiquark density $\rho_\gamma(\lambda, Q^2, R, z)$ of the photon defined in (8.97), and the forward amplitude $T_{\text{dip}}(W^2, \mathbf{R}, z)$ for scattering a quark-antiquark dipole on a proton:

$$T_{\gamma p}(\lambda, W^2, Q^2)\big|_{t=0} = \int d^2R \int_0^1 dz\, \rho_\gamma(\lambda, Q^2, R, z)\, T_{\text{dip}}(W^2, \mathbf{R}, z) \qquad (9.4)$$

where $\mathbf{R}$ is the distance vector from the quark to the antiquark. Using the optical theorem (3.1), we replace the imaginary part of the forward scattering amplitude with the total cross section and average over all directions of the quark-antiquark pair. A possible z dependence of the dipole-target cross section is generally neglected. This is justified in the model described in section 8.5, as can be seen from figure 9.2, which plots the z-dependent

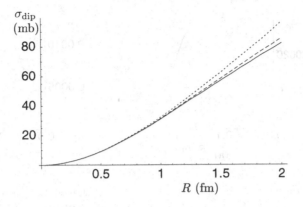

Figure 9.2. Dipole cross section[435] $\sigma_{\text{dip}}(W^2, R, z)$ for various values of z, the longitudinal momentum fraction of the quark, at $W \approx 20$ GeV. The solid curve is for $z = 0.5$, the dashed curve for $z = 0.25$ and the dotted curve for $z = 0$.

dipole cross section defined as

$$\sigma_{\text{dip}}(W^2, R, z) = \frac{1}{W^2} \int_0^{2\pi} \frac{d\phi_T}{2\pi} \, \text{Im} \, T_{\text{dip}}(W^2, \mathbf{R}, z). \tag{9.5}$$

This leads to

$$\sigma^{\gamma^* p}(\lambda, W^2, Q^2) = \int dR \, \tilde{\rho}_\gamma(\lambda, Q^2, R) \, \sigma_{\text{dip}}(W^2, R) \tag{9.6}$$

where the dipole cross section is

$$\sigma_{\text{dip}}(W^2, R) = \int_0^1 dz \, \sigma_{\text{dip}}(W^2, R, z) \tag{9.7}$$

and the planar density of the photon is

$$\tilde{\rho}_\gamma(\lambda, Q^2, R) = 2\pi R \int_0^1 dz \, \rho_\gamma(\lambda, Q^2, R, z) \tag{9.8}$$

of which (8.107) is a special case. Equations (9.4) and (9.6) have been made plausible starting from the parton picture and QCD factorisation[480], which is valid for high photon virtualities. This factorisation should not be confused with factorisation in the Regge sense (2.32). A more general derivation is given in terms of the functional approach as described in section 8.5.

For longitudinal photons of high virtuality the perturbative result for the photon wave function (8.95) is a good approximation and therefore one can

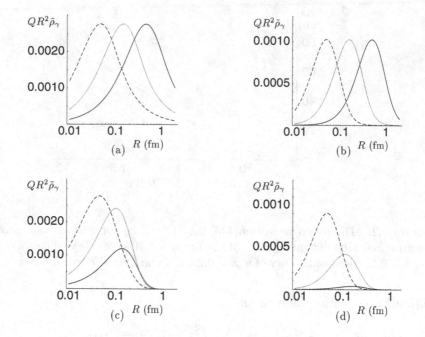

Figure 9.3.   The dimensionless quantity $QR^2\tilde{\rho}_\gamma(\lambda, Q^2, R)$ for $Q^2 = 1$ GeV^2 (thick line), 10 GeV^2 (thin line), and 100 GeV^2 (dashed line). (a) u and d-quark contribution, transverse photon; (b) u and d-quark contribution, longitudinal photon; (c) charm-quark contribution, transverse photon; (d) charm-quark contribution, longitudinal photon.

extract from the longitudinal structure function the dipole-proton cross section for small quark-antiquark separations in a model-independent way. One finds[480]

$$\sigma_{\mathrm{dip}}(Q^2(1/x - 1), R) = \tfrac{1}{3}\alpha_s(Q)\, xg(x, Q^2)\, R^2 \qquad (9.9)$$

where $x\,g(x, Q^2)$ is the gluon density of the proton. The dipole size R can be related to Q^2 in the following way. According to (9.6) and the relations between the structure functions and the cross sections given in section 7.1, the structure functions can be expressed as integrals over dipole sizes. The longitudinal structure function is

$$F_L(x, Q^2) = \frac{Q^2}{4\pi^2\alpha} \int dR\, \sigma_{\mathrm{dip}}(Q^2(1/x - 1), R)\, \tilde{\rho}_\gamma(0, Q^2, R). \qquad (9.10)$$

An analogous expression holds for the structure function $F_2(x, Q^2)$:

$$F_2(x, Q^2) = \frac{Q^2}{4\pi^2\alpha} \int dR \, \sigma_{\text{dip}}(Q^2(1/x - 1), R) \left(\tilde{\rho}_\gamma(0, Q^2, R) + \tilde{\rho}_\gamma(1, Q^2, R) \right).$$

(9.11)

For small values of R we can use (9.9) and obtain for the integrand of (9.10)

$$\tfrac{1}{3} Q^2 \alpha_s(Q) \, xg(x, Q^2) \, R^2 \tilde{\rho}_\gamma(0, Q^2, R)$$

(9.12)

and similarly for (9.11). The dimensionless quantities $QR^2\tilde{\rho}_\gamma(\lambda, Q^2, R)$ are plotted in figure 9.3. For small R they have the behaviour of the integrand (9.12). They show pronounced maxima at

$$R_{\text{max}} = \frac{C}{\sqrt{Q^2 + 4m_f^2}}$$

(9.13)

with $C \approx 0.4$ fm GeV ≈ 2.

The main contribution to the integral is concentrated around $R \approx R_{\text{max}}$. The relation between Q^2 and R^2 in (9.9) is therefore

$$Q^2 \approx \frac{C^2}{R^2} - 4m_f^2.$$

(9.14)

For transverse photons and massless quarks the function $QR^2\tilde{\rho}_\gamma(\pm 1, Q^2, R)$ has a tail that extends to large distances. This tail comes from values of z near 0 or 1. It gives a finite contribution to the structure function also in the limit of very large Q^2, as can be inferred from (8.98). Therefore the value $C^2 \approx 4$ quoted above should be regarded rather as a lower limit for C^2. Some authors, for instance[481], prefer for that reason values near $C^2 = 10$.

Since the energy dependence of the structure function is conveniently expressed as a function of x it is quite common to express the dipole cross section also as a function of x rather than of W. However this seems inconsistent, since the Q^2 dependence has been explicitly eliminated from the dipole cross section. Furthermore it is inconvenient, since an interesting feature of the dipole formalism is the extension into the fully nonperturbative domain, that is also to reactions of real photons and light hadrons. In that case x is no longer the most appropriate variable. It is more suitable[482] to use the dimensionless quantity $\zeta = W^2 R^2$ as the variable for the dipole cross section. For high values of Q^2 we can relate R^2 to Q^2, and obtain from (9.14)

$$\zeta = C^2 \frac{1 - x}{x}.$$

(9.15)

We shall use this relation quite generally to compare different models for the dipole cross section.

The separation of photon and hadron cross sections into a part describing the parton dynamics of scattering and a part describing the structure of the asymptotic states has advantages. It allows different scattering processes to be related using the same dipole cross section and gives the possibility of incorporating nonperturbative effects both in the wave function and in the dipole cross section. On the other hand, in most practical cases this separation introduces a model dependence, since neither the wave functions nor the parton scattering amplitudes are accessible to experiment nor can they yet be treated from first principles. In section 9.3 we shall see that different approaches do indeed lead to very different dipole cross sections.

9.2 Production processes

We consider such processes as exclusive electroproduction of vector mesons or more generally reactions of the type $ap \to bp$. The dipole approach relates these processes to deep inelastic scattering. The density ρ_γ of the photon in (9.4) has to be replaced by the overlap integral of the wave functions of the incoming and produced particles, yielding

$$T_{ap \to bp}(W^2, t)\Big|_{t=t_{\min}} = \int d^2R \int_0^1 dz \, \psi_b^*(\mathbf{R}, z) \, \psi_a(\mathbf{R}, z) \, T_{\mathrm{dip}}(W^2, \mathbf{R}, z)$$

(9.16)

for the forward amplitude, where $t_{\min}$ is the value of t for forward production. If the overlap integral depends at most weakly on the planar angle ϕ_T we can perform the integration over it and, neglecting the real part of T_{dip}, obtain the forward differential cross section analogously to (9.6):

$$\frac{d}{dt}\sigma(W^2, t)\Big|_{t=t_{\min}} \approx \frac{1}{16\pi}\left| \int dR \, \tilde{\rho}_{ab}(R) \, \sigma_{\mathrm{dip}}(W^2, R)\right|^2$$

(9.17)

where $\tilde{\rho}_{ab}$ is the planar overlap

$$\tilde{\rho}_{ab}(R) = R \int_0^{2\pi} d\phi_T \int_0^1 dz \, \psi_b^*(\mathbf{R}, z) \, \psi_a(\mathbf{R}, z).$$

(9.18)

If the mass or virtuality of the incoming state is different from that of the outgoing one there must be a small momentum transfer in the longitudinal direction. This means that the longitudinal-momentum fraction of the partons changes. For high-energy electroproduction of a vector meson with mass m_V the change in the longitudinal-momentum fraction of the parton is of order $(Q^2 + m_V^2)/W^2$. In the dipole approach it is generally neglected.

If one wants to calculate σ_{dip} in terms of the gluon distribution, as in (9.9), then $g(x, Q^2)$ has to be given by a more general skewed (or off-diagonal) gluon distribution[483–485]. We should also remember the caveat in section 8.6: an integral involving the product of two different wave functions is much more sensitive to details of the wave functions than is an integral involving the density.

If the outgoing particle b is the same as the incoming one, (9.16) describes the elastic forward scattering amplitude and from that the total cross section for the reaction $ap \to X$ can be obtained using the optical theorem:

$$\sigma_{ap}^{\text{Tot}}(W^2) = \int dR \, \tilde{\rho}_a(R) \, \sigma_{\text{dip}}(W^2, R) \tag{9.19}$$

where $\tilde{\rho}_a(R)$ is given by (9.18) with b replaced with a.

Comparison of (9.6) and (5.7) allows some insight into the mechanisms and limitations of the vector-meson dominance model discussed in section 5.2. The transverse vector-meson–proton cross section σ_n^{T} in (5.7) can be expressed in the dipole approach as

$$\sigma_n^{\text{T}} = \int dR \, \tilde{\rho}_n^T(R) \, \sigma_{\text{dip}}(W^2, R) \tag{9.20}$$

where $\tilde{\rho}_n^T(R)$ is the planar quark-antiquark density of the transverse vector meson V_n:

$$\tilde{\rho}_n^T(R) = R \int_0^1 dz \int_0^{2\pi} d\phi_T \, |\psi_n^T(\mathbf{R}, z)|^2. \tag{9.21}$$

It is instructive to concentrate on the lowest-lying resonances contributing to (5.6). Comparison of (5.7), (9.6) and (9.20) yields

$$\tilde{\rho}_\gamma(\pm 1, Q^2, R) = \sum_{V=\rho,\omega} \frac{4\pi\alpha}{\gamma_V^2} \left(\frac{m_V^2}{m_V^2 + Q^2} \right)^2 \rho_V^T(R) + \text{higher resonances.} \tag{9.22}$$

Figure 8.10 shows the combined u and d-quark contribution to $\tilde{\rho}_\gamma(\pm 1, 0, R)$ for a real photon and the contribution of the ρ and the ω to the right hand side of (9.22). A Gaussian form (8.108) was chosen as the vector-meson wave function with transverse radius $R_h = 0.68$ fm. In a simple quark model this wave function reproduces the observed leptonic decay widths, yielding $(4\pi/\gamma_\rho^2 + 4\pi/\gamma_\omega^2) = 0.55 = 0.004/\alpha$. In the region $R \approx R_h$ where the integral (9.6) receives its main contributions the two curves agree reasonably well, which explains the success of the vector-dominance model in its most simple form. For $Q^2 > 0$ the approximation by the lowest-lying

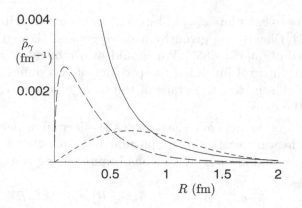

Figure 9.4. Combined u and d-quark contribution to the planar density $\tilde{\rho}_\gamma(\lambda, Q^2, R)$ of a transverse photon (solid line) and a longitudinal photon with virtualities $Q^2 = m_\rho^2$ (long dashed line), and the combined ρ and ω contribution to the right hand side of (9.22) at $Q^2 = m_\rho^2$ (short dashed line).

resonances of course becomes much worse. The quark-antiquark density of the vector meson is only multiplied by the factor $m_V^2/(Q^2 + m_V^2)$ whereas the photon wave function changes its form, as can be seen in figure 9.4 where the solid line is the photon wave function for $Q^2 = m_\rho^2$ and the short-dashed line the ρ and ω contribution to the right hand side of (9.22). Also, the short distance part becomes more and more important with increasing Q^2. This part can be approximated well only if many high-lying resonances are taken into account, and to reproduce the singularity of the photon density one even needs an infinite sum. Another important feature of the photon density $\tilde{\rho}_\gamma$ is the shrinkage of the mean square radius with increasing Q^2; see (9.11). In the vector-meson dominance model it can be reproduced only by taking non-diagonal terms[486] into account; see (5.6).

The simplest prescription for obtaining a longitudinal photon-proton amplitude in the vector-meson dominance model is to multiply the transverse amplitude by the factor $(Q^2/m_V^2)^2$ as in (5.5). In this prescription the longitudinal contribution at $Q^2 = m_\rho^2$ is equal to the transverse one, which is the short dashed curve in figure 9.4. There is little similarity to the longitudinal photon density $\tilde{\rho}_\gamma(0, m_\rho^2, R)$ indicated by long dashes in figure 9.4. Indeed the longitudinal cross section is severely overestimated if the above prescription is applied, as it can be seen from figure 7.30b that the ratio σ_L/σ_T increases much more slowly than Q^2.

The dipole cross section $\sigma_{\mathrm{dip}}(W^2, R)$ can be used to give at least a lower bound on the inclusive diffractive cross section for the reaction $a\,p \to X\,p$

in the forward direction with a large rapidity gap between the final proton and the hadronic state X. We imagine that the system X arises from the production of a quark-antiquark pair and that there is unit probability of this pair materialising into a system of hadrons. So

$$\frac{d}{dt}\sigma_{ap\to Xp}(W^2,t)|_{t=t_{\min}} =$$

$$\frac{1}{16\pi W^4}\int d^2R \int_0^1 dz\, |\psi_a(\mathbf{R},z)|^2\, |T_{\mathrm{dip}}(W^2,\mathbf{R},z)|^2. \qquad (9.23)$$

We neglect the z dependence of the dipole-scattering amplitude and apply the inequality

$$\int_0^{2\pi} d\phi_T\, |F(\phi_T)|^2 \ge \left|\int_0^{2\pi} d\phi_T\, F(\phi_T)\right|^2 \qquad (9.24)$$

to the integration over azimuthal angle. Thus we obtain

$$\frac{d}{dt}\sigma_{ap\to Xp}(W^2,t)|_{t=t_{\min}} \ge \frac{1}{16\pi}\int d^2R \int_0^1 dz\, |\psi_a(\mathbf{R},z)|^2\, (\sigma_{\mathrm{dip}}(W^2,R))^2. \tag{9.25}$$

In most applications it is assumed that $T_{\mathrm{dip}}(W^2,\mathbf{R})$ depends only weakly on ϕ_T and (9.25) is taken as an equality.

The ϕ_T dependence of the dipole cross section $T_{\mathrm{dip}}(W^2,\mathbf{R})$ can be investigated in models. In the model discussed in section 8.5 it turns out that the only appreciable variation in ϕ_T is proportional to $\cos(2\phi_T)$:

$$T_{\mathrm{dip}}(W^2,\mathbf{R}) \approx \sigma_{\mathrm{dip}}(W^2,R) + \sigma_2(W^2,R)\cos(2\phi_T). \qquad (9.26)$$

As can be seen from figure 9.5, the ϕ_T-dependent term is indeed small compared with $\sigma_{\mathrm{dip}}(R)$ in the relevant range of R.

In (9.23) it is assumed that one may calculate quark-antiquark pair production and that this gives the cross section for hadron production; this corresponds to global quark-hadron duality. The assumption of local quark-hadron duality is much stronger. This requires that a colour-singlet quark-antiquark state of invariant mass M approximates the hadronic contribution at the same mass. Evidently a certain averaging has to be performed if M is still in the resonance region, as in sections 4.2 and 4.3. In the approximation of local duality one can calculate the diffractive structure function, introduced in section 7.10, in the forward direction. The squared invariant mass of the quark-antiquark pair is

$$M^2 = \frac{\mathbf{p}_T^2 + m^2}{z(1-z)} \qquad (9.27)$$

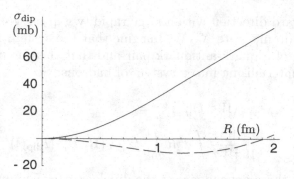

Figure 9.5.   Dipole cross section $\sigma_{\text{dip}}(W^2, R)$ (solid line) and the second Fourier coefficient $\sigma_2(W^2, R)$ of (9.26) (dashed line) at $W = 20$ GeV

where $\mathbf{p}_T$ is the relative momentum of the pair. The variables of the diffractive structure functions can be expressed through x, z and $\mathbf{p}_T^2$ as

$$\beta = \frac{\epsilon^2}{\mathbf{p}_T^2 + \epsilon^2} \qquad \xi = x/\beta \tag{9.28}$$

where $\epsilon = \sqrt{z(1-z)Q^2 + m_f^2}$ as in (8.96). Complete expressions for the forward diffractive structure function in terms of the dipole cross section may be found in [404].

9.3 Different approaches to dipole cross sections

In this section we discuss some typical models for the dipole cross section, emphasising their most distinct features. For details and motivation we refer to the original literature. We start with the lowest-order perturbative contribution, namely the two-gluon contribution to the total cross section of two colourless quark-antiquark pairs at high energies as first discussed by Low[3] and Nussinov[4]. These results play a role in an alternative approach [487,389] to the BFKL pomeron, in which the proton is considered as a superposition of many small dipoles. It is also instructive to see that the perturbative result can be extended into the nonperturbative domain, with seemingly-reasonable results. In the high-energy limit the perturbative results for dipole-dipole scattering simplify considerably and are reduced to an integral over the internal transverse momentum $\mathbf{k}_T$. In this limit the scattering cross section of two dipoles with sizes R_1 and R_2 is

$$\sigma(R_1, R_2) = \frac{32}{9}\alpha_s^2 \int d^2 k_T \left(1 - J_0(k_T R_1)\right)\left(1 - J_0(k_T R_2)\right)\left(\Delta(k_T)\right)^2 \tag{9.29}$$

where $\Delta(k_T)$ is the gluon propagator. For the free propagator $1/k_T^2$ one can perform the integration analytically and obtain

$$\sigma(R_1, R_2) = \frac{16\pi}{9} \alpha_s^2 R_<^2 \left(1 + \log \frac{R_>}{R_<}\right) \qquad (9.30)$$

with $R_<$ and $R_>$ the smaller and larger of R_1, R_2. The larger dipole cuts off the lower frequencies and therefore the result is infrared-finite and one obtains reasonable numbers even if R_1 and R_2 are of hadron size. But the perturbative result can be trusted only if this cutoff is well below the confinement radius; that is both dipoles need to be small. If a gluon mass m_g is introduced into the propagator $\Delta(k_T)$ in order to suppress contributions from small k_T and if this mass is larger than the inverse radius of the larger dipole, it determines the infrared cutoff and the perturbative contribution to the cross section saturates at a correspondingly lower value. If one of the radii, say R_1, is small the perturbative expression yields the correct power behaviour $\sigma \sim R_1^2$ (see (9.9)) but the coefficient proportional to the gluon density of the proton is of course determined by nonperturbative QCD. If both R_1 and R_2 are large the contributions of the oscillating Bessel functions in (9.29) become small and the asymptotic value of the cross section is

$$\lim_{R_1, R_2 \to \infty} \sigma(R_1, R_2) = 32\pi\alpha_s^2/(9m_g^2). \qquad (9.31)$$

From lattice results[257] for the gluon correlator one can deduce a value $m_g \approx 1$ GeV and, with a frozen value $\alpha_s = 0.6$, one obtains for the asymptotic perturbative cross section the numerical value 1.6 mb, which is only a small fraction of a typical hadronic cross section. If, however, both radii are small, a situation which can be realized in $\gamma^*\gamma^*$ scattering (see section 7.6), the perturbative expressions are dominant.

The space-time picture in deep-inelastic scattering was revived[478] mainly to explain nuclear-shadowing effects. The first investigation[478] started with a perturbative expression of the type (9.29). This expression was later taken as a phenomenological nonperturbative and energy-independent input. To it were added contributions which take into account the occurrence of higher Fock states and which are solutions of a BFKL-type equation[488,489]. They appear as a series of Regge poles which lead to a strong rise with energy and are dominant at very high energies. These dipole cross sections have been applied to deep-inelastic scattering[478,488], production of vector mesons[490,491] and inclusive diffraction[492].

The two-pomeron approach[194] described in chapter 7 has motivated a particular way to incorporate the energy dependence into the dipole cross section. The hard pomeron, which is dominant for processes involving large virtualities, is associated with small dipoles, the soft pomeron with large dipoles. Therefore in [482] the dipole cross section is assumed to consist of two parts, a hard part dominant at small distances which increases like $(W^2R^2)^{0.38}$, and a soft part dominant at large distances, which increases like $(W^2R^2)^{0.06}$. The forms of the dipole cross sections and the power of the hard part were fitted in such a way as best to describe the structure functions. Another approach[442,493] is based on the dipole cross section obtained from the nonperturbative formalism using the functional methods described in section 8.5. The contributions from small dipoles are assumed to contribute to the hard pomeron with an intercept of 1.4, and the large dipoles to the soft pomeron. We shall describe this approach in more detail in section 9.5.

The dipole cross section of [494,495] is also based on functional methods. But in contrast to the approach discussed in section 8.5 the proton is not resolved into partons but considered as a colour background field consisting of many regions with uncorrelated colour field strengths. This leads to a saturation of the dipole cross section at a radius proportional to the inverse square root of the number of colour domains in the proton. The result is used as an input to a DGLAP evolution of the proton structure function starting from a low scale $Q_0^2 \approx 1.2$ GeV^2 corresponding to a separation of the quark and antiquark of about 0.4 fm. The strong increase of the structure function with $1/x$ at large values of Q^2 in this approach is a consequence partly of the x dependence of the input, partly of the DGLAP evolution.

The starting point of [481] is the relation (9.9) of the dipole cross section to the gluon density of the proton. This relation is implemented with suitably-chosen values of Q^2 and x for $R < R_p$, where the cross section given by (9.9) is smaller than half the pion-nucleon cross section. In this region the x dependence is assumed to be obtained purely from the DGLAP evolution and the initial conditions. For values $R \geq R_\pi \approx 0.65$ fm a nonperturbative ansatz is made. In the intermediate-R region perturbation theory is assumed to be modified by saturation and there is a smooth interpolation between the perturbative and nonperturbative domains.

A simple phenomenological expression for the dipole cross section has been presented[403,404] which gives a good description of the data for deep-inelastic scattering:

$$\sigma_{\mathrm{dip}}(x, R) = \sigma_0 \left(1 - \exp\left(- \frac{R^2}{R_0^2\, x^{0.288}} \right) \right) \tag{9.32}$$

with $\sigma_0 = 23$ mb and $R_0 = 0.73$ fm. The essential feature of this model is the saturation of the dipole cross section at σ_0 at large $1/x$. The value of x

at which saturation sets in depends strongly on R; the larger R the earlier the saturation. This simple formula gives a good quantitative description of the structure function F_2 and a reasonable description of the diffractive structure function $F_2^{D(3)}$, but fails when confronted with the diffractive dijet data[399].

For an application of the dipole model to reactions of photons with low values of Q^2, say $Q^2 \leq 1$ GeV^2, the choice of the photon wave function has a large influence. All applications are based on the perturbative expressions (8.95) with a constituent mass. An additional enhancement of the photon wave function around 1 fm was introduced in [482].

In some approaches [488,403,481] the dipole cross section increases monotonically with the size of the dipole, in others[442,482,493] it can also decrease with size in the transition region from the hard to the soft pomeron. In yet others[403,481] the cross section saturates with increasing $1/x$. In figure 9.6 we plot all the dipole cross sections of the models discussed here. The dependence on $\zeta = W^2 R^2$ in [482,493] has been converted into an x dependence according to (9.15). The big differences in the nonperturbative region are not surprising, since in this region the dipole cross sections are either guesses or model-dependent. But as can be seen from figure 9.6 there are considerable differences even in the perturbative region. The cross sections of [481] and [494] differ by about a factor 2, though both approaches start from the same relation (9.9). In [494] the relevant scale is fixed to $Q_0^2 \approx 1.2$ GeV^2 and the result is used as a starting point for a DGLAP evolution whereas in [481] the scale for the gluon distribution in the proton is adjusted to the distance R through $Q^2 = C^2/R^2$, with $C^2 \approx 10$. The different scales lead to the considerable difference in the dipole cross section. But the very different dipole cross sections nevertheless lead to similar results for the structure function. The reason is the following. In the approach of [494] only about 20% of the contribution to the structure function at $Q^2 \approx 1.2$ GeV^2 comes from the region below 0.4 fm, whereas for [481] the contribution from this region is more than 40%. This explains the difference of the dipole cross section obtained from different approaches even in a seemingly-perturbative region. It can be compensated easily by the behaviour of the cross section in the nonperturbative domain. This corroborates our statement made at the end of section 9.1 that the concept of a dipole cross section is model-dependent. This model dependence is natural in the nonperturbative domain, but extends its effects even to distances well below 0.4 fm.

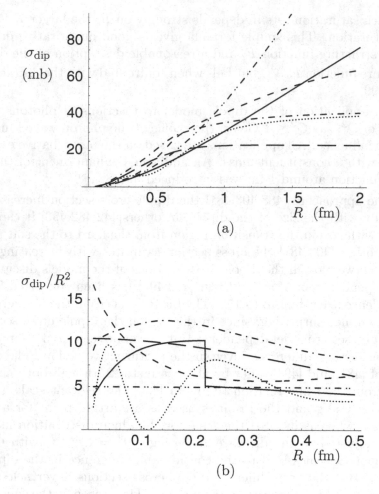

$\sigma_{\rm dip}$ (mb)

R (fm)

(a)

$\sigma_{\rm dip}/R^2$

R (fm)

(b)

Figure 9.6.   Dipole cross section $\sigma_{\rm dip}$ and $\sigma_{\rm dip}/R^2$ at $x = 10^{-3}$ for different models; short dashes[496]; solid line[493]; long dashes[403]; dotted line[482]; dot-dashes[481]; dash-double-dots[494,495]

9.4 Saturation

The strong increase with $1/x$ of the $\gamma^* p$ cross section observed at HERA gives rise to the question of whether this increase stops or slows down significantly, and if so where. Since there are no unitarity limits for photon-induced processes there are no constraints on the $1/x$ dependence of structure functions. But if a significant change occurred at large values of Q^2, at which confinement effects should be negligible, it would indicate a new regime of QCD. High-density effects in the gluon distribution may lead to a new kind of nonperturbative phenomenon. In large nuclei the projected gluon density in transverse space is high. This has been exploited in the saturation model[497] for deep-inelastic scattering on nuclei. The arguments have been extended[498] to deep-inelastic scattering on protons where the high gluon density is attributed to small x.

In the treatment of large nuclei[497,499] the valence quarks are considered as sources which give rise to classical gluon radiation through the Weizsäcker-Williams mechanism. In this semiclassical treatment the gluon density $xg(x, Q^2)$ tends to a constant for small x. The distribution of gluons with transverse distance $\mathbf{R}$ is obtained by averaging over the transverse components of the matrix-valued gluon potentials (see (6.4)) and integrating over the impact parameter $\mathbf{b}$ inside the nucleus:

$$N(\mathbf{R}) = \frac{2}{\pi}\mathrm{Tr}\int d^2b\, \langle \mathbf{A}_i^{\mathrm{cl}}(\mathbf{R}+\mathbf{b})\mathbf{A}_i^{\mathrm{cl}}(\mathbf{b})\rangle. \qquad (9.33)$$

In a covariant gauge the gluon fields are supposed to fall off fast enough outside the nucleons for it to be justified to assume that the gluon field of a nucleus is additive in the nucleons. A transformation is then made to the light-cone gauge and the interactions are absorbed into the gluon distribution. For a nucleus with large radius R_A the result is[499]

$$N(\mathbf{R}) = \int d^2b\, \frac{N_c^2-1}{\pi^2\alpha_s N_c R^2}\left(1 - e^{-R^2Q_s^2/4}\right) \qquad (9.34)$$

where

$$Q_s^2 = \frac{8\pi^2\alpha_s N_c}{N_c^2-1}\sqrt{R_A^2 - b^2}\,\rho_A(\mathbf{b})\,x\,g(x, Q^2). \qquad (9.35)$$

Here ρ_A is the nucleon density of the nucleus, $xg(x, Q^2)$ is the gluon density of the nucleon and Q^2 is related to the transverse distance through $Q^2 = C^2/R^2$, $C^2 \geq 4$; see (9.14).

In a thin slice of the nucleus the interaction of a colourless gluon pair separated by distance $\mathbf{R}$ is proportional to $|\mathbf{R}|^2 \rho_A\, xg(x, Q^2)$ and summing repeated interactions gives the exponential form $(1 - \exp(-R^2Q_s^2/4))$. This is analogous to the eikonal discussed in section 2.4. The quantity Q_s plays the role of a saturation momentum. Using $Q^2 \approx C^2/R^2$ we see that for $Q^2 \gg Q_s^2$ only the leading term of the exponential contributes significantly and the gluon density of the nucleus is proportional to that of the nucleon. For values $Q^2 < Q_s^2$ saturation effects due to the exponentiation play an important role. For large nuclear radii Q_s will be large and saturation effects can become important at values of Q^2 where confinement effects should play no role.

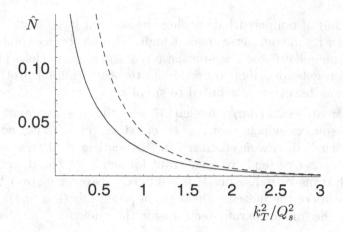

Figure 9.7.   Gluon density $\hat{N}(\mathbf{b}, \mathbf{k}_T)$ as a function of k_T^2/Q_s^2 for $\alpha_s = 0.2$. Solid line: full expression (9.37); dashed line: asymptotic expansion valid for large values of k_T^2/Q_s^2, where Q_s^2 is defined in (9.35).

To obtain differential gluon densities we introduce first the differential gluon density in impact parameter as the integrand of (9.34):

$$\bar{N}(\mathbf{b}, \mathbf{R}) = \frac{N_c^2 - 1}{\pi^2 \alpha_s N_c R^2} \left(1 - e^{-R^2 Q_s^2/4} \right) \tag{9.36}$$

The differential gluon density in impact parameter and transverse momentum is obtained by Fourier transformation:

$$\hat{N}(\mathbf{b}, \mathbf{k}_T) = \int \frac{d^2 R}{(2\pi)^2} e^{i\mathbf{k}_T \cdot \mathbf{R}} \bar{N}(\mathbf{b}, \mathbf{R})$$

$$= \frac{N_c^2 - 1}{4\pi^3 \alpha_s N_c} \Gamma\left(0, \frac{k_T^2}{Q_s^2}\right). \tag{9.37}$$

We see that, at fixed α_s, $\hat{N}(\mathbf{b}, \mathbf{k}_T)$ is a function just of k_T^2/Q_s^2, where Q_s^2 is related to b and Q^2 in (9.35). In principle, the relation involves also x, but not in the semiclassical approximation. The incomplete Γ-function has standard expansions[9] The full expression (9.37) and its expansion, valid only for large values values of k_T^2/Q_s^2, are shown in figure 9.7. The suppression of small values of k_T^2 compared with the large-k_T expansion is evident.

For values $k_T^2 \ll Q_s^2$ expansion of the incomplete Γ-function yields

$$\hat{N}(\mathbf{b}, \mathbf{k}_T) \approx \frac{N_c^2 - 1}{4\pi^3 \alpha_s N_c} \log\left(\frac{Q_s^2}{k_T^2}\right). \tag{9.38}$$

.

Lowest order radiative corrections have been calculated [501] and a $\log(1/x)$ correction to the semiclassical result has been obtained. This logarithmic correction is due to the increase of the longitudinal phase space with increasing energy[500]. It is argued that the increasing volume in longitudinal momentum space leads to similar effects to those induced by increasing longitudinal position space in a large nucleus. It is also argued[498,501] that BFKL dynamics (see section 7.3), which for sufficiently large Q^2 leads to a power behaviour of the cross section like $(1/x)^\kappa$, also induces a saturation momentum $Q_s^2 \propto (1/x)^\kappa$. The differential gluon density in impact parameter and transverse momentum is

$$x\hat{g}(x, \mathbf{b}, \mathbf{k}_T) = K \frac{N_c^2 - 1}{4\pi^3 \alpha_s N_c} \log\left(\frac{Q_s^2(x)}{k_T^2}\right). \tag{9.39}$$

This has the same structure as the corresponding equation (9.38), where the saturation scale is a function of the nuclear radius. The factor K and the power κ are related directly to BFKL dynamics; they are not determined reliably[501].

These considerations predict a new domain of QCD where confinement effects play at most a minor role, but where perturbative QCD is not applicable. The domains of QCD for diffractive reactions induced by virtual photons are thus

- $Q^2 < \Lambda$ nonperturbative, confinement
- $Q^2 > Q_s^2 \gg \Lambda^2$ perturbative
- $Q_s^2 > Q^2 \gg \Lambda^2$ nonperturbative, high density.

The high-density domain is a consequence of a high gluon density in the transverse phase space of $\mathbf{b}$ and $\mathbf{k}_T$. This high density can be reached either in large nuclei or at very small x in nucleons.

The dipole approach predicts a regime of nonperturbative high density too. Consider a small quark-antiquark separation in the dipole, where confinement effects are not yet important. Relation (9.9) between the gluon density and the dipole cross section suggests a similar relation between the differential quantities in impact parameter. We introduce the impact-parameter differential dipole cross section $\bar{\sigma}_{\text{dip}}(x, R, \mathbf{b})$ and gluon density $\bar{g}(x, \mathbf{b}, Q^2)$:

$$\sigma_{\text{dip}}(x, R) = \int d^2 b \, \bar{\sigma}_{\text{dip}}(x, R, \mathbf{b})$$

$$g(x, Q^2) = \int d^2 b \, \bar{g}(x, \mathbf{b}, Q^2)$$

$$= \int d^2 b \int_{|\mathbf{k}_T| \leq Q} d^2 k_T \, \hat{g}(x, \mathbf{b}, \mathbf{k}_T). \tag{9.40}$$

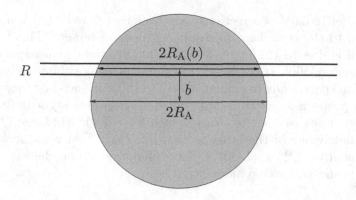

Figure 9.8.   Scattering of a small dipole on a large nucleus with radius R_A at fixed impact parameter b

Compare (9.39). They are related by

$$\bar{\sigma}_{\text{dip}}(x, R, \mathbf{b}) = \tfrac{1}{3}\alpha_s(Q^2)x\bar{g}(x, \mathbf{b}, Q^2)R^2 \qquad (9.41)$$

where again, up to logarithmic terms, we have $Q^2 = C^2/R^2$, $C^2 \geq 4$.

In the dipole approach the cross sections for photon and hadron reactions are expressed through the same dipole cross sections σ_{dip}. The dipole cross section with small radius R can be realised approximately through heavy-quarkonium–proton scattering. Purely-hadronic reactions such as quarkonium-proton scattering are subject to unitarity constraints which then also apply to the dipole cross sections. Therefore $\bar{\sigma}_{\text{dip}}(x, R, \mathbf{b})$ is subject to unitarity constraints. If the cross section is fully absorptive, corresponding to purely-imaginary phase shifts, then in the limit of small x the bound for the dimensionless density in impact parameter is

$$\bar{\sigma}_{\text{dip}}(x, R, \mathbf{b}) \leq 1. \qquad (9.42)$$

Otherwise

$$\bar{\sigma}_{\text{dip}}(x, R, \mathbf{b}) \leq 2. \qquad (9.43)$$

For large nuclei the saturation scale Q_s in (9.34) can be understood very intuitively in the dipole picture. Consider scattering of a small dipole on a large nucleus in a semiclassical approach; see figure 9.8. The absorption probability in a thin slice of the nucleus with width Δ is proportional to $R^2\rho_A xg(x, Q^2)\Delta$, where $xg(x, Q^2)$ is the gluon density of the nucleon and ρ_A the nucleon density of the nucleus. Large dipoles with large cross section are absorbed near the surface, small ones travel through the whole nucleus. We therefore have a saturation scale $c\rho_A x g(x, Q^2) R_A(b)$ with $R_A(b) = \sqrt{R_A^2 - b^2}$ and some constant c. If $c\rho_A x g(x, Q^2) R_A(b) R^2 \ll 1$ the cross section is proportional to R^2; if $c\rho_A x g(x, Q^2) R_A(b) R^2 \gg 1$ it is independent of R. This means that $c\rho_A x g(x, Q^2) R_A(b)$ plays the role of Q_s defined in (9.35). The equivalence of this approach to the original one[497] has been shown[495].

There remains the important question: how small must a dipole be so that we may neglect confinement effects? Since we do not understand confinement from first principles this question cannot be answered in a model-independent way. But if we assume that vacuum properties are essential for confinement, the gauge-invariant gluon correlator discussed in section 6.4 may give us some quantitative hint. As can be seen from figure 6.7 the non-perturbative term is dominant for distances above 0.2 fm for $D_\parallel$ and above 0.3 fm for $D_\perp$. Therefore we should consider only quark-antiquark pairs with spatial separation well below 0.2 fm in order to be free of major nonperturbative effects. This scale, which corresponds to about 1 GeV, appears also in other contexts of nonperturbative QCD, for example continuum thresholds in QCD sum rules[250]. If we express the distance in terms of the virtuality of the photon, $R \ll 0.2$ fm translates to $Q^2 \gg 4$ GeV^2.

The dipole model of (9.32) contains[403,404] more stringent saturation than implied by (9.39). There only the differential quantity saturates and the cross section can still rise with R if the size in impact parameter space increases, whereas (9.32) saturates for any value of R if x is small enough. Also the model of [481] assumes saturation of the dipole cross section and not its differential in impact parameter. In other approaches ([502–504] and the literature quoted there) saturation is assumed to take place at fixed impact parameter. A convenient way to ensure the unitarity constraint (9.42) is to write the eikonal form of (2.49) for the dipole cross section:

$$\bar{\sigma}_{\text{dip}}(x, R, \mathbf{b}) = 1 - \exp(-\bar{\chi}(x, R, b). \tag{9.44}$$

If $\bar{\chi}(x, R, b) = R^2 \chi(x, b)$, the relation (9.41) can be fulfilled for small R. The function $\chi(x, b)$ is closely related to the saturation scale Q_s^2 introduced in (9.34). If these models are extended to soft high-energy scattering, the x dependence of $\chi(x, b)$ has to be rather strong, since for large values of $\chi(x, b)$ the exponentiation moderates the increase with $1/x$. If pomeron dominance is assumed, the intercept has to be chosen to be about 1.2. For hard reactions the influence of the exponentiation becomes less important because of the decreasing effective value of R, and therefore the increase with energy becomes faster, in accordance with experiment. For values of

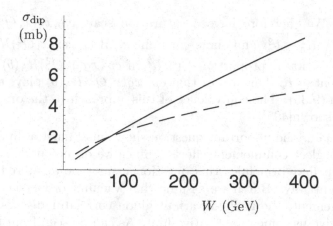

Figure 9.9. Dipole cross section at $R = 0.15$ fm for the non-saturating model of [493] (solid line) and for the saturating model of [403] (dashed line)

Q^2 larger than a few GeV^2 the resulting gluon distribution can be used as input for the DGLAP evolution; see section 7.2. In this way not only the ordinary structure functions but also the diffractive structure functions have been described successfully.

There are other approaches without saturation such as the two-pomeron approach[194], and dipole models based on this approach[442,482,493], which fit the data equally well. In the dipole cross section given through the stochastic-vacuum model there is a deviation from a pure-R^2 behaviour which becomes noticeable at around 0.2 fm. This is a consequence of the string formation which leads asymptotically to a linearly-rising dipole cross section and which in the model is a direct consequence of confinement.

We therefore have the problem that several quite-contradictory models explain the data. The charm structure function offers the best chance to distinguish between models in the perturbative region, since the charm density of the photon scans the dipole cross section in a relatively narrow band around $R = 1/\sqrt{Q^2/4 + m_c^2}$. In figure 9.9 we plot the dipole cross section as a function of W for $R = 0.15$ fm for two extreme models, a non-saturating one[493] and a saturating one[403]. The conversion from the variable x to W^2 in [403] has been made using (9.15). It is clear that precise charm data in the energy range of HERA could discriminate between the two models.

In conclusion, high gluon densities may lead to a new domain of QCD in which confinement effects are unimportant but nevertheless perturbation theory cannot be applied. Colour-glass-condensate models[505] provide an example. This situation can be realised in deep-inelastic scattering of electrons on large nuclei. The high density in transverse phase space in that case is obtained by integration over a large region in the longitudinal direction of the nucleus. It is argued that a similar situation can occur in small-x physics. Here the high density in transverse space is a consequence of the large phase space in the longitudinal momentum direction. From present HERA data one cannot deduce evidence for saturation effects, since models with and without saturation describe the data equally well.

9.5 Two-pomeron dipole model

The nonperturbative approach described in sections 8.5, 8.7 and 8.8 gives an explicit expression for the dipole-dipole scattering amplitude. The total cross section averaged over all dipole directions can be numerically approximated to an accuracy of better than 10% by the factorising form

$$\sigma_{\text{dip}}(R_1, R_2) = 0.67 \frac{1}{4\pi^2} (\langle g^2 FF \rangle a^4)^2 \, R_1 \left(1 - e^{-R_1/(3.1a)}\right) R_2 \left(1 - e^{-R_2/(3.1a)}\right) \tag{9.45}$$

where the parameters $\langle g^2 FF \rangle$ and a are given in (8.112). This expression is valid at centre-of-mass energies $W \approx 20$ GeV. It can be adapted to include the hard pomeron introduced in chapter 7 and applied to the study of structure functions. The new ingredient is to assume[442,493] that small dipoles couple to the hard pomeron and large dipoles couple to the soft pomeron. To be economical with parameters a sharp cut is introduced such that only the soft pomeron couples if both dipoles are larger than a certain value R_c, whereas the hard pomeron couples if at least one of the dipoles is smaller than R_c. Energy dependence is introduced by hand into the dipole cross section by dividing the amplitude into a soft part and a hard part with coefficients σ_s and σ_h:

$$T_{ab \to cb}(W) = iW^2 \left(\sigma_s \, (W/W_0)^{2\epsilon_s} + \sigma_h \, (W/W_0)^{2\epsilon_h}\right) \tag{9.46}$$

with

$$W_0 = 20 \text{ GeV} , \quad \epsilon_s = 0.08, \quad \epsilon_h = 0.42 . \tag{9.47}$$

With (8.91) the soft-pomeron contribution is given by

$$\sigma_s = \int_{R_c}^{\infty} 2\pi dR_1 \, R_1 \int_{R_c}^{\infty} 2\pi dR_2 \, R_2 \int_0^1 dz_1 \, dz_2 \, \psi_c^*(\vec{R}_1, z_1) \psi_a(\vec{R}_1, z_1)$$

$$\times \, |\psi_b(\vec{R}_2, z_2)|^2 \, \sigma_{\text{dip}}(R_1, R_2). \tag{9.48}$$

For the hard pomeron it has been argued [482] that the appropriate dimensionless variable is RW. Highly-virtual photons have a hadronic radius

$R \propto 1/Q$, so to ensure scaling behaviour for the dimensionless quantity $Q^2 \sigma_{\text{dip}}(R, W)$ the W dependence should enter in the combination $W^2 R^2$, which corresponds to the inverse of the Bjorken variable x; see the discussion after (9.9). If one dipole is small, say $R_1 \leq R_c$, the hard contribution should depend on $(R_1 W)^2$, if both dipoles are small on $(R_1 R_2 W^2)$. Since the factor $W^{2\epsilon_h}$ has been extracted in (9.46), the coefficient σ_h is

$$
\begin{aligned}
\sigma_h = & \int_0^{R_c} 2\pi dR_1 \, R_1 \int_0^{R_c} 2\pi dR_2 \, R_2 \int_0^1 dz_1 \, dz_2 \, \psi_c^*(\vec{R}_1, z_1) \psi_a(\vec{R}_1, z_1) \\
& \times |\psi_b(\vec{R}_2, z_2)|^2 \, \sigma_{\text{dip}}(R_1, R_2)(R_1 R_2 / R_c^2)^{\epsilon_h} \\
+ & \int_0^{R_c} 2\pi dR_1 \, R_1 \int_{R_c}^{\infty} 2\pi dR_2 \, R_2 \int_0^1 dz_1 \, dz_2 \, \psi_c^*(\vec{R}_1, z_1) \psi_a(\vec{R}_1, z_1) \\
& \times |\psi_b(\vec{R}_2, z_2)|^2 \, \sigma_{\text{dip}}(R_1, R_2)(R_1 / R_c)^{2\epsilon_h} \\
+ & \int_0^{R_c} 2\pi dR_2 \, R_2 \int_{R_c}^{\infty} 2\pi dR_1 \, R_1 \int_0^1 dz_1 \, dz_2 \, \psi_c^*(\vec{R}_1, z_1) \psi_a(\vec{R}_1, z_1) \\
& \times |\psi_b(\vec{R}_2, z_2)|^2 \, \sigma_{\text{dip}}(R_1, R_2)(R_2 / R_c)^{2\epsilon_h}. \qquad (9.49)
\end{aligned}
$$

Although the stochastic-vacuum model is a model for the infrared behaviour of QCD and was originally applied only to hadron-hadron scattering, it gives reasonable results[435,433,441] for photon-induced processes for photon virtualities Q^2 up to about 10 GeV^2. For higher values of Q^2 the model over-estimates the cross sections. For consistency of the model with low-energy theorems the strong coupling in the infrared domain must have the frozen value[275] $\alpha_s \approx 0.57$. It is plausible that upon introduction of a hard scale through a highly-virtual photon the coupling of the gluons to the corresponding dipole is governed by that hard scale. Therefore the dipole cross section should be rescaled by the factor

$$
\frac{\alpha_s(Q)}{\alpha_s(0)} = \frac{1}{0.57} \frac{4\pi}{11 \log(Q^2/Q_0^2 + 7.42)} \qquad (9.50)
$$

with $Q_0^2 = 1$ GeV^2. This corresponds to a running coupling $\alpha_s(Q)$ in leading order for zero quark flavours, $n_f = 0$, adjusted to give $\alpha_s(0) = 0.57$; see (6.19) and (6.20).

The photon and hadron wave functions and the fixing of their parameters are discussed in sections 8.6 and 8.7. With these known, the only free parameter is R_c, which can be determined from the total $\gamma^* p$ cross section $\sigma_{\gamma^* p}^{\text{Tot}}$. Fitting the cross section for $W > 50$ GeV and $Q^2 < 150$ GeV^2 yields[493] $R_c = 0.22$ fm. The model is then completely determined and can be applied to any dipole-dipole reaction for which the wave functions of the participating particles are known. The model has been applied successfully to the proton's charm structure function $F_2^c(x, Q^2)$ and its longitudinal structure function $F_2^L(x, Q^2)$, J/ψ photoproduction, deep virtual-Compton scattering $\gamma^* p \to \gamma p$, the real photon-proton total cross section $\sigma^{\gamma p}(s)$, the real photon-photon total cross section $\sigma^{\gamma\gamma}(s)$, and the photon structure function $F_2^\gamma(x, Q^2)$.

Both for the proton's charm structure function $F_2^c(x, Q^2)$ and for its longitudinal structure function $F_L(x, Q^2)$, for $Q^2 \geq 5 \text{ GeV}^2$ the photon wave function is concentrated at small distances. In the case of F_2^c this is a consequence of the mass of the charm quark occurring in the argument ϵ of (8.96). For F_2^L it is a consequence of the factor $z(1 - z)$ in the wave function (8.95) of the longitudinal photon. This factor suppresses contributions from small values of ϵ, which correspond to large distances. Thus in both these cases the hard pomeron is already dominant at moderate energies. Figure 9.10 shows the soft and hard-pomeron contributions to structure functions at various values of x. The first row of figure 9.10 contains the contributions to the proton structure function F_2, and reproduces the results of [194,313] well. Comparison of the first and second rows of figure 9.10 shows that for the longitudinal structure function the increase of the short-range (hard) part with increasing Q^2 is not as strong as for the transverse structure function, a consequence of the less-singular behaviour of the Bessel functions at the origin in the relevant photon wave function (8.95). Nevertheless as the long-range (soft) part of the longitudinal structure function is even more suppressed at large Q^2 relative to its contribution to the transverse structure function, the hard pomeron is dominant sooner in $F_L(x, Q^2)$ than in $F_2(x, Q^2)$. The third row gives the contributions to the proton charm structure function $F_2^c(x, Q^2)$. The strong suppression of the soft pomeron relative to the hard pomeron in $F_2^c(x, Q^2)$, which is purely a wave-function effect, is notable and provides an explanation for the almost-complete flavour blindness of the hard pomeron commented on in section 7.4. The last row gives the soft and hard-pomeron contributions to the photon structure function F_2^γ/α. This shows a remarkable similarity to the proton structure function, in conformity with the conclusions of section 7.6.

The quantitative agreement of the model with the data for $F_2^c(x, Q^2)$, J/ψ photoproduction, $\sigma_{\gamma p}^{\text{Tot}}(s)$, $\sigma_{\gamma\gamma}^{\text{Tot}}(s)$ and $F_2^\gamma(x, Q^2)$ is as good as the fits shown in chapter 7, and so is not shown here. The comparison of the model with data[328] for the longitudinal structure function $F_L(x, Q^2)$ of the proton is made in figure 9.11 and with the H1 data[506] for deep virtual-Compton scattering, $\gamma^* p \to \gamma p$, in figure 9.12. The agreement is clearly very satisfactory.

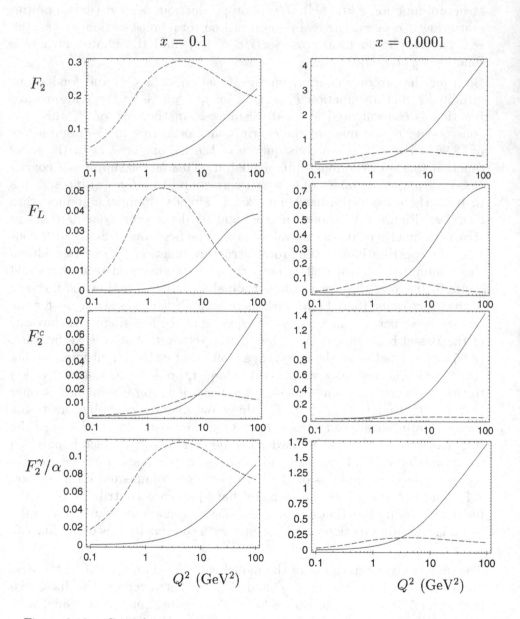

Figure 9.10.   Contributions to structure functions at two values of x from the hard pomeron (solid lines) and the soft pomeron (dashed lines)

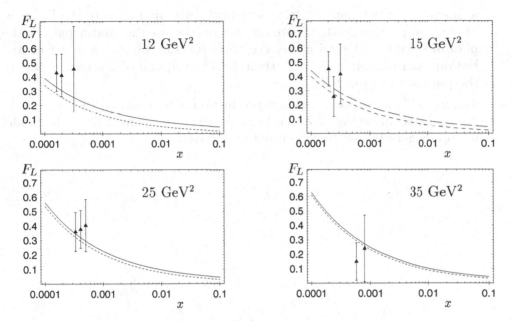

Figure 9.11. The longitudinal proton structure function[328] $F_L(x, Q^2)$ for $Q^2 = 12, 15, 25$ and 35 GeV^2. The solid line is the full result and the dashed line the hard-pomeron contribution.

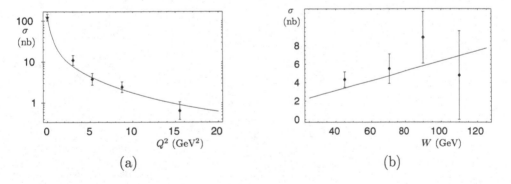

Figure 9.12. Integrated cross section for the reaction $\gamma^* p \to \gamma p$. (a) at $\langle W \rangle = 75$ GeV; (b) at $\langle Q^2 \rangle = 4.5$ GeV^2. The data are from [506].

We saw in section 7.8 that, although the soft pomeron does not contribute significantly to $F_2^c(x, Q^2)$, it is an important component of J/ψ photoproduction at comparable energies. This can be easily understood in the dipole-model approach. The virtual $c\bar{c}$ pair in the photon wave function has a transverse extension of about $1/m_c$, but the J/ψ wave function has a much larger tail, ranging from a typical hadronic radius to the Bohr radius of order $1/(\alpha_s m_c)$. Therefore the overlap of the charm part of the photon wave function with the J/ψ wave function obtains a larger contribution from distances $R > R_c$ than does the square of the charm part of the photon wave function.

The model fails to describe charm production in $\gamma\gamma$ interactions, $\gamma\gamma \to c\bar{c}X$. As we found in section 7.7, the hard-pomeron component from the model is approximately half that required by the data.

10

Questions for the future

The experimental and theoretical study of diffractive phenomena has been very fruitful over the last two decades. With the ongoing experiments at HERA, the Tevatron and RHIC, and the start-up of the LHC, this field of research will remain of high topical interest. It is clear that the description of diffractive phenomena through QCD is rather complicated. Thus, what can we hope to learn by studying diffractive phenomena? For testing the basic principles of QCD in experiment and for determining the fundamental parameters of QCD we should choose simple reactions where perturbation theory is without doubt applicable. In the nonperturbative domain we should choose quantities which can be calculated reliably on the lattice. Diffractive phenomena are unlikely to be of direct benefit here, until theorists make significant progress with their calculational methods. Diffraction can be considered as an area for "applied QCD", where theorists have been developing various approximation schemes in order to understand the observed phenomena. For this a continual interplay of experiment and theory has been essential, and in the future will still be so. It has been the experimental possibility of studying diffractive phenomena under conditions very different from elastic hadron-hadron scattering which has brought diffractive phenomena to the fore and stimulated so much theoretical development. In this final chapter we review briefly the outstanding issues.

Regge theory, introduced more than forty years ago, remains remarkably successful in providing a phenomenological description of soft high-energy hadronic and photon-induced processes. As Regge theory is the theory of exchanges, for example of the ρ, ω, f_2 and a_2 families of trajectories, it is natural to think of the soft pomeron in the same terms. Is the soft pomeron then simply the exchange of a glueball trajectory? The assumption that the exchange of a single pomeron dominates, that it has a linear trajectory and couples to single quarks, describes high-energy pp and $\bar{p}p$ elastic scattering data extraordinarily well, as described in chapter 3.

We saw in chapter 3 that, to a good approximation, all total cross sections rise at high energy with the same effective power of the energy, which we may associate with the intercept of an effective pomeron trajectory. At extremely large energies, fixed-power behaviour is in conflict with the Froissart bound, so the the observed behaviour of total cross sections can only be an effective power, which must reduce as the energy increases. While it is agreed that this reduction is caused by multiple exchanges, there are two divergent views about it. One is that multiple-scattering corrections are small at present-day energies, so that the effective-pomeron intercept is close to that of the "bare" pomeron. Alternatively it has been argued that multiple-scattering corrections are large at present-day energies and hence the intercept of the bare pomeron is very different from that of the effective pomeron. The implications of which of these two divergent views is correct are far-reaching. For example, they affect how one views the additive-quark rule, which is found to be in rather good agreement with experiment. The simplest interpretation is that it is a fundamental property of the single exchange; an alternative is that more than one quark in each hadron is involved and the apparent additivity is a more-or-less accidental property associated with multiple exchanges and related to hadron sizes.

Regge theory is a mathematical formalism that associates exchanges with singularities of amplitudes in the complex angular-momentum plane. It allows a huge amount of data to be correlated, even if one does not understand the dynamical origin of the singularities. The singularity associated with the soft pomeron might not be a simple pole, but experimental data are not in conflict with assuming that it is. The fact that the total cross section and the parameters describing $d\sigma/dt$, for instance the slope parameter b, are dimensioned parameters suggests considering the pomeron as originating from nonperturbative QCD. In this approach, described in chapter 8, soft hadronic scattering results from the quarks in the hadrons feeling the nonperturbative fluctuations of the gluon fields in the vacuum. High-energy-scattering data can be used to extract the parameters of the vacuum, that is the gluon condensate, the correlation length and the non-Abelian parameter, and relate them to the string tension. The results compare well with lattice calculations. In its basic form this approach leads to constant total cross sections, but it may be that other structures in the vacuum, such as instantons, play an important role for high-energy scattering and are responsible for the rising total cross sections. How do we bring together the nonperturbative-QCD approach and the Regge picture of the soft pomeron? Novel pictures and calculational techniques for understanding the pomeron in QCD have been proposed, including considering diffractive phenomena as being similar to critical phenomena in statistical physics or to glass phenomena. Also, the pomeron may be related to a surface in the curved space of a gravitational black hole in 10-dimensional anti-de Sitter space. How do these relate to the more conventional phenomenological and nonperturbative-QCD approaches?

In chapter 3 we described the rather complicated story associated with the data for diffraction dissociation, $pp \to pX$, where the final-state proton has lost only a small fraction ξ of its initial momentum. The energy dependence of the data is not compatible with the straightforward application of conventional theory, and even the assumption of strong multiple-scattering corrections cannot naturally resolve the discrepancy. There are both experimental and theoretical issues here. New measurements at the LHC should help to resolve the experimental problems and improve our theoretical understanding.

A potential problem for theory is the failure to observe any contribution to elastic scattering at small t from the odderon, the conjectured $C = -1$ partner of the pomeron which is discussed in chapters 3, 7 and 8. Mechanisms to suppress the odderon contribution at small t in elastic scattering can be envisaged, and other reactions in which such mechanisms do not operate have been proposed. However there is as yet no evidence for the existence of the odderon. It is important to continue the search. The continuing non-observation of the odderon would have a major impact on our understanding of diffractive phenomena.

We have explained in chapter 7 that the data for hard collisions show clearly that the soft pomeron is not enough, and suggest the existence of a second pomeron, the hard pomeron. Regge theory should be applicable to such processes when the appropriate variable x is sufficiently small, and the data agree well with this hypothesis. However, perturbative QCD should also be applicable and we are only just beginning to learn how these two apparently disparate approaches can be reconciled. We have explained that our understanding of both the BFKL and the DGLAP approaches presents very difficult unsolved problems, and it is important to try and solve them. Whether the solution lies purely in the realm of perturbative QCD is not known, and probably the theory is too difficult to be worked out without the help of additional accurate data at even smaller values of x than are so far available. A related question is: does the soft pomeron become progressively harder as the hard scale of a reaction increases, or is the hard pomeron a separate object? One way of resolving this issue is to determine whether the hard pomeron is present in $\sigma^{\gamma p}$ and $\sigma^{\gamma\gamma}$ already at $Q^2 = 0$. Present-day data indicate that this may be the case, but are not definitive. If the presence of the hard pomeron is confirmed, it would demonstrate that the existence of a hard-pomeron effect in hard collisions is not attributable to perturbative evolution, only its increasing strength as the hard scale increases. If a hard-pomeron component is present in $\sigma^{\gamma p}$ and $\sigma^{\gamma\gamma}$ one might attribute this to the presence of a pointlike component in the real photon. But occasionally the quarks in a proton are very close together too, so although we have shown that a hard-pomeron component need not be present in σ^{pp}, it is plausible that it may be. Will it be large enough to be identified at the LHC?

As we have discussed in chapter 7, it has become possible to probe the pomeron at very short distance scales, either in diffractive processes in deep-inelastic scattering or in hard diffractive hadron-hadron interactions. Considering the pomeron as a quasi-particle which can be described in terms of partons, quarks and gluons, as is usual for hadrons, has been an extremely useful and successful concept. In this framework it is possible to ask about the parton content of the pomeron, what fractions of gluons and quarks does it contain and how are they distributed in momentum space. The answers obtained so far are controversial and confusing. The two HERA experiments, H1 and ZEUS, agree that the gluons provide 80 to 90% of the momentum of the pomeron. The two Tevatron experiments, CDF and D0, conclude that the gluon momentum fraction is no more than 50%, and they find a dramatic breakdown in factorisation. However, there are large theoretical uncertainties. If a diffractive event is defined as one with a very fast proton the theory is relatively well defined. But in practice much of the data is rather obtained by requiring just a large rapidity gap, for which the theory is much less certain. Until it becomes possible to compare data from HERA and the Tevatron where both triggers are on a very fast final-state proton (or antiproton) there is a worry that we may not be comparing like with like. Again, there are uncertainties in extracting the gluon component of the pomeron structure function; the HERA experiments have obtained rather different outputs using different analyses. Further, one should not expect factorisation. Whatever the explanation is, whether it be BFKL, a combination of hard and soft pomerons, important screening effects, or any other, all would agree that factorisation should break down. Nevertheless, it is surprising that the breakdown is so dramatic. For example, the Tevatron cross section for W production accompanied by a very fast proton or antiproton is nearly an order of magnitude smaller than predicted, and diffractive dijet production has similar problems. More data are needed to try and gain a better understanding.

It has been suggested that exclusive Higgs production, $pp \rightarrow pHp$, with both final-state protons very fast, provides a clean environment for discovering the Higgs boson. We have seen in chapter 3 that various calculations claim this to be an experimentally-feasible process. Other calculations disagree. The difference lies in the estimate of the screening correction to the basic production process. At present there is no consensus on how this can be reliably estimated.

The whole question of screening has been a controversial one. Many people believe that the rapid rise with increasing $1/x$ discovered in the HERA measurements of the proton structure function will be significantly moderated by so-called unitarity corrections when smaller values of x become accessible. In chapter 7 we explained that unitarity imposes no such constraint. It will be interesting to see whether in fact the rapid rise persists at smaller x or whether saturation sets in, as discussed in chapter 9. That chapter describes the dipole model, which is a natural bridge between perturbative and nonperturbative calculations of QCD. As yet though, there is no unique dipole model, because different assumptions about wave functions lead to very different dipole cross sections, as figure 9.2 quite dramatically shows.

Issues for experiment

- Will the pp total cross section at the LHC follow the extrapolation given by the soft-pomeron theory? If it comes out higher, this may point to a hard pomeron being present in hadronic collisions. If it comes out significantly smaller, screening effects must be very large.

- Are there glueballs and are some of them on pomeron trajectories?

- What is the cross section for diffraction dissociation at the LHC?

- Is a hard-pomeron contribution already present at $Q^2 = 0$ in the γp and $\gamma \gamma$ total cross sections?

- Will $d\sigma/dt$ for pp elastic scattering at large t be the same at LHC energies as at ISR energies?

- Will $pp \to p + \text{Higgs} + p$ be large enough to be seen?

- Is there an odderon at small t? What are the best reactions to study the odderon? If it does exist, are there soft and hard-odderon phenomena as for pomerons?

- How do diffractive phenomena respond to polarisation of the incoming particles?

Issues for theory

- Can diffractive phenomena be calculated in a reliable way starting from the QCD Lagrangian?

- Are the hard and soft pomeron distinct objects? What is their dynamical origin?

- Can hard diffractive reactions be understood from perturbative QCD? Can we calculate gap survival probabilities?

- Can we learn better how to combine perturbative and nonperturbative concepts – evolution and analyticity?

- Can we complete the construction of the theory of perturbative evolution at small x?

- Are there saturation effects in hard processes and can they be calculated reliably?

- Is screening in soft processes large or small?

- Can screening corrections in $pp \to p + \text{Higgs} + p$ be calculated reliably?

Appendix A

Sommerfeld-Watson transform

In this appendix we show how to define two signatured amplitudes $A^{\pm}(s,t)$ so as to allow the derivation of the Sommerfeld-Watson transforms (2.12). We then explain how Mandelstam[507] modified the background integrals, which in (2.12) are along the line of integration Re $l = -\frac{1}{2}$, so that when they are moved further to the left in the complex l-plane they become as small as one wishes. Consequently, the Regge representation (2.15) of the high-energy scattering amplitude should include the contributions from all Regge poles, not just those whose trajectories satisfy Re $\alpha_i^{\pm}(t) \geq -\frac{1}{2}$.

In (2.10) we defined the two signatured amplitudes $A^{\pm}(s,t)$ as partial-wave series and then in (2.11) we wrote the two series as integrals. In order to deform the contour of the l integration in these integrals so as to give (2.12), and then the Regge asymptotic form (2.15), the Regge amplitudes $A^{\pm}(l,t)$ must have appropriate behaviour as $l \to \infty$ with $-\pi \leq \arg l \leq \pi$. There are infinitely many choices that satisfy the requirement (2.8) that for even physical values of l, $A^{+}(l,t)$ coincides with the ordinary partial wave amplitude $A_l(t)$, and similarly $A^{-}(l,t)$ for odd values. But the constraint on the asymptotic behaviour picks out unique choices.

These choices are based on the t-channel Froissart-Gribov formula for the t-channel partial-wave amplitude, the s-channel version of which is given in (1.54):

$$32\pi^2 i A_l(t) = \int_1^{\infty} dz_t'\, D_s(s, t(s, z_t'))\, Q_l(z_t')$$

$$+ \int_{-1}^{-\infty} dz_t'\, D_u(s, u(s, z_t'))\, Q_l(z_t'). \quad \text{(A.1)}$$

We use $Q_l(-z) = (-1)^l Q_l(z)$ to rewrite this as

$$16\pi A_l(t) = \frac{1}{2\pi i} \int_{z_0}^{\infty} dz'_t \left(D_s(t, s(t, z'_t)) + (-1)^l D_u(t, u(t, -z'_t)) \right) Q_l(z'_t).$$

(A.2)

The term $(-1)^l$ makes this unsuitable for analytic continuation: if we write it as $e^{i\pi l}$ it causes a problem when Im $l \to -\infty$, while $e^{-i\pi l}$ would cause a problem when Im $l \to +\infty$. This is why we introduce the signatured amplitudes

$$16\pi A^{\pm}(l, t) = \frac{1}{2\pi i} \int_{z_0}^{\infty} dz'_t \left(D_s(t, s(t, z'_t)) \pm D_u(t, u(t, -z'_t)) \right) Q_l(z'_t) \quad \text{(A.3)}$$

and split the partial-wave series into two sums (2.10), one over even l and the other over odd l. $A^+(l, t)$ agrees with $A_l(t)$ for $l = 0, 2, 4, \ldots$ and $A^-(l, t)$ agrees with $A_l(t)$ for $l = 1, 3, 5, \ldots$. As we shall show, the definitions (A.3) satisfy our requirements for continuation to complex values of l, and do so uniquely.

As we explained in section 1.6, the integrals (A.3) converge only if Re l is sufficiently large, in fact larger than the position of the rightmost singularity of each amplitude $A^{\pm}(l, t)$ in the complex l-plane. So we start with the representations (A.3) for values of l where they converge and continue them analytically to regions where the representations no longer converge. Suppose for illustration that the rightmost singularities of $A^{\pm}(l, t)$ are as is shown in figure 2.7. Then the representation for $A^+(l, t)$ no longer converges when it is continued as far as $l = 2$, because there is a pole to the right of that point. Nevertheless it might be that its continuation to there is equal to the partial-wave amplitude $A_2(t)$. A corresponding statement applies to its continuation to $l = 0$ and to the continuation of $A^-(l, t)$ to $l = 1$. If this is not true, then we must add to the representation (2.12)

$$16\pi \left(A_0(t) - A^+(0, t) + (A_1(t) - A^-(1, t))P_1(z_t) + (A_2(t) - A^+(2, t))P_2(z_t) \right).$$

(A.4)

In consequence, the large-s behaviour (2.15) of $A(s, t)$ would have additional terms proportional to s^2, s and a constant. This is assumed not to be the case.

In order to satisfy ourselves that the definition (A.3) is the appropriate one, we need the integrand

$$\frac{(2l + 1)A^{\pm}(l, t)(P_l(-z_t) \pm P_l(z_t))}{\sin(\pi l)}$$

(A.5)

of the Watson-Sommerfeld transform (2.12) to vanish when $l \to \infty$ in such a way that we may neglect the contribution to the integral from the infinite-semicircular part of the integration contour shown in figure 2.5, and then move the contour further to the left as in figures 2.6 and 2.7. That is, we need the integrand to go to zero faster than $1/l$ when $l \to \infty$ in the directions $-\pi \le \arg l \le \pi$. This is for physical values of $z_t = \cos\theta_t$ and t above the t-channel threshold; it is only after we have deformed the contour and discarded the part of it at infinity that we make the analytic continuation to the s-channel physical region, with s large.

For physical values of $z_t = \cos\theta_t$, the large-l behaviour of $P_l(z)$ is[9]

$$P_l(z) = \sqrt{\frac{2}{\pi l \sin\theta_t}} \cos((l + \tfrac{1}{2})\theta_t + \tfrac{1}{4}\pi)(1 + O(l^{-1})). \tag{A.6}$$

Because (A.5) contains also $P_l(-z_t)$ we need too a similar expression with θ_t replaced with $\pi - \theta_t$. The large-l behaviour of $(P_l(-z_t) \pm P_l(z_t))/\sin(\pi l)$ is, up to a constant factor,

$$\sqrt{\frac{1}{l \sin\theta_t}} e^{-\bar{\theta}_t |\mathrm{Im}\, l|} \tag{A.7}$$

where $\bar{\theta}_t = \mathrm{Min}(\theta_t, \pi - \theta_t)$.

In the t-channel physical region, $z_t' > 1$ in the integration in the definition (A.3) of $A^{\pm}(l,t)$. The behaviour of $Q_l(z_t')$ as $l \to \infty$ is then given, up to a constant factor, by (1.55). Hence $A^{\pm}(l,t) \to 0$ exponentially as $l \to \infty$ in directions in the complex l-plane for which $-\pi < \arg l < \pi$. Thus the integrands of the Sommerfeld-Watson integrals (2.11) do have the right large-l behaviour to allow the infinite-semicircular part of the integration contour shown in figure 2.4 to be neglected, and the integration contour may then be moved to the left as shown in figures 2.6 and 2.7. Having achieved this, we may now continue the Sommerfeld-Watson integrals analytically from the t-channel physical region to the s-channel physical region and go to the large-s limit.

A theorem known as Carlson's theorem ensures that the definitions (A.3) of $A^{\pm}(l,t)$ are unique; that is, they are the only definitions that agree with the partial-wave amplitude $A_l(t)$ at the appropriate physical values of l, as in (2.8), and have the required large-l behaviour. In order to be able to perform the desired analytic continuation to the s-channel physical region, we must start with an integral that correctly reproduces the t-channel amplitude for some continuous interval of physical values of θ_t. A very small interval would suffice, and (A.7) tells us that the least stringent constraint on $A^{\pm}(l,t)$ obtains if we choose an interval in the neighbourhood of $\theta_t = \tfrac{1}{2}\pi$. Then we can allow $A^{\pm}(l,t)$ to become large for large l, provided that it does not diverge as fast as $e^{\frac{1}{2}\pi l}$. Carlson's theorem says that, if $f(z)$ is regular and diverges no faster than $e^{k|z|}$ for large z with Re $z \geq 0$, where $k < \pi$, and $f(z) = 0$ for $z = 0, 1, 2, 3, \ldots$, then $f(z) = 0$ everywhere. We identify $z = \frac{1}{2}l$ and apply the theorem, so concluding that we cannot add a term to $A^{\pm}(l, t)$ that preserves (2.8) while maintaining the required asymptotic behaviour.

Note that, according to (1.33) and (1.34), once we have continued to the s-channel physical region the two discontinuity functions $D_s((t, s(t, z'_t))$ and $D_u(t, u(t, -z'_t))$ are pure imaginary, so that $A^{\pm}(l, t)$ defined by (A.3) are real for real l. It is assumed that this is achieved by the positions of their poles, and the residues at these poles, each being real. Then the phase of the contribution from a Regge pole to the large-s behaviour of the amplitudes is given by the signature factor $\xi_\alpha^\pm = 1 \pm e^{i\pi\alpha}$ of (2.18).

So far, then, we have derived the Sommerfeld-Watson transforms (2.12) for the two signatured amplitudes, with background integrals taken along Re $l = -\frac{1}{2}$. It is immediate from what we have said that we may push the integration contour further to the left, picking up contribution from any Regge poles and cuts that we may encounter on the way. However, when we have pushed the integration contour to Re $l = L$, with $L < -\frac{1}{2}$, the integrands of the background integrals behave as s^{-L-1}, according to (2.13). That is, the background integral now apparently dominates over the contributions from the poles and cuts that have been exposed. We now explain how Mandelstam[507] modified the background integrals so that they could be seen to become steadily smaller as he pushed the integration contour further to the left in the complex l-plane, so that all the poles to the right of them should be included in the Regge representation (2.15) and dominate over the background at large s.

We start by pushing the background integrals back just a little to the right, so that the integrals are along Re $l = L$ with L just greater than $-\frac{1}{2}$. We use the relation[508]

$$\pi P_l(z) = \tan(\pi l) \left(Q_l(z) - Q_{-l-1}(z) \right) \tag{A.8}$$

so that the background integrals in (2.12) become

$$8i \int_{L-i\infty}^{L+i\infty} dl \, (2l+1) A^{\pm}(l, t) \frac{Q_l(-z_t) \pm Q_l(z_t) - Q_{-l-1}(-z_t) \mp Q_{-l-1}(z_t)}{\cos(\pi l)}. \tag{A.9}$$

Because we have analytically continued to the large-s physical region, z_t is large. For large z,

$$Q_l(z) \sim \sqrt{\pi} \, \frac{\Gamma(l+1)}{\Gamma(l+\frac{3}{2})} (2z)^{-l-1}. \tag{A.10}$$

Hence, $Q_l(-z_t)$ and $Q_l(z_t)$ are exponentially small. This means that, for these two terms, we may reverse the deformation of the contour that took us from the contour of figure 2.4a to Re $l = L$. In the process, of course we pick up contributions from any Regge poles or cuts of $A^\pm(l, t)$ we may encounter. The integrand now has poles at the zeros $l = l' - \frac{1}{2}$ of $\cos(\pi l)$, and we pick up the residues of those at $l' = 1, 2, \ldots$. So the Q_l terms give us

$$16\pi \sum_i (2\alpha_i^\pm(t) + 1)\beta_i^\pm(t) \frac{Q_{\alpha_i^\pm(t)}(-z_t) \pm Q_{\alpha_i^\pm(t)}(z_t)}{\cos(\pi\alpha_i^\pm(t))}$$

$$+ 16 \sum_{l'=1}^{\infty} l' A^\pm(l' - \tfrac{1}{2}, t) \left(Q_{l'-\frac{1}{2}}(-z_t) \pm Q_{l'-\frac{1}{2}}(z_t)\right). \quad (A.11)$$

As usual, we have not explicitly written the contribution from any Regge cuts. The first sum in (A.11) extends over the same Regge poles as in the original Sommerfeld-Watson representation (2.12). Because of the identity (A.8), the two sums together are

$$16\pi \sum_i (2\alpha_i^\pm(t) + 1)\beta_i^\pm(t) \frac{Q_{-\alpha_i^\pm(t)-1}(-z_t) \pm Q_{-\alpha_i^\pm(t)-1}(z_t)}{\cos(\pi\alpha_i^\pm(t))}. \quad (A.12)$$

For the two Q_{-l-1} terms in (A.9), we push the contour to the left in the complex l-plane, making L as large negative as we like. As we do so, we pick up the residues of the poles of $1/\cos(\pi l)$ that we cross. These are at $l = -l' - \frac{1}{2}$, with $l' = 0, 1, 2, \ldots$, though the factor $(2l+1)$ in the numerator removes the pole at $l = -\frac{1}{2}$. We also pick up contributions from any further Regge poles (as well as cuts, which again we do not include explicitly). So the Q_{-l-1} terms give us

$$16\pi \sum_j (2\alpha_j^\pm(t) + 1)\beta_j^\pm(t) \frac{Q_{-\alpha_j^\pm(t)-1}(-z_t) \pm Q_{-\alpha_j^\pm(t)-1}(z_t)}{\cos(\pi\alpha_j^\pm(t))}$$

$$- 16 \sum_{l'=0}^{N-1} l' A^\pm(-l' - \tfrac{1}{2}, t) \left(Q_{l'-\frac{1}{2}}(-z_t) \pm Q_{l'-\frac{1}{2}}(z_t)\right) \quad (A.13)$$

together with (A.9), where now L is between $-N \pm \frac{1}{2}$, with N an integer greater than 1.

For integer n, $Q_{n-\frac{1}{2}}(z) = Q_{-n-\frac{1}{2}}(z)$ so that the definition (A.3) of $A^\pm(l, t)$ gives

$$A^\pm(n - \tfrac{1}{2}, t) = A^\pm(-n - \tfrac{1}{2}, t) \qquad n \text{ integer.} \quad (A.14)$$

Thus the sums over l' in (A.11) and (A.13) combine in a simple way to give

$$16 \sum_{l'=N}^{\infty} l' A^{\pm}(l' - \tfrac{1}{2}, t) \left(Q_{l'-\frac{1}{2}}(-z_t) \pm Q_{l'-\frac{1}{2}}(z_t) \right). \tag{A.15}$$

Because of (A.10), the sum (A.15) converges rapidly for large z and is dominated by its first term, which behaves as z^{-N-1}; so we may make it as small as we like by choosing N to be large enough. The background integral has become

$$-8i \int_{L-i\infty}^{L+i\infty} dl \ (2l+1) A^{\pm}(l, t) \frac{Q_{-l-1}(-z_t) \pm Q_{-l-1}(z_t)}{\cos(\pi l)} \tag{A.16}$$

with $L < -N - \tfrac{1}{2}$. From (A.10) again, its integrand is smaller than z^{-N}, so the background integral too can be made as small as we like. Apart from Regge-cut contributions, we are left with the Regge-pole terms in (A.12) and (A.13). These have the same form; that is, the sum (A.12) should extend over all Regge poles whose trajectories satisfy $\operatorname{Re} \alpha_i^{\pm}(t) > -N + \tfrac{1}{2}$. Because of (A.10), we obtain from each Regge pole the same large-s behaviour as is recorded in (2.15).

Appendix B
The group $SU(3)$

The group $SU(3)$ is the group of all unitary 3×3 matrices $\mathbf{U}$ with determinant 1:

$$\mathbf{U}\mathbf{U}^\dagger = 1$$

$$\det \mathbf{U} = 1. \tag{B.1}$$

The infinitesimal transformations are:

$$\mathbf{U} = \mathbf{1} + i\,\delta\varphi_a\,\mathbf{t}_a$$

$$\mathbf{t}_a = \tfrac{1}{2}\lambda_a \tag{B.2}$$

where we make use of the summation convention. The $\delta\varphi_a$ are infinitesimal parameters and the $\lambda_a(a = 1,\ldots,8)$ are the Gell-Mann matrices:

$$\lambda_1 = \begin{pmatrix} 0 & 1 & 0 \\ 1 & 0 & 0 \\ 0 & 0 & 0 \end{pmatrix} \quad \lambda_2 = \begin{pmatrix} 0 & -i & 0 \\ i & 0 & 0 \\ 0 & 0 & 0 \end{pmatrix} \quad \lambda_3 = \begin{pmatrix} 1 & 0 & 0 \\ 0 & -1 & 0 \\ 0 & 0 & 0 \end{pmatrix}$$

$$\lambda_4 = \begin{pmatrix} 0 & 0 & 1 \\ 0 & 0 & 0 \\ 1 & 0 & 0 \end{pmatrix} \quad \lambda_5 = \begin{pmatrix} 0 & 0 & -i \\ 0 & 0 & 0 \\ i & 0 & 0 \end{pmatrix} \quad \lambda_6 = \begin{pmatrix} 0 & 0 & 0 \\ 0 & 0 & 1 \\ 0 & 1 & 0 \end{pmatrix}$$

$$\lambda_7 = \begin{pmatrix} 0 & 0 & 0 \\ 0 & 0 & -i \\ 0 & i & 0 \end{pmatrix} \quad \lambda_8 = \frac{1}{\sqrt{3}}\begin{pmatrix} 1 & 0 & 0 \\ 0 & 1 & 0 \\ 0 & 0 & -2 \end{pmatrix}. \tag{B.3}$$

The λ_a satisfy

$$\operatorname{tr}\lambda_a = 0 \tag{B.4}$$

$$\operatorname{tr}(\lambda_a\lambda_b) = 2\delta_{ab} \tag{B.5}$$

$$[\lambda_a, \lambda_b] = 2if_{abc}\lambda_c \tag{B.6}$$

$$\{\lambda_a, \lambda_b\} = \tfrac{4}{3}\delta_{ab} + 2d_{abc}\lambda_c \tag{B.7}$$

(a,b,c)	f_{abc}	(a,b,c)	d_{abc}
123	1	118	$1/\sqrt{3}$
147	$1/2$	146	$1/2$
156	$-1/2$	157	$1/2$
246	$1/2$	228	$1/\sqrt{3}$
257	$1/2$	247	$-1/2$
345	$1/2$	256	$1/2$
367	$-1/2$	338	$1/\sqrt{3}$
458	$\sqrt{3}/2$	344	$1/2$
678	$\sqrt{3}/2$	355	$1/2$
		366	$-1/2$
		377	$-1/2$
		448	$-1/(2\sqrt{3})$
		558	$-1/(2\sqrt{3})$
		668	$-1/(2\sqrt{3})$
		778	$-1/(2\sqrt{3})$
		888	$-1/\sqrt{3}$

Table B1. The independent nonvanishing components of f_{abc} and d_{abc}

where f_{abc} is totally antisymmetric, and d_{abc} is totally symmetric with values for the independent components as given in Table B1.

The following relations hold:

$$f_{abr}f_{rcs} + f_{bcr}f_{ras} + f_{car}f_{rbs} = 0 \tag{B.8}$$

$$f_{abr}d_{rcs} + f_{cbr}d_{ras} = d_{acr}f_{rbs} \tag{B.9}$$

$$f_{ars}f_{brs} = 3\delta_{ab} \tag{B.10}$$

$$d_{aab} = 0 \tag{B.11}$$

$$d_{ars}d_{brs} = \tfrac{5}{3}\delta_{ab} \tag{B.12}$$

$$f_{abc}\lambda_b\lambda_c = 3i\lambda_a \tag{B.13}$$

$$\lambda_a\lambda_a = 16/3 \tag{B.14}$$

$$\lambda_a\lambda_b\lambda_a = -\tfrac{2}{3}\lambda_b \tag{B.15}$$

$$(\lambda_a)_{\alpha\beta}(\lambda_a)_{\alpha'\beta'} = -\tfrac{2}{3}\delta_{\alpha\beta}\delta_{\alpha'\beta'} + 2\delta_{\alpha\beta'}\delta_{\alpha'\beta}. \tag{B.16}$$

It is often convenient to consider a generalization of QCD in which the colour-group $SU(3)$ is replaced with $SU(N)$ $(N \geq 2)$, the group of all unitary $N \times N$ matrices $\mathbf{U}$ with determinant 1. The infinitesimal transformations of $SU(N)$ are as in (B.2) except that we have now $N^2 - 1$ matrices

$\lambda_a(a = 1, \ldots, N^2 - 1)$ which again satisfy (B.4) and (B.5). Also (B.6) is still valid but with different structure constants for each N.

The infinitesimal generators in the fundamental representation (B.2) are $\mathbf{t}_a = \lambda_a/2$, in the adjoint representation $(\mathbf{T}_a)_{bc} = f_{abc}/i$. We have then

$$\mathbf{t}_a\mathbf{t}_a = \tfrac{1}{4}\lambda_a\lambda_a = C_F \mathbf{1}$$

$$C_F = \frac{N^2 - 1}{2N} \tag{B.17}$$

$$\operatorname{tr}(\mathbf{T}_a\mathbf{T}_b) = f_{acd}f_{bcd} = C_A\delta_{ab}$$

$$C_A = N. \tag{B.18}$$

Appendix C
Feynman rules of QCD

In this Appendix we present the Feynman rules of QCD. The derivation of these rules from the Lagrange density (6.1) is best achieved using the path-integral formalism. Here we restrict ourselves to giving only the results.

The Feynman diagrams of QCD are constructed from three types of lines: quark lines, gluon lines, and so-called Faddeev-Popov ghost lines. Within QCD it is not possible that lines corresponding to different quark flavours are joined together. The ghosts are to be treated like a complex scalar particle with zero mass. Thus ghost and antighost are distinct. The corresponding lines, like the quark lines, are given an arrow. The ghosts transform according to the adjoint representation of the colour group $SU(3)$ and thus carry a colour index a $(a = 1, \ldots, 8)$. We shall make use of the following scheme to relate physical situations, parts of diagrams, and algebraic expressions:

quark in initial state   incoming quark line

$u(p)$

quark in final state   outgoing quark line

$\bar{u}(p)$

antiquark in initial state   outgoing quark line

$\bar{v}(p)$

antiquark in final state

$v(p)$

incoming quark line

$$\xrightarrow{\quad p \quad}\;\bullet$$

ghost in initial or antighost in final state

1

incoming ghost line

$$- - \blacktriangleright\!- \;\!-\bullet$$

ghost in final or antighost in initial state

1

outgoing ghost line

$$\bullet\!- \;- \;\blacktriangleright\!- \;-$$

virtual quark with flavour j

$$\frac{i}{\gamma.p - m_j + i\epsilon}$$

internal quark line

$$\bullet\!\xrightarrow{\quad p \quad}\!\bullet$$

gluon in initial state

ϵ^μ

incoming gluon line

gluon in final state

$\epsilon^{*\mu}$

outgoing gluon line

virtual gluon

$$i\delta^{ab}\left(\frac{-g_{\mu\nu}}{k^2 + i\epsilon} + \frac{(1-\xi)k_\mu k_\nu}{(k^2 + i\epsilon)^2}\right)$$

internal gluon line

$$\bullet\!\underset{\quad}{\overset{k}{\text{e e e e e}}}\!\bullet$$

virtual ghost

$$i\frac{\delta^{ab}}{p^2 + i\epsilon}$$

internal ghost line

$$\bullet\!- \;-\overset{p}{\blacktriangleright}\!- \;-\bullet$$

The elementary processes of QCD correspond to the vertices shown on the next page.

Appendix C

three-gluon vertex $\quad (\sum_{j=1}^{3} k_j = 0$, all momenta incoming)

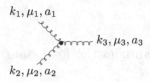

$$-g f_{a_1 a_2 a_3} \Big((k_1 - k_2)_{\mu_3} g_{\mu_1 \mu_2} + (k_2 - k_3)_{\mu_1} g_{\mu_2 \mu_3} + (k_3 - k_1)_{\mu_2} g_{\mu_3 \mu_1} \Big)$$

four-gluon vertex $\quad (\sum_{j=1}^{4} k_j = 0$, all momenta incoming)

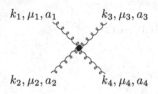

$$ig^2 f_{a_1 a_2 b} f_{a_3 a_4 b} \left(g_{\mu_2 \mu_3} g_{\mu_1 \mu_4} - g_{\mu_1 \mu_3} g_{\mu_2 \mu_4} \right) + \text{ cyclic permutations of } 1, 2, 3$$

ghost-gluon vertex $\quad (p, k$ incoming; p' outgoing; $p + k = p')$

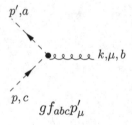

$$g f_{abc} p'_\mu$$

quark-gluon vertex $\quad (p, k$ incoming; p' outgoing; $p + k = p')$

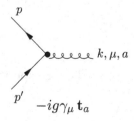

$$-i g \gamma_\mu \mathbf{t}_a$$

The parameters that appear in the Feynman rules are the coupling constant g, the quark masses $m_j (j = 1, \ldots, n_f)$, and the arbitrary gauge parameter ξ which must drop out in the expressions obtained for observable quantities. It is usually most convenient to set $\xi = 1$; this is the so-called Feynman gauge. For every closed fermion loop and every ghost loop a factor of (-1) must be included. For closed gluon loops one must take the so-called statistical factor into account. Loop momenta are to be integrated over with a measure given for each loop by

$$\int \frac{d^4 l}{(2\pi)^4}$$

Putting together all these factors for diagrams with given number of incoming and outgoing lines gives i times the transition matrix element T_{fi} for the reaction.

When calculating the polarisation sums for gluons in the initial or final states there are likewise a few points that should be noted. It is always possible to do the calculation noncovariantly, that is to consider only the transverse polarisation degrees of freedom of the gluons. Alternatively, one can work with covariant quantities; in Feynman gauge one may replace the polarisation sum of a gluon in the initial or final state by

$$\sum_{\text{spins}} \epsilon_\mu \epsilon_\nu^* = -g_{\mu\nu}.$$

In this case, however, it is also necessary to take into account processes of the same order with incoming and outgoing ghost lines, treating ghosts and antighosts formally as if they were extra degrees of freedom for the gluons to be considered in their spin sums. (Of course only two external gluon lines can be replaced at a time by ghost lines since an odd number of external ghost lines is impossible, as is an odd number of external fermion lines.) Amplitudes corresponding to diagrams which are identical except for permutations of the labels of external ghost lines receive extra sign factors relative to each other in the same way as for fermions. In calculating transition probabilities one has furthermore to include a factor $(-1)^{n/2}$, where n is the number of external ghost lines in the diagram. Thus some ghost processes, for instance the production of a single ghost-antighost pair ($n = 2$), are assigned a negative probability (see for instance [509]).

These rules are sufficient for the calculation of all tree diagrams, that is diagrams without loops. For the regularisation and renormalisation procedures to be applied beyond this level the reader is referred to the textbooks, for instance [254,510,222].

Appendix D
Pion-nucleon amplitudes

In this appendix we give the connection between the invariant amplitudes, s-channel helicity amplitudes and t-channel helicity amplitudes for pion-nucleon scattering and pion photoproduction on nucleons

Pion-nucleon scattering amplitudes

Let the four-momenta of the incident and outgoing pion be denoted by k_1 and k_2 respectively, and those of the initial and final nucleons by p_1 and p_2. Energy-momentum conservation means that only three of these are independent, and it is conventional to take $p = \frac{1}{2}(p_1 + p_2)$, $k = \frac{1}{2}(k_1 + k_2)$, and $\kappa = \frac{1}{2}(k_1 - k_2)$ as the three independent four vectors. Note that $t = -\frac{1}{4}\kappa^2$.

The complete invariant pion-nucleon scattering amplitude may be written as[511]

$$\bar{u}(p_2)T^{\beta\alpha}u(p_1) = \bar{u}(p_2)\Big(- A_1^{\beta\alpha}(\nu, t) + i\gamma.\kappa\, A_2(\nu, t)^{\beta\alpha}\Big)u(p_1) \qquad \text{(D.1)}$$

where $\nu = \frac{1}{4}(s - u)$. In (D.1) the isospin index of the incoming pion is denoted by α and that of the outgoing pion by β. There are two possible independent isospin combinations

$$\begin{aligned}
I_{\beta\alpha}^+ &= \tfrac{1}{2}(\tau_\beta\tau_\alpha + \tau_\alpha\tau_\beta) = \delta_{\beta\alpha} \\
I_{\beta\alpha}^- &= \tfrac{1}{2}(\tau_\beta\tau_\alpha - \tau_\alpha\tau_\beta) = \tfrac{1}{2}[\tau_\beta, \tau_\alpha]\,.
\end{aligned} \qquad \text{(D.2)}$$

where τ is the isospin-matrix operator. The combinations in (D.2) are chosen so as to be hermitian and anti-hermitian. Thus the isospin decomposition of the $A_i^{\beta\alpha}$ is

$$A_i^{\beta\alpha} = \delta_{\beta\alpha} \, A_i^{(+)} + \tfrac{1}{2}[\tau_\beta, \tau_\alpha] \, A_i^{(-)}, \tag{D.3}$$

$A_i^{(+)}$ corresponds to isospin-0 exchange in the t-channel and $A_i^{(-)}$ corresponds to isospin-1 exchange in the t-channel. The $A_i^{(\pm)}$ are given in terms of the amplitudes $A^{\frac{1}{2}}$ and $A^{\frac{3}{2}}$ of specific isospin by

$$A_i^{(+)} = \tfrac{1}{3}(A^{\frac{1}{2}} + 2A^{\frac{3}{2}})$$
$$A_i^{(-)} = \tfrac{1}{3}(A^{\frac{1}{2}} - A^{\frac{3}{2}}), \tag{D.4}$$

and the physical amplitudes are given by

$$A_i(\pi^- p \to \pi^- p) = A_i^{(+)} + A_i^{(-)}$$
$$A_i(\pi^+ p \to \pi^+ p) = A_i^{(+)} - A_i^{(-)}$$
$$A_i(\pi^- p \to \pi^0 n) = -\sqrt{2}A_i^{(-)} \tag{D.5}$$

The amplitudes $A_i(\nu, t)$ are convenient because of their simple analytic and crossing properties. $A_1^{(+)}$ and $A_2^{(-)}$ are even functions of ν, while $A_1^{(-)}$ and $A_2^{(+)}$ are odd functions of ν. Because of these properties one can write down fixed-t dispersion relations for the $A_i^{\pm}$.

The connection to s-channel partial wave amplitudes is made via functions f_1 and f_2 defined by

$$f_1 = \frac{E + m_p}{8\pi\sqrt{s}}\left(A_1 + (\sqrt{s} - m_p)A_2\right)$$
$$f_2 = \frac{E - m_p}{8\pi\sqrt{s}}\left(-A_1 + (\sqrt{s} + m_p)A_2\right) \tag{D.6}$$

where E is the centre-of-mass energy of the proton and m_p the proton mass. The f_i are given in terms of the partial-wave amplitudes f_{l+} and f_{l-} corresponding to orbital angular momentum l and total angular momentum $J = l \pm \frac{1}{2}$ by

$$f_1 = \sum_{l=0}^{\infty} f_{l+} P'_{l+1}(\cos\theta_s) - \sum_{l=2}^{\infty} f_{l-} P'_{l-1}(\cos\theta_s)$$
$$f_2 = \sum_{l=1}^{\infty}(f_{l-} - f_{l+})P'_l(\cos\theta_s). \tag{D.7}$$

In (D.7) the $f_{l\pm}$ are defined by

$$f_{l\pm} = \frac{e^{i\delta_{l\pm}}\sin\delta_{l\pm}}{|\mathbf{p}_1|} \tag{D.8}$$

which differs from the definition (1.47) by a factor $\frac{1}{2}\sqrt{s}$.

There are two independent s-channel helicity amplitudes: the non-helicity-flip amplitude A_{++} and the helicity-flip amplitude A_{+-}. As the pion has zero spin we have not written its helicity label and we have also omitted the $\frac{1}{2}$ for the nucleon helicity, retaining only the sign. With the amplitudes normalised in the same way as for the unpolarised case, as in (1.19), the differential cross section is given by

$$\frac{d\sigma}{dt} = \frac{1}{64\pi|\mathbf{p}_1|^2 s}\left(|A_{++}|^2 + |A_{+-}|^2\right) \qquad \text{(D.9)}$$

For convenience of writing, in the following we denote $|\mathbf{p}_1|$ by q. The s-channel helicity amplitudes are related to the invariant amplitudes A_i by

$$A_{++} = \left(1 + \frac{t}{4q^2}\right)^{\frac{1}{2}}\left(2m_p A_1 + (s - m_p^2 - m_\pi^2)A_2\right)$$

$$A_{+-} = \left(\frac{-t}{4q^2}\right)^{\frac{1}{2}}\left(2E\,A_1 + (s - m_p^2 - m_\pi^2)A_2\right) \qquad \text{(D.10)}$$

where m_π is the pion mass.

The crossing matrix for obtaining the t-channel helicity amplitudes $A_{++}^{(t)}$ and $A_{+-}^{(t)}$ from the s-channel helicity amplitudes is[512]

$$\begin{pmatrix} \sin\chi & \cos\chi \\ -\cos\chi & \sin\chi \end{pmatrix} \qquad \text{(D.11)}$$

where

$$\cos\chi = \frac{s - m_p^2 - m_\pi^2}{2q\sqrt{s}}\left(\frac{-t}{4m_p^2 - t}\right)^{\frac{1}{2}}$$

$$\sin\chi = \frac{m}{q}\left(\frac{4q^2 + t}{4m_p^2 - t}\right)^{\frac{1}{2}}. \qquad \text{(D.12)}$$

The resulting t-channel helicity amplitudes are then given in terms of the A_i by[513]

$$A_{++}^{(t)} = \left(4m_p^2 - t\right)^{\frac{1}{2}}\left(A_1 + \frac{2m_p(s - m_p^2 - m_\pi^2 + \frac{1}{2}t)}{4m_p^2 - t}A_2\right)$$

$$A_{+-}^{(t)} = (-t)^{\frac{1}{2}}\left(\frac{(s + m_p^2 - m_\pi^2)^2}{4m_p^2 - t} - s\right)^{\frac{1}{2}}A_2. \qquad \text{(D.13)}$$

A complete experimental determination of pion-nucleon elastic scattering can be achieved by analysing the final polarisation $\mathbf{P}_f$ of the protons recoiling from a target whose initial polarisation $\mathbf{P}_i$ is in the scattering plane. Let $\hat{\mathbf{k}}_i$, $\hat{\mathbf{k}}_f$ and $\hat{\mathbf{p}}_f$ be unit vectors in the direction of the incident pion, scattered pion and recoil proton in the laboratory system, and define $\hat{\mathbf{n}} = (\hat{\mathbf{k}}_i \times \hat{\mathbf{k}}_f)/|\hat{\mathbf{k}}_i \times \hat{\mathbf{k}}_f|$. Then the polarisation P and the spin-rotation parameters R and A are defined by[514]

$$\mathbf{P}_f.(\hat{\mathbf{p}}_f \times \hat{\mathbf{n}}) = \frac{R\,\mathbf{P}_i.(\hat{\mathbf{k}}_i \times \hat{\mathbf{n}}) - A\,\mathbf{P}_i.\hat{\mathbf{k}}_i}{1 + P\,\mathbf{P}_i.\hat{\mathbf{n}}}$$

$$\mathbf{P}_f.\hat{\mathbf{p}}_f = \frac{A\,\mathbf{P}_i.(\hat{\mathbf{k}}_i \times \hat{\mathbf{n}}) + R\,\mathbf{P}_i.\hat{\mathbf{k}}_i}{1 + P\,\mathbf{P}_i.\hat{\mathbf{n}}}$$

$$\mathbf{P}_f.\hat{\mathbf{n}} = \frac{P + \mathbf{P}_i.\hat{\mathbf{n}}}{1 + P\,\mathbf{P}_i.\hat{\mathbf{n}}} \,. \tag{D.14}$$

With the normalisation of the s-channel helicity amplitudes defined by (D.9), P, R and A are given by

$$P\frac{d\sigma}{dt} = \frac{-1}{32\pi|\mathbf{p}_1|^2 s}\,\mathrm{Im}\ A_{++}A_{+-}^*$$

$$R\frac{d\sigma}{dt} = \frac{-1}{64\pi|\mathbf{p}_1|^2 s}\left(\left(|A_{++}|^2 - |A_{+-}|^2\right)\cos\theta_p - 2\,\mathrm{Re}\ A_{++}A_{+-}^*\sin\theta_p\right)$$

$$A\frac{d\sigma}{dt} = \frac{1}{64\pi|\mathbf{p}_1|^2 s}\left(\left(|A_{++}|^2 - |A_{+-}|^2\right)\sin\theta_p - 2\,\mathrm{Re}\ A_{++}A_{+-}^*\cos\theta_p\right)$$

$$\tag{D.15}$$

where $\cos\theta_p = \hat{\mathbf{k}}_i.\hat{\mathbf{p}}_f$.

Pion-photoproduction amplitudes

Let the four momenta of the incident photon and outgoing pion be denoted by q and k respectively, and those of the initial and final nucleons by p_1 and p_2. Energy-momentum conservation means that only three of these are independent, and it is conventional to take $p = \frac{1}{2}(p_1 + p_2)$ together with q and k as the three independent four vectors.

The complete invariant pion photoproduction matrix element may be written as[515]

$$\bar{u}(p_2)T^\beta u(p_1) = \bar{u}\Big(i\gamma_5\,\gamma.\epsilon\,\gamma.q\,A_1^\beta(\nu,t)$$

$$+\,2i\gamma_5(p.\epsilon\,q.k - p.q\,k.\epsilon)\,A_2^\beta(\nu,t)$$

$$+\,\gamma_5(\gamma.\epsilon\,q.k - \gamma.q\,k.\epsilon)\,A_3^\beta(\nu,t)$$

$$+\,2\gamma_5(\gamma.\epsilon\,p.q - \gamma.q\,p.\epsilon - im_p\gamma.\epsilon\,\gamma.q)\,A_4^\beta(\nu,t)\Big)u(p)$$

$$\text{(D.16)}$$

where m_p is the nucleon mass, ϵ is the photon polarisation vector and $\nu = \frac{1}{4}(s-u)$. The factors of 2 and i were introduced for convenience in subsequent calculations[515] and the superscript β is an isospin label. Denoting the isospin index of the outgoing pion by β and the isospin matrix operator by τ, there are three independent isospin combinations possible:

$$I_\beta^+ = \tfrac{1}{2}(\tau_\beta\tau_3 + \tau_3\tau_\beta) = \delta_{\beta 3}$$

$$I_\beta^- = \tfrac{1}{2}(\tau_\beta\tau_3 - \tau_3\tau_\beta) = \tfrac{1}{2}[\tau_\beta, \tau_3]$$

$$I_\beta^0 = \tau_\beta.$$

$$\text{(D.17)}$$

These particular combinations are chosen so as to be either hermitian or anti-hermitian. Thus the isospin decomposition of each of the A_i^β is

$$A_i^\beta = \delta_{\beta 3}\,A_i^{(+)} + \tfrac{1}{2}[\tau_\beta, \tau_3]\,A_i^{(-)} + \tau_3\,A_i^{(0)}.$$

$$\text{(D.18)}$$

The physical amplitudes are then given by

$$A_i(\gamma p \to \pi^0 p) = A_i^{(+)} + A_i^{(0)}$$

$$A_i(\gamma n \to \pi^0 n) = A_i^{(+)} - A_i^{(0)}$$

$$A_i(\gamma p \to \pi^+ n) = \sqrt{2}(A_i^{(-)} + A_i^{(0)})$$

$$A_i(\gamma n \to \pi^- p) = -\sqrt{2}(A_i^{(-)} - A_i^{(0)}).$$

$$\text{(D.19)}$$

The amplitudes $A_i(\nu, t)$ are convenient because of their simple analytic and crossing properties. From (D.16) and (D.18) we see that $A_{1,2,4}^{(+,0)}$ and $A_3^{(-)}$ are even functions of ν, while $A_3^{(+,0)}$ and $A_{1,2,4}^{(-)}$ are odd functions of ν. Because of these properties of the A_i one can write down fixed-t dispersion relations and finite-energy sum rules for them.

For discussing forward high-energy reactions it is more convenient to use a set of amplitudes F_i rather than the amplitudes A_i. The F_i are defined by

$$F_1 = A_1 - 2m_p A_4 \qquad F_2 = A_1 + tA_2$$

$$F_3 = 2m_p A_1 - tA_4 \qquad F_4 = A_3.$$

$$\text{(D.20)}$$

These amplitudes also have simple crossing and analytic properties. F_1 and F_2 are respectively natural and unnatural parity t-channel amplitudes, and although F_3 and F_4 are of mixed parity, at large ν they become respectively natural and unnatural parity t-channel amplitudes also. Further, the amplitudes $F_2(\nu, t)$ and $F_3(\nu, t)$ satisfy a constraint equation at $t = 0$:

$$F_3(\nu, 0) = 2m_p F_2(\nu, 0). \tag{D.21}$$

This can be derived immediately from (D.20) as the amplitudes A_2 and A_4 do not have poles at $t = 0$.

Although the A_i and F_i are appropriate for dispersion relations and sum rules, they are not so suitable for the application of Regge theory. For this purpose it is much more useful to use helicity amplitudes.

Denote the s-channel helicity amplitudes by $T_{\lambda_N \lambda_N'}^{\lambda_\gamma}$, where λ_γ is the photon helicity and λ_N, λ_N' are the helicities of the initial and final nucleons. The helicity amplitudes satisfy the conditions

$$N \equiv T_{+-}^1 = T_{-+}^{-1} \qquad S_1 \equiv T_{--}^1 = T_{++}^{-1}$$
$$S_2 \equiv T_{++}^1 = T_{--}^{-1} \qquad D \equiv T_{-+}^1 = -T_{+-}^{-1}. \tag{D.22}$$

In (D.22) we have used the usual convention of denoting the nucleon helicities only by their sign as there is no ambiguity. Because of the relations (D.22) it is also conventional to use the helicity amplitudes with $\lambda_\gamma = 1$ and omit the photon-helicity label. As the pion has zero helicity there is automatic helicity flip at the photon-pion-reggeon vertex, so the net helicity flip is defined by the nucleon helicities. Then T_{-+} is a non-flip amplitude, T_{++} and T_{--} are both single-flip amplitudes and T_{+-} is double flip. This explains the notation N, S_1, S_2 and D which we shall now use.

If the amplitudes are normalised such that

$$\frac{d\sigma}{dt}\bigg|_{\text{unpolarised}} = |N|^2 + |S_1|^2 + |S_2|^2 + |D|^2 \tag{D.23}$$

then to leading order in s the relation between the helicity amplitudes and the amplitudes F_i is given by

$$\begin{bmatrix} F_1 \\ F_2 \\ F_3 \\ F_4 \end{bmatrix} \sim \frac{1}{\sqrt{-t}} \begin{bmatrix} 2m & \sqrt{-t} & -\sqrt{-t} & 2m \\ 0 & \sqrt{-t} & \sqrt{-t} & 0 \\ t & 2m\sqrt{-t} & -2m\sqrt{-t} & t \\ 1 & 0 & 0 & -1 \end{bmatrix} \begin{bmatrix} T_{++} \\ T_{-+} \\ T_{+-} \\ T_{--} \end{bmatrix}. \tag{D.24}$$

We can see immediately from (D.24) that F_1 and F_4 are mainly s-channel helicity flip and F_2 and F_3 a combination of no-flip and double-flip.

The cross section for a polarised photon incident on a polarised target has the form

$$\frac{d\sigma}{dt}(\mathbf{P}, P_T, \phi, P_\odot) = \frac{d\sigma}{dt}\Big|_{\text{unpolarised}} \Big(1 - P_T \cos 2\phi \, \Sigma$$
$$+ P_x(-P_T \sin 2\phi \, H + P_\odot F) - P_y(-T + P_T \cos 2\phi \, R)$$
$$- P_z(-P_T \sin 2\phi \, G + P_\odot E)\Big). \tag{D.25}$$

Here Σ, R, T, E, F, G and H are functions of ν and t. They are defined in terms of the helicity amplitudes by

$$\Sigma \, d\sigma/dt = 2 \operatorname{Re}(S_1 S_2^* - N D^*)$$
$$T \, d\sigma/dt = 2 \operatorname{Im}(S_1 N^* - S_2 D^*)$$
$$R \, d\sigma/dt = 2 \operatorname{Im}(S_2 N^* - S_1 D^*)$$
$$E \, d\sigma/dt = |S_2|^2 - |S_1|^2 - |D|^2 + |N|^2$$
$$F \, d\sigma/dt = 2 \operatorname{Re}(S_2 D^* + S_1 N^*)$$
$$G \, d\sigma/dt = -2 \operatorname{Im}(S_1 S_2^* + N D^*)$$
$$H \, d\sigma/dt = -2 \operatorname{Re}(S_1 D^* + S_2 N^*). \tag{D.26}$$

In terms of the t-channel amplitudes F_i we have asymptotically

$$\frac{d\sigma}{dt} = \frac{-1}{32\pi} \frac{1}{t - 4m_p^2} \Big(- t|F_1|^2 + |F_3|^2 - (t - 4m_p^2)(|F_2|^2 - t|F_4|^2)\Big)$$

$$\Sigma \frac{d\sigma}{dt} = \frac{-1}{32\pi} \frac{1}{t - 4m_p^2} \Big(- t|F_1|^2 + |F_3|^2 + (t - 4m_p^2)(|F_2|^2 - t|F_4|^2)\Big)$$

$$T \frac{d\sigma}{dt} = \frac{-1}{16\pi} \frac{\sqrt{-t}}{t - 4m_p^2} \operatorname{Im}\Big(F_1 F_3^* - (t - 4m_p^2) F_4 F_2^*\Big)$$

$$R \frac{d\sigma}{dt} = \frac{-1}{16\pi} \frac{\sqrt{-t}}{t - 4m_p^2} \operatorname{Im}\Big(F_1 F_3^* + (t - 4m_p^2) F_4 F_2^*\Big)$$

$$E \frac{d\sigma}{dt} = \frac{-1}{16\pi} \frac{1}{t - 4m_p^2} \operatorname{Re}\Big(t F_4(2m_p F_1^* - F_3^*) + F_2(-t F_1^* + 2m_p F_3^*)\Big)$$

$$G \frac{d\sigma}{dt} = \frac{-1}{16\pi} \frac{1}{t - 4m_p^2} \operatorname{Im}\Big(t F_4(2m_p F_1^* - F_3^*) + F_2(-t F_1^* + 2m_p F_3^*)\Big)$$

$$F \frac{d\sigma}{dt} = \frac{-1}{16\pi} \frac{\sqrt{-t}}{t - 4m_p^2} \operatorname{Re}\Big(F_2(2m_p F_1^* - F_3^*) + F_4(-t F_1^* + 2m_p F_3^*)\Big)$$

$$H \frac{d\sigma}{dt} = \frac{-1}{16\pi} \frac{\sqrt{-t}}{t - 4m_p^2} \operatorname{Im}\Big(F_2(2m_p F_1^* - F_3^*) + F_4(-t F_1^* + 2m_p F_3^*)\Big).$$

$$\tag{D.27}$$

Plane-polarised photon beams allow a separation of natural and unnatural-parity exchange: if the beam is polarised in the production plane ($\phi = 0$) then $d\sigma_\parallel/dt$ selects unnatural-parity exchange; and if the beam is polarised perpendicular to the production plane ($\phi = \frac{1}{2}\pi$) then $d\sigma_\perp/dt$ selects natural-parity exchange. This follows immediately from (D.25) on noting that $d\sigma_\parallel/dt \propto 1 - \Sigma$ and $d\sigma_\perp/dt \propto 1 + \Sigma$ and using the second line of (D.27).

Appendix E

The density matrix of vector mesons

In this appendix we give formulae for the two-body-decay angular distributions of vector mesons in electroproduction, using $\rho \to \pi\pi$ for definiteness.

Let k, q, p be the four momenta of the electron, the virtual photon γ^*, and the proton. The differential cross section for $\gamma^* N \to \rho N$ is conventionally written as

$$\frac{d\sigma_{\gamma^* N \to \rho N}}{dt d\psi} = \frac{1}{128\pi^2 |\mathbf{q}|^2 s} \tfrac{1}{2} \operatorname{tr}\left(T \rho(\gamma) T^\dagger\right) \tag{E.1}$$

where T represents the helicity amplitudes $T_{\lambda_\rho \lambda'_N ; \lambda_\gamma \lambda_N}(\theta_\rho)$, $\rho(\gamma)$ is the photon density matrix[195], ψ is the angle between the normals to the lepton scattering plane and the hadron production plane, λ_γ, λ_ρ are the helicities of the photon and the ρ, and $\lambda_{N'}$ and λ_N are the helicities of the initial and final proton.

The normalised vector-meson density matrix is given by

$$\rho(\rho) = \frac{\tfrac{1}{2}\left(T \rho(\gamma) T^\dagger\right)}{\frac{1}{2\pi} \int d\psi \, \tfrac{1}{2} \operatorname{tr}\left(T \rho(\gamma) T^\dagger\right)} \tag{E.2}$$

where summation over the nucleon helicities is understood. The complete derivation of the ρ density matrix can be found in [195]. Here we give only the results relevant for our discussion.

The exclusive electroproduction of ρ mesons, at given s, Q^2, t and $m_{\pi\pi}$, is described by three angles θ, ϕ and ψ. The angle ψ is defined in the $\gamma^* p$ centre-of-mass frame as the angle between the lepton scattering plane and the plane containing the ρ and the scattered proton. In the s-channel helicity frame, that is the frame in which the z-axis is the ρ^0 direction in the $\gamma^* p$ centre-of-mass frame, the ρ decay is described by the polar angle θ and the azimuthal angle ϕ of the positive pion in the ρ rest frame.

The angular distribution $W(\cos\theta, \phi, \psi)$ is parametrised by spin-density matrix elements ρ^α_{ik}. The subscripts $i, k = -1, 0, 1$, refer to the values of the ρ meson helicity λ_ρ entering the matrix element. The superscript $\alpha = 0, 1, 2, 3, 4, 5, 6, 7$ and denotes the contributions from different photon polarisation states. For an unpolarised lepton beam the convention is: transverse photons (0); linearly polarised transverse photons (1,2); longitudinally polarised photons (4); and the interference of the longitudinal and transverse amplitudes (5,6). The superscripts 3, 7, 8 arise with a longitudinally polarised lepton beam.

For unpolarised leptons the ρ-decay angular distribution is given by[195]

$$
\begin{aligned}
W(\cos\theta, \phi, \psi) = {} & \frac{1}{1 + \epsilon R} \frac{3}{4\pi} \\
& \times \Big(\tfrac{1}{2}(1 - \rho^0_{00}) + \tfrac{1}{2}(3\rho^0_{00} - 1)\cos^2\theta - \sqrt{2}\,\mathrm{Re}\,\rho^0_{10}\sin 2\theta \cos\phi \\
& \quad - \rho^0_{1-1}\sin^2\theta \cos 2\phi \\
& \quad - \epsilon \cos 2\psi\,(\rho^1_{11}\sin^2\theta + \rho^1_{00}\cos^2\theta - \sqrt{2}\,\mathrm{Re}\,\rho^1_{10}\sin 2\theta \cos\phi \\
& \qquad - \rho^1_{1-1}\sin^2\theta \cos 2\phi) \\
& \quad - \epsilon \sin 2\psi\,(\sqrt{2}\,\mathrm{Im}\,\rho^2_{10}\sin 2\theta \sin\phi + \mathrm{Im}\,\rho^2_{1-1}\sin^2\theta \sin 2\psi) \\
& \quad + \epsilon R(\tfrac{1}{2}(1 - \rho^4_{00}) + \tfrac{1}{2}(3\rho^4_{00} - 1)\cos^2\theta \\
& \qquad - \sqrt{2}\,\mathrm{Re}\,\rho^4_{10}\sin 2\theta \cos\phi - \rho^4_{1-1}\sin^2\theta \cos 2\phi) \\
& \quad + \sqrt{2\epsilon R(1 + \epsilon)}\,\cos\psi\,(\rho^5_{11}\sin^2\theta + \rho^5_{00}\cos^2\theta \\
& \qquad - \sqrt{2}\,\mathrm{Re}\,\rho^5_{10}\sin 2\theta \cos\phi - \rho^5_{1-1}\sin^2\theta \cos 2\phi) \\
& \quad + \sqrt{2\epsilon R(1 + \epsilon)}\,\sin\psi\,(\sqrt{2}\,\mathrm{Im}\,\rho^6_{10}\sin 2\theta \sin\phi \\
& \qquad + \mathrm{Im}\,\rho^6_{1-1}\sin^2\theta \sin 2\phi) \Big).
\end{aligned}
$$

$$(E.3)$$

Here $R = \sigma^L_{\gamma^* p} / \sigma^T_{\gamma^* p}$, with $\sigma^L_{\gamma^* p}$ and $\sigma^T_{\gamma^* p}$ the cross sections for ρ production from longitudinal and transverse photons, and ϵ is the ratio of the longitudinal to transverse photon fluxes, given by

$$
\epsilon = \frac{1 - y}{1 - y - \tfrac{1}{2}y^2} \tag{E.4}
$$

where $y = p.q/p.k$, and in the proton rest frame is equal to the fraction of the lepton energy transferred to the proton.

If the lepton has longitudinal polarisation P_L, then in addition to (E.3) one has

$$W_P(\cos\theta, \phi, \psi) = \pm\frac{3}{4\pi}P_L$$
$$\times \left(\sqrt{1-\epsilon^2}(\sqrt{2}\,\mathrm{Im}\,\rho_{10}^3\sin 2\theta\sin\phi + \mathrm{Im}\,\rho_{1-1}^3\sin^2\theta\sin 2\phi)\right.$$
$$+ \sqrt{2\epsilon(1-\epsilon)}\cos\psi\,(\sqrt{2}\,\mathrm{Im}\,\rho_{10}^7\sin 2\phi$$
$$+ \mathrm{Im}\,\rho^{1-1}\sin^2\theta\sin 2\phi)$$
$$+ \sqrt{2\epsilon(1-\epsilon)}\sin\psi\,(\rho_{11}^8\sin^2\theta + \rho_{00}^8\cos^2\theta$$
$$\left. - \sqrt{2}\,\mathrm{Re}\,\rho_{10}^8\sin 2\theta\cos\phi - \rho_{1-1}^8\sin^2\theta\cos 2\phi)\right) \qquad (E.5)$$

where the $\pm$ specifies whether the lepton polarisation is along or against the lepton momentum.

If it is not possible to separate σ^T and σ^L then the matrix elements which can be determined are

$$r_{ik}^{04} = \frac{\rho_{ik}^0 + \epsilon R\rho_{ik}^4}{1 + \epsilon R}$$

$$r_{ik}^\alpha = \frac{\rho_{ik}^\alpha}{1 + \epsilon R} \qquad \alpha = 1 \text{ to } 3$$

$$r_{ik}^\alpha = \frac{\sqrt{R}\rho_{ik}^\alpha}{1 + \epsilon R} \qquad \alpha = 5 \text{ to } 8. \qquad (E.6)$$

The spin density matrix elements are bilinear combinations of the s-channel helicity amplitudes $T_{\lambda_\rho\lambda_{N'};\lambda_\gamma\lambda_N}$. If the helicities of the initial and final proton are not measured it is normal to omit $\lambda_{N'}$, λ_N and simply label the helicity amplitudes by λ_ρ, λ_γ. Defining the normalisation factors N^T and N^L by[195]

$$N^T = \frac{1}{2}\sum_{\lambda_\rho\lambda_{N'}\lambda_N}\left(|T_{\lambda_\rho\lambda_{N'},1\lambda_N}|^2 + |T_{\lambda_\rho\lambda_{N'},-1\lambda_N}|^2\right)$$

$$N^L = \sum_{\lambda_\rho\lambda_{N'}\lambda_N}|T_{\lambda_\rho\lambda_{N'},0\lambda_N}|^2 \qquad (E.7)$$

then

$$\rho_{\lambda\lambda'}^0 = \frac{1}{N^T}\sum_{\lambda_\gamma=\pm 1}T_{\lambda\lambda_\gamma}T_{\lambda'\lambda_\gamma}^*$$

$$\rho_{\lambda\lambda'}^1 = \frac{1}{N^T}\sum_{\lambda_\gamma=\pm 1}T_{\lambda-\lambda_\gamma}T_{\lambda'\lambda_\gamma}^*$$

$$\rho_{\lambda\lambda'}^{2} = \frac{i}{N^T} \sum_{\lambda_\gamma = \pm 1} \lambda_\gamma T_{\lambda - \lambda_\gamma} T_{\lambda'\lambda_\gamma}^{*}$$

$$\rho_{\lambda\lambda'}^{3} = \frac{1}{N^T} \sum_{\lambda_\gamma = \pm 1} \lambda_\gamma T_{\lambda\lambda_\gamma} T_{\lambda'\lambda_\gamma}^{*}$$

$$\rho_{\lambda\lambda'}^{4} = \frac{1}{N^L} T_{\lambda 0} T_{\lambda' 0}^{*}$$

$$\rho_{\lambda\lambda'}^{5} = \frac{1}{\sqrt{2N^T N^L}} \sum_{\lambda_\gamma = \pm 1} \frac{1}{2}\lambda_\gamma (T_{\lambda 0} T_{\lambda'\lambda_\gamma}^{*} + T_{\lambda\lambda_\gamma} T_{\lambda' 0}^{*})$$

$$\rho_{\lambda\lambda'}^{6} = \frac{i}{\sqrt{2N^T N^L}} \sum_{\lambda_\gamma = \pm 1} \frac{1}{2} (T_{\lambda 0} T_{\lambda'\lambda_\gamma}^{*} - T_{\lambda\lambda_\gamma} T_{\lambda' 0}^{*})$$

$$\rho_{\lambda\lambda'}^{7} = \frac{1}{\sqrt{2N^T N^L}} \sum_{\lambda_\gamma = \pm 1} \frac{1}{2} (T_{\lambda 0} T_{\lambda'\lambda_\gamma}^{*} + T_{\lambda\lambda_\gamma} T_{\lambda' 0}^{*})$$

$$\rho_{\lambda\lambda'}^{8} = \frac{i}{\sqrt{2N^T N^L}} \sum_{\lambda_\gamma = \pm 1} \frac{1}{2}\lambda_\gamma (T_{\lambda 0} T_{\lambda'\lambda_\gamma}^{*} - T_{\lambda\lambda_\gamma} T_{\lambda' 0}^{*}). \qquad \text{(E.8)}$$

The decay angular distributions also allow a separation of the natural and unnatural-parity t-channel exchange amplitudes. This is possible because of a further symmetry of the helicity amplitudes which is valid at high energy[196]:

$$T_{-\lambda_\rho, -\lambda_\gamma} = \pm(-1)^{\lambda_\rho - \lambda_\gamma} T_{\lambda_\rho, \lambda_\gamma} \qquad \text{(E.9)}$$

with the nucleon helicities unchanged. The overall positive sign applies to natural-parity exchange and the negative sign to unnatural-parity exchange. Thus at high energy we have

$$T_{\lambda_\rho, \lambda_\gamma}^{\binom{N}{U}}(\theta_\rho) = \frac{1}{2}\left(T_{\lambda_\rho, \lambda_\gamma}(\theta_\rho) \pm (-1)^{\lambda_\rho - \lambda_\gamma} T_{-\lambda_\rho, -\lambda_\gamma}(\theta_\rho)\right) \qquad \text{(E.10)}$$

where the superscripts N and U go with the upper and lower sign of $\pm$ on the right-hand side of the equation. Correspondingly we can write

$$\rho_{\lambda\lambda'}^{\alpha} = \rho_{\lambda\lambda'}^{\alpha N} + \rho_{\lambda\lambda'}^{\alpha U}. \qquad \text{(E.11)}$$

The $\rho^{\alpha\binom{N}{U}}$ are given by[195]

$$\rho_{\lambda\lambda'}^{0\binom{N}{U}} = \frac{1}{2}(\rho_{\lambda\lambda'}^{0} \mp (-1)^{\lambda'} \rho_{\lambda - \lambda'}^{1})$$

$$\rho_{\lambda\lambda'}^{1\binom{N}{U}} = \frac{1}{2}(\rho_{\lambda\lambda'}^{1} \mp (-1)^{\lambda'} \rho_{\lambda - \lambda'}^{0})$$

$$\rho_{\lambda\lambda'}^{2\binom{N}{U}} = \frac{1}{2}(\rho_{\lambda\lambda'}^{2} \pm i(-1)^{\lambda'} \rho_{\lambda - \lambda'}^{3})$$

$$\rho_{\lambda\lambda'}^{3\binom{N}{U}} = \tfrac{1}{2}(\rho_{\lambda\lambda'}^3 \mp i(-1)^{\lambda'}\rho_{\lambda-\lambda'}^2)$$

$$\rho_{\lambda\lambda'}^{4\binom{N}{U}} = \tfrac{1}{2}(\rho_{\lambda\lambda'}^4 \pm (-1)^{\lambda'}\rho_{\lambda-\lambda'}^4)$$

$$\rho_{\lambda\lambda'}^{5\binom{N}{U}} = \tfrac{1}{2}(\rho_{\lambda\lambda'}^5 \pm (-1)^{\lambda'}\rho_{\lambda-\lambda'}^5)$$

$$\rho_{\lambda\lambda'}^{6\binom{N}{U}} = \tfrac{1}{2}(\rho_{\lambda\lambda'}^6 \pm i(-1)^{\lambda'}\rho_{\lambda-\lambda'}^7)$$

$$\rho_{\lambda\lambda'}^{7\binom{N}{U}} = \tfrac{1}{2}(\rho_{\lambda\lambda'}^7 \mp i(-1)^{\lambda'}\rho_{\lambda-\lambda'}^6)$$

$$\rho_{\lambda\lambda'}^{8\binom{N}{U}} = \tfrac{1}{2}(\rho_{\lambda\lambda'}^8 \pm (-1)^{\lambda'}\rho_{\lambda-\lambda'}^8). \tag{E.12}$$

The cross sections for natural and unnatural-parity exchange, σ^N and σ^U are then given by

$$\sigma_T^{\binom{N}{U}} = \tfrac{1}{2}(1 \pm (2\rho_{1-1}^1 - \rho_{00}^1))\sigma_T$$

$$\sigma_L^{\binom{N}{U}} = \tfrac{1}{2}(1 \mp (2\rho_{1-1}^4 - \rho_{00}^4))\sigma_L \tag{E.13}$$

The parity asymmetry, P_σ, defined as

$$P_\sigma = \frac{\sigma^N - \sigma^U}{\sigma^N + \sigma^U} \tag{E.14}$$

can then be written for transverse and longitudinal photons as

$$P_\sigma^T \sigma_T = 2\rho_{1-1}^1 - \rho_{00}^1$$

$$P_\sigma^L \sigma_L = -(2\rho_{1-1}^4 - \rho_{00}^4). \tag{E.15}$$

References

[1] P D B Collins, *An Introduction to Regge Theory* Cambridge University Press (1977)

[2] A Donnachie and P V Landshoff, Physics Letters 123B (1983) 345

[3] F E Low, Physical Review D12 (1975) 163

[4] S Nussinov, Physical Review Letters 34 (1975) 1286

[5] H Cheng and T T Wu, *Expanding Protons* MIT Press (1987)

[6] J R Forshaw and D A Ross, *Quantum Chromodynamics and the Pomeron* Cambridge University Press (1997)

[7] P V Landshoff and O Nachtmann, Zeitschrift für Physik C35 (1987) 405

[8] G Ingelman and P Schlein, Physics Letters B152 (1985) 256

[9] A Erdélyi ed, *Higher Transcendental Functions* volume 1 McGraw-Hill (1953)

[10] R J Eden, P V Landshoff, D I Olive and J C Polkinghorne, *The Analytic S-Matrix* Cambridge University Press (1966)

[11] H Lehmann, Il Nuovo Cimento 10 (1958) 579

[12] M Froissart, Physical Review 123 (1961) 1053

[13] A Martin, Il Nuovo Cimento 42A (1966) 930

[14] L Lukaszuk and A Martin, Il Nuovo Cimento 47A (1967) 265

[15] A Martin in *Proceedings of the First International Workshop on Diffractive and Elastic Scattering*, B Nicolescu and J Tranh Thanh Van eds, Editions Frontières (1985)

[16] I Y Pomeranchuk, Soviet Physics: Journal of Experimental and Theoretical Physics 7 (1958) 499

[17] R J Eden, Physical Review Letters 16 (1966) 39

[18] G Grunberg and T N Truong, Physical Review Letters 31 (1973) 63

[19] J S Loos, U E Kruse and E L Goldwasser, Physical Review 173 (1968) 1330

[20] H W Atherton *et al*, Physics Letters 30B (1969) 494

[21] T Regge, Il Nuovo Cimento 14 (1959) 951

[22] T Regge, Il Nuovo Cimento 18 (1960) 947

[23] A Bottino, A M Longhoni and T Regge, Il Nuovo Cimento 23 (1962) 954

[24] A Sommerfeld, *Partial Differential Equations in Physics* Academic Press (1949)

[25] G N Watson, Proceedings of The Royal Society A95 (1919) 83

[26] G N Watson, Proceedings of The Royal Society A95 (1919) 546

[27] Particle Data Group, European Physical Journal C15 (2000) 1

[28] G F Chew and S C Frautschi, Physical Review Letters 8 (1962) 41

[29] J K Storrow, Physics Reports 103 (1984) 317

[30] R J Eden and P V Landshoff, Physical Review 136B (1964) 1817

[31] S Mandelstam, Il Nuovo Cimento 30 (1963) 1127

[32] P V Landshoff and J C Polkinghorne, Physical Review 81 (1969) 1989

[33] P V Landshoff and J C Polkinghorne, Physics Reports 5C (1972) 1

[34] D Branson, Physical Review 179 (1969) 1608

[35] L M Jones and P V Landshoff, Nuclear Physics B94 (1975) 145

[36] V N Gribov, Soviet Physics: Journal of Experimental and Theoretical Physics 26 (1968) 414

[37] M Moshe, Physics Reports 37C (1978) 255

[38] B Renner, *Current Algebra and its Applications* Pergamon (1968)

[39] J B Bronzan *et al*, Physical Review Letters 18 (1967) 32

[40] J B Bronzan *et al*, Physical Review 157 (1967) 1448

[41] V Singh, Physical Review Letters 17 (1967) 340

[42] S Fubini, Il Nuovo Cimento 43 (1966) 475

[43] D Gross and H Pagels, Physical Review Letters 20 (1968) 961

[44] H G Dosch and D Gordon, Il Nuovo Cimento 57A (1968) 82

[45] M Jacob and G C Wick, Annals of Physics 7 (1959) 404

[46] A R Edmonds, *Angular Momentum in Quantum Mechanics* Princeton University Press (1966)

[47] L Lukaszuk and A Martin, Il Nuovo Cimento 52 (1967) 122

[48] Particle Data Group, Physical Review D45 (1992) 1

[49] J R Cudell *et al*, Physical Review D61 (1998) 034019

[50] P D B Collins and F Gault, Physics Letters 73B (1978) 330

[51] A Donnachie and P V Landshoff, Physics Letters B296 (1992) 227

[52] N A Amos *et al*, E710 Collaboration, Physical Review Letters 63 (1989) 2784

[53] F Abe *et al*, CDF Collaboration, Physical Review D50 (1994) 5550

[54] U Dersch *et al*, SELEX Collaboration, Nuclear Physics B579 (2000) 277

[55] J Rosner, C Rebbi and R Slansky, Physical Review 188 (1969) 2367

[56] A Capella *et al*, European Physical Journal C5 (1998) 111

[57] E Gotsman *et al*, European Physical Journal C10 (1999) 689

[58] H G Dosch, E Ferreira and A Krämer, Physics Letters B289 (1992) 153

[59] V Kundrat, M Lokajicek and D Krupa, hep-ph/0001047 (2000)

[60] P V Landshoff and J C Polkinghorne, Nuclear Physics B28 (1971) 225

[61] N H Buttimore *et al*, Physical Review D59 (1999) 114010

[62] N A Amos *et al*, Nuclear Physics B262 (1985) 689

[63] A Breakstone *et al*, Nuclear Physics B248 (1984) 253

[64] E Nagy *et al*, Nuclear Physics B150 (1979) 221

[65] N A Amos *et al*, E710 Collaboration, Physics Letters B247 (1990) 127

[66] G A Jaroskiewicz and P V Landshoff, Physical Review D10 (1974) 170

[67] A Donnachie and P V Landshoff, Nuclear Physics B231 (1984) 189

[68] C J Bebek *et al*, Physical Review D17 (1978) 1693

[69] S R Amendolia *et al*, Physics Letters B146 (1984) 116

[70] C W Akerlof *et al*, Physical Review D14 (1976) 2864

[71] P Desgrolard *et al*, European Physical Journal C16 (200) 499

[72] J Pumplin, Physical Review D8 (1973) 2899

[73] N A Amos *et al*, E710 Collaboration, Physical Review Letters 68 (1992) 2433

[74] C Augier *et al*, UA4/2 Collaboration, Physics Letters B316 (1993) 448

[75] P Gauron, E Leader and B Nicolescu, Physical Review Letters 52 (1984) 1952

[76] F Pereira and E Ferreira, Physical Review D59 (1999) 014008

[77] A Donnachie and P V Landshoff, Nuclear Physics B267 (1986) 690

[78] T Akesson *et al*, AFS Collaboration, Physics Letters B152 (1985) 140

[79] T T Chou and C N Yang, Physical Review D17 (1989) 1889

[80] C Bourrely, J Soffer and T T Wu, Zeitschrift für Physik C37 (1988) 369

[81] A Breakstone *et al*, Physical Review Letters 54 (1985) 1985

[82] W Faissler *et al*, Physical Review D23 (1981) 33

[83] A Donnachie and P V Landshoff, Zeitschrift für Physik C2 (1979) 55

[84] A Donnachie and P V Landshoff, Physics Letters B387 (1996) 637

[85] P Desgrolard *et al*, European Physical Journal C16 (2000) 499

[86] P V Landshoff and J C Polkinghorne, Physics Letters 44B (1973) 63

[87] S J Brodsky and G R Farrar, Physical Review Letters 31 (1973) 1153

[88] N H Buttimore, E Gotsman and E Leader, Physical Review D18 (1978) 694

[89] H J Lipkin and F Scheck, Physical Review Letters 16 (1966) 71

[90] E Leader, Physical Review 166 (1968) 1599

[91] N Akchurin *et al*, Physical Review D48 (1993) 3026

[92] B Z Kopeliovich and L I Lapidus, Soviet Journal of Nuclear Physics 19 (1974) 114

References

[93] A H Mueller, Physical Review D2 (1970) 2963

[94] J Coster and H P Stapp, Journal of Mathematical Physics 16 (1975) 1288

[95] A Donnachie and P V Landshoff, Nuclear Physics B244 (1984) 322

[96] P V Landshoff, Nuclear Physics B15 (1970) 284

[97] D P Roy and R G Roberts, Nuclear Physics B77 (1974) 240

[98] C Adloff *et al*, H1 Collaboration, Nuclear Physics B497 (1997) 3

[99] K Goulianos, Physics Letters B358 (1995) 379

[100] S Erhan and P E Schlein, Physics Letters B481 (2000) 177

[101] M G Albrow *et al*, CHLM Collaboration, Nuclear Physics B108 (1976) 1

[102] M Bozzo *et al*, UA4 Collaboration, Physics Letters B136 (1984) 217

[103] E Gotsman, E M Levin and U Maor, Physical Review D49 (1994) 4321

[104] A M Smith *et al*, R608 Collaboration, Physics Letters B163 (1985) 267

[105] T Henkes *et al*, R608 Collaboration, Physics Letters B283 (1992) 155

[106] A Schäfer, O Nachtmann and R Schöpf, Physics Letters B249 (1990) 331

[107] A Bialas and P V Landshoff, Physics Letters B256 (1991) 540

[108] V A Khoze, A D Martin and M G Ryskin, European Physical Journal C14 (2000) 525

[109] A H Mueller, Il Nuovo Cimento 37 (1965) 731

[110] N Bali, G F Chew and A Pignotti, Physical Review 163 (1967) 1572

[111] J-R Cudell and O F Hernandez, Nuclear Physics B471 (1996) 471

[112] M G Albrow and A Rostovtsev, hep-ph/0009336 (2000)

[113] J Pumplin, Physical Review D52 (1995) 1477

[114] C E DeTar, S D Ellis and P V Landshoff, Nuclear Physics B87 (1975) 176

[115] E Gotsman, E M Levin and U Maor, Physics Letters B353 (1995) 526

[116] T Arens *et al*, Zeitschrift für Physik C74 (1997) 651

[117] F E Close and G Schuler, Physics Letters B458 (1999) 127

[118] F E Close and G Schuler, Physics Letters B464 (1999) 279

[119] D Barberis *et al*, WA102 Collaboration, Physics Letters B397 (1997) 339

[120] A Kirk, Physics Letters B489 (2000) 29

[121] D Robson, Nuclear Physics B310 (1977) 328

[122] F E Close, Reports on Progress in Physics 51 (1988) 833

[123] F E Close, A Kirk and G Schuler, Physics Letters B477 (2000) 13

[124] D Barberis *et al*, WA102 Collaboration, Physics Letters B479 (2000) 59

[125] F E Close and A Kirk, Physics Letters B483 (2000) 345

[126] C Michael, M Foster and C McNeile, Nuclear Physics Proceedings Supplement 83 (2000) 185

[127] G S Bali *et al*, Physical Review D62 (2000) 054503

[128] S Abatzis *et al*, WA91 Collaboration, Physics Letters B324 (1994) 509

[129] D R O Morrison, Physical Review 165 (1968) 1699

[130] V N Gribov, Soviet Journal of Nuclear Physics 5 (1967) 138

[131] L Wang, Physical Review 153 (1967) 1664

[132] S D Drell and K Hiida, Physical Review Letters 7 (1961) 199

[133] R T Deck, Physical Review Letters 13 (1964) 169

[134] L Stodolsky, Physical Review Letters 18 (1967) 973

[135] G Ascoli *et al*, Physical Review D8 (1973) 3894

[136] M G Bowler *et al*, Nuclear Physics B97 (1975) 227

[137] M G Bowler, Journal of Physics G5 (1979) 203

[138] L Lukaszuk and B Nicolescu, Lettere Nuovo Cimento 8 (1973) 405

[139] D Joynson *et al*, Il Nuovo Cimento 30A (1975) 345

[140] M M Block and R N Cahn, Reviews of Modern Physics 57 (1985) 563

[141] P Desgrolard, M Giffon and E Predazzi, Zeitschrift für Physik C63 (1994) 241

[142] E Leader and T L Trueman, Physical Review D61 (2000) 077504

[143] A Schäfer, L Mankiewicz and O Nachtmann, Physics Letters B272 (1991) 419

[144] E R Berger *et al*, European Physical Journal C14 (2000) 673

[145] A Schäfer, L Mankiewicz and O Nachtmann in *Workshop on Physics at HERA*, W Buchmüller and G Ingelman eds, DESY (1992)

[146] W Kilian and O Nachtmann, European Physical Journal C5 (1998) 317

[147] E R Berger *et al*, European Physical Journal C9 (1999) 491

[148] J Czyzewski *et al*, Physics Letters B398 (1997) 400

[149] R Engel *et al*, European Physical Journal C4 (1998) 93

[150] S J Brodsky, J Rathsman and C Merino, Physics Letters B461 (1999) 114

[151] R J Glauber in *Lectures in Theoretical Physics*, W E Britten ed, Interscience (1959)

[152] W Czyz and L C Maximon, Annals of Physics 52 (1969) 59

[153] A Capella *et al*, Physics Reports 236 (1994) 225

[154] X M Wang, Physics Reports 280 (1997) 287

[155] S Bondarenko *et al*, Nuclear Physics A683 (2001) 649

[156] V A Abramovsky, V N Gribov and V Kancheli, Soviet Journal of Nuclear Physics 18 (1974) 308

[157] K Werner, Physics Reports C232 (1993) 87

[158] B Andersson, *The Lund Model* Cambridge University Press (1998)

[159] M Hladik *et al*, Physical Review Letters 86 (2001) 3506

[160] K Igi, Physical Review Letters 9 (1962) 76

[161] R Dolen, D Horn and C Schmid, Physical Review Letters 19 (1967) 402

[162] K Igi and S Matsuda, Physical Review Letters 18 (1967) 625

[163] A A Logunov, L D Soloviev and A N Tavkhelidze, Physics Letters 24B (1967) 181

[164] C Schmid, Proceedings of the Royal Society 318A (1970) 257

[165] V Barger and R J N Phillips, Physical Review 187 (1969) 2210

[166] H Harari and Y Zarmi, Physical Review 187 (1969) 2230

[167] G Veneziano, Il Nuovo Cimento 57A (1968) 190

[168] P H Frampton, *Dual Resonance Models* Benjamin (1974)

[169] S Mandelstam, Physics Reports 23C (1976) 245

[170] G Brandenburg *et al*, Physics Letters 58B (1975) 367

[171] G Cozzika *et al*, Physics Letters B40 (1972) 281

[172] G Abbiendi *et al*, OPAL Collaboration, European Physical Journal C14 (2000) 199

[173] M Acciari *et al*, L3 Collaboration, Physics Letters B519 (2001) 33

[174] J J Sakurai, Annals of Physics 11 (1960) 1

[175] M Gell-Mann and F Zachariasen, Physical Review 124 (1961) 953

[176] A Donnachie and G Shaw in *Electromagnetic Interactions of Hadrons* volume 2, A Donnachie and G Shaw eds, Plenum Press (1978)

[177] S Godfrey and N Isgur, Physical Review D32 (1985) 189

[178] D Aston *et al*, Nuclear Physics B209 (1982) 56

[179] A Donnachie and P V Landshoff, Physics Letters B348 (1995) 213

[180] A Donnachie and P V Landshoff, Physics Letters B478 (2000) 146

[181] G Gounaris and J J Sakurai, Physical Review Letters 21 (1968) 244

[182] F M Renard, Nuclear Physics B15 (1970) 267

[183] J Breitweg *et al*, ZEUS Collaboration, European Journal of Physics C1 (1998) 81

[184] T J Chapin *et al*, Physical Review D31 (1985) 17

[185] P Söding, Physics Letters 19 (1966) 702

[186] S D Drell, Review of Modern Physics 33 (1961) 458

[187] D W G S Leith in *Electromagnetic Interactions of Hadrons* volume 1, A Donnachie and G Shaw eds, Plenum Press (1978)

[188] M G Ryskin and Y M Shabelski, Physics of Atoms and Nuclei 61 (1998) 89

[189] R Ross and L Stodolsky, Physical Review 149 (1966) 1173

[190] S Okubo, Physics Letters 5 (1963) 165

[191] G Zweig, CERN report TH-412 (1964)

[192] G Zweig in *Developments in the Quark Theory of Hadrons* volume 1, D B Lichtenberg and S P Rosen eds, Hadronic Press (1980)

[193] J Iizuka, Progress in Theoretical Physics Supplement 37/38 (1966) 21

[194] A Donnachie and P V Landshoff, Physics Letters B437 (1998) 408

[195] K Schilling and G Wolf, Nuclear Physics B61 (1973) 381

[196] K Schilling, P Seyboth and G Wolf, Nuclear Physics B15 (1971) 397

[197] J Ballam *et al*, Physical Review D7 (1973) 3150

[198] A A Savin, ZEUS Collaboration, Nuclear Physics Proceedings Supplement 79 (1999) 333

[199] B Clerbaux, H1 Collaboration, Nuclear Physics Proceedings Supplement 79 (1999) 327

[200] C Adloff *et al*, H1 Collaboration, Zeitschrift für Physik C74 (1997) 221

[201] J K Storrow in *Electromagnetic Interactions of Hadrons* volume 1, A Donnachie and G Shaw eds, Plenum Press (1978)

[202] R L Anderson *et al*, Physical Review D4 (1971) 1937

[203] I S Barker, A Donnachie and J K Storrow, Nuclear Physics B95 (1975) 347

[204] I S Barker and J K Storrow, Nuclear Physics B137 (1978) 413

[205] I S Barker, A Donnachie and J K Storrow, Nuclear Physics B79 (1974) 431

[206] A M Boyarski *et al*, Physical Review Letters 20 (1968) 300

[207] P K Williams, Physical Review D1 (1970) 1312

[208] G C Fox and C Quigg, Annual Review of Nuclear Science 23 (1973) 219

[209] R Worden, Nuclear Physics B37 (1972) 253

[210] P A M Dirac, Physikalische Zeitschrift der Sowjetunion 3 (1933) 64

[211] R P Feynman, Reviews of Modern Physics 20 (1948) 367

[212] G 't Hooft and M Veltman, Nuclear Physics B44 (1972) 189

[213] C G Bollini and J J Gianbiagi, Physics Letters 40B (1972) 566

[214] K G Wilson, Physical Review D10 (1974) 2445

[215] W A Bardeen *et al*, Physical Review D18 (1978) 3998

[216] T van Ritbergen, J A M Vermaseren and S A Larin, Physics Letters B400 (1997) 379

[217] H D Politzer, Physical Review Letters 30 (1973) 1346

[218] D Gross and F Wilczek, Physical Review Letters 30 (1973) 1343

[219] S Coleman and E Weinberg, Physical Review D7 (1973) 1888

[220] W Bernreuther, Annals of Physics 151 (1983) 127

[221] W Marciano, Physical Review D29 (1984) 580

[222] R K Ellis, W J Stirling and B R Webber, *QCD and Collider Physics* Cambridge University Press (1996)

[223] P Lepage in *Perturbative and Nonperturbative Aspects of Quantum Field Theory*, H Latal and W Schweiger eds, Springer (1997)

[224] H J Rothe, *Lattice Gauge Theories, an Introduction* World Scientific (1992)

[225] G Altarelli, Physics Reports 81 (1982) 1

[226] J L Cardy and G A Winbow, Physics Letters B52 (1974) 95

[227] H D Politzer, Nuclear Physics B129 (1977) 301

[228] C T Sachrajda, Physics Letters B73 (1978) 185

[229] D Amati, R Petronzio and G Veneziano, Nuclear Physics B140 (1978) 54

[230] D Amati, R Petronzio and G Veneziano, Nuclear Physics B146 (1978) 29

[231] R K Ellis *et al*, Nuclear Physics B152 (1979) 285

[232] S B Libby and G Sterman, Physical Review D18 (1978) 3252

[233] S Gupta and A H Mueller, Physical Review D20 (1979) 118

[234] J C Collins, D E Soper and G Sterman in *Perturbative QCD*, A H Mueller ed, World Scientific (1989)

[235] G Sterman in *Perturbative and Nonperturbative Aspects of Quantum Field Theory*, H Latal and W Schweiger eds, Springer (1997)

[236] J Ellis, M K Gaillard and W J Zakrzewski, Physics Letters B81 (1979) 224

[237] R Doria, J Frenkel and J C Taylor, Nuclear Physics B168 (1980) 93

[238] O Nachtmann and A Reiter, Zeitschrift für Physik C24 (1984) 283

[239] S Falciano *et al*, NA10 Collaboration, Zeitschrift für Physik C31 (1986) 513

[240] J S Conway *et al*, Physical Review D39 (1989) 92

[241] A Brandenburg, E Mirkes and O Nachtmann, Zeitschrift für Physik C60 (1993) 697

[242] W Heisenberg, Zeitschrift für Physik 133 (1952) 65

[243] G Giacomelli and M Jacob, Physics Reports 55 (1979) 1

[244] O Nachtmann in *Perturbative and Nonperturbative Aspects of Quantum Field Theory*, H Latal and W Schweiger eds, Springer (1997)

[245] J Ambjørn and P Olesen, Nuclear Physics B170 (1980) 60

[246] J C Maxwell, Philosophical Magazine 21 (1861) 281

[247] E V Shuryak, Physics Reports C115 (1984) 151

[248] H M Fried and B Müller eds, *QCD Vacuum Structure* World Scientific (1993)

[249] G K Saviddy, Physics Letters 71B (1977) 133

[250] M A Shifman, A I Vainshtein and V I Zakharov, Nuclear Physics B147 (1979) 385

[251] S Narison, *QCD Spectral Sum Rules* World Scientific (1989)

[252] M A Shifman, *Vacuum Structure and QCD Sum Rules* North Holland (1992)

[253] K G Wilson, Physical Review D3 (1971) 1818

[254] C Itzykson and J Zuber, *Quantum Field Theory* McGraw Hill (1980)

[255] S Narison, Nuclear Physics Proceedings Supplement 54A (1997) 238

[256] V A Novikov *et al*, Nuclear Physics B237 (1984) 525

[257] M D'Eglia, A Di Giacomo and E Meggiolaro, Physics Letters B408 (1997) 315

[258] H G Dosch, M Eidemüller and M Jamin, Physics Letters B452 (1999) 379

[259] H G Dosch, E Ferreira and A Krämer, Physical Review D50 (1994) 1992

[260] E Meggiolaro, Physics Letters B451 (1999) 414

[261] F Wegner, Journal of Mathematical Physics 12 (1971) 2259

[262] I Y Aref'eva, Teoreticheskaya i Matematicheskaya Fizika 43 (1980) 111

[263] N E Bralic, Physical Review D22 (1980) 3090

[264] Y A Simonov, Yadernaya Fizika 48 (1988) 1381

[265] H G Dosch, Physics Letters 190B (1987) 177

[266] H G Dosch and Y A Simonov, Physics Letters 205B (1988) 339

[267] H G Dosch, Progress in Particle and Nuclear Physics 33 (1994) 121

[268] H G Dosch in *Hadron Physics 96*, E Ferreira *et al* eds, World Scientific (1997)

[269] Y A Simonov, Physics Uspekhi 39 (1996) 313

[270] N G Van Kampen, Physics Reports C24 (1976) 172

[271] W Buchmüller and S-H H Tye, Physical Reviw D24 (1981) 132

[272] S Aoki *et al*, CP-PACS Collaboration, Nuclear Physics Proceedings Supplement 73 (1999) 216

[273] G S Bali, Nuclear Physics Proceedings Supplement 83 (1999) 422

[274] V I Shevchenko and Y A Simonov, Physical Review Letters 85 (2000) 1811

[275] M Rueter and H G Dosch, Zeitschrift für Physik C66 (1995) 245

[276] H G Dosch, O Nachtmann and M Rueter, hep-ph/9503386 (1995)

[277] G S Bali, K Schilling and C Schlichter, Physical Review D51 (1995) 5165

[278] M Beneke, Physics Reports C137 (1999) 1

[279] A H Mueller, Nuclear Physics B250 (1985) 327

[280] B R Webber, Nuclear Physics Proceedings Supplement 71 (1999) 66

[281] Y L Dokshitzer, hep-ph/9911299 (1999)

[282] G Källen and W Pauli, Kgl Danske Videnskab Selskab, Mat-Fys Medd 30 (1955) 3

[283] F J Dyson, Physical Review 85 (1952) 631

[284] L D Landau, *Quantum theory of fields*, in *Niels Bohr and the Development of Physics*, W Pauli ed, Pergamon Press (1955)

[285] M Göckeler *et al*, Physical Review Letters 80 (1998) 4119

[286] J D Bjorken, Physical Review 179 (1969) 1547

[287] R G Roberts, *The Structure of the Proton* Cambridge University Press (1990)

[288] L N Hand, Physical Review 129 (1963) 1834

[289] F J Gilman, Physical Review 167 (1967) 1365

[290] C Adloff *et al*, H1 Collaboration, European Physical Journal C19 (2001) 269

[291] C Adloff *et al*, H1 Collaboration, Physics Letters B520 (2001) 183

[292] A D Martin *et al*, European Physical Journal C4 (1998) 463

[293] H L Lai *et al*, Physical Review D55 (1997) 1280

[294] M Glück, E Reya and A Vogt, European Physical Journal C5 (1998) 461

[295] R D Ball and S Forte, Physics Letters B335 (1994) 77

[296] A De Rujula *et al*, Physical Review D10 (1974) 1649

[297] E A Kuraev, L N Lipatov and V S Fadin, Soviet Physics: Journal of Experimental and Theoretical Physics 45 (1977) 199

[298] Y Y Balitskii and L N Lipatov, Soviet Journal of Nuclear Physics 28 (1978) 822

[299] J C Collins and P V Landshoff, Physics Letters B276 (1992) 196

[300] J Bartels, H Lotter and M Vogt, Physics Letters B373 (1996) 215

[301] J Bartels, A De Roeck and H Lotter, Physics Letters B389 (1996) 742

[302] J Brodsky, F Hautmann and D E Soper, Physical Review D56 (1997) 6957

[303] S Fadin and L N Lipatov, Physics Letters B429 (1998) 127

[304] G Camici and M Ciafaloni, Physics Letters B412 (1997) 396

[305] M Ciafaloni, D Colferai and G P Salam, Journal of High Energy Physics 9910 (1999) 017

[306] G Altarelli, R D Ball and S Forte, Nuclear Physics B575 (1999) 313

[307] R S Thorne, Nuclear Physics Proceedings Supplement 79 (1999) 210

[308] T Jaroszewicz, Physics Letters 116B (1982) 291

[309] R S Thorne, Physical Review D60 (1999) 054031

[310] M Ciafaloni, D Colferai and G P Salam, Journal of High Energy Physics 0007 (2000) 054

[311] G Altarelli, R D Ball and S Forte, Nuclear Physics B599 (2001) 383

[312] R S Thorne, Physical Review D64 (2001) 074005

[313] A Donnachie and P V Landshoff, Physics Letters B518 (2001) 63

[314] S J Brodsky and G R Farrar, Physical Review Letters 31 (1973) 1153

[315] V A Matveev, R M Murddyan and A N Tavkhelidze, Lettere Nuovo Cimento 7 (1973) 719

[316] J J Aubert *et al*, EMC Collaboration, Nuclear Physics B293 (1987) 74

[317] G Altarelli, R D Ball and S Forte, hep-ph/0104246 (2001)

[318] J Breitweg *et al*, ZEUS Collaboration, European Physical Journal C12 (2000) 35

[319] L V Gribov, E M Levin and M G Ryskin, Physics Reports 100 (1983) 1

[320] A Donnachie and P V Landshoff, hep-ph/0111427 (2001)

[321] J-R Cudell, A Donnachie and P V Landshoff, Physics Letters B448 (1999) 281

[322] Z Sullivan and P M Nadolsky, hep-ph/0111358 (2001)

[323] C Adloff *et al*, H1 Collaboration, European Physical Journal C21 (2001) 33

[324] A Cooper-Sarkar, in *International Europhysics Conference on High Energy Physics Budapest 2001*, D Horvath, P Levai and A Patkos eds, JHEP (http://jhep.sissa.it/) Proceedings Section, PrHEP-hep2001/009 (2001)

[325] A D Martin *et al*, European Physics Journal C18 (2000) 117

[326] A D Martin *et al*, European Physics Journal C23 (2002) 73

[327] H L Lai *et al*, European Physics Journal C12 (2000) 375

[328] C Adloff *et al*, H1 Collaboration, European Physical Journal C21 (2001) 33

[329] V M Budnev *et al*, Physics Reports 15C (1975) 181

[330] K Ackerstaff *et al*, OPAL Collaboration, Physics Letters B411 (1997) 387

[331] K Ackerstaff *et al*, OPAL Collaboration, Physics Letters B412 (1997) 225

[332] G Abbiendi *et al*, OPAL Collaboration, European Physical Journal C18 (2000) 15

[333] M Acciarri *et al*, L3 Collaboration, Physics Letters B436 (1998) 403

[334] M Acciarri *et al*, L3 Collaboration, Physics Letters B447 (1999) 147

[335] D Barate *et al*, ALEPH Collaboration, Physics Letters B458 (1999) 152

[336] M Acciarri *et al*, L3 Collaboration, Physics Letters B514 (2001) 19

[337] M Acciarri *et al*, L3 Collaboration, Physics Letters B514 (2001) 19

[338] S Frixione, E Laenen and M Krämer, Nuclear Physics B571 (2000) 169

[339] M Glück, E Reya and I Schienbein, Physical Review D60 (1999) 054019

[340] M Glück, E Reya and M Stratman, European Physical Journal C2 (1998) 159

[341] M Glück, E Reya and I Schienbein, European Physical Journal C10 (1999) 313

[342] P Achard *et al*, L3 Collaboration, hep-ex/0111012 (2001)

[343] G Abbiendi *et al*, OPAL Collaboration, hep-ex/0110006 (2001)

[344] M Boonekamp *et al*, Nuclear Physics B555 (1999) 540

[345] A Donnachie, H G Dosch and M Rueter, European Physical Journal C13 (2000) 141

[346] E Gotsman *et al*, European Physical Journal C14 (2000) 511

[347] J Kwiecinski and L Motyka, European Physical Journal C18 (2000) 343

[348] C Adloff *et al*, H1 Collaboration, Physics Letters B483 (2000) 23

[349] J Breitweg *et al*, ZEUS Collaboration, European Physical Journal C1 (1998) 81

[350] J Breitweg *et al*, ZEUS Collaboration, European Physical Journal C14 (2000) 213

[351] J Busenitz *et al*, Phys Rev D40 (1989) 1

[352] M Derrick *et al*, ZEUS Collaboration, Physics Letters B377 (1996) 259

[353] J Breitweg *et al*, ZEUS Collaboration, European Physical Journal C6 (1999) 603

[354] S Aid *et al*, H1 Collaboration, Nuclear Physics B463 (1996) 3

[355] S Aid *et al*, H1 Collaboration, Nuclear Physics B468 (1996) 3

[356] J Breitweg *et al*, ZEUS Collaboration, European Physical Journal C2 (1998) 247

[357] A Donnachie, J Gravelis and G Shaw, Physical Review D63 (2001) 114013

[358] V V Anisovich *et al*, Physical Review D60 (1999) 074011

[359] E L Berger and D Jones, Physical Review D27 (1981) 1521

[360] S J Brodsky *et al*, Physical Review D50 (1994) 3134

[361] M G Ryskin, Zeitschrift für Physik C57 (1993) 89

[362] M G Ryskin *et al*, Zeitschrift für Physik C76 (1997) 231

[363] C Adloff *et al*, H1 Collaboration, European Physical Journal C10 (1999) 373

[364] C Adloff *et al*, H1 Collaboration, European Physical Journal C13 (2000) 371

[365] L Frankfurt, M McDermott and M Strikman, Journal of High Energy Physics 3 (2001) 045

[366] A G Shuvaev *et al*, Physical Review D60 (1999) 014015

[367] L Frankfurt, W Koepf and M Strikman, Physical Review D54 (1996) 319

[368] L Frankfurt, W Koepf and M Strikman, Physical Review D57 (1998) 513

[369] J C Collins, L Frankfurt and M Strikman, Physical Review D56 (1997) 2982

[370] A Donnachie, J Gravelis and G Shaw, European Physical Journal C18 (2001) 539

[371] I Royon and J-R Cudell, Nuclear Physics B545 (1999) 505

[372] M Diehl, Zeitschrift für Physik C66 (1995) 181

[373] J Breitweg *et al*, ZEUS Collaboration, European Physical Journal C12 (2000) 393

[374] D Y Ivanov and R Kirschner, Physical Review D58 (1998) 114026

[375] E V Kuraev, N N Nikolaev and B G Zakharov, Journal of Experimental and Theoretical Physics Letters 68 (1998) 696

[376] J R Forshaw and M G Ryskin, Zeitschrift für Physik C68 (1995) 137

[377] J Bartels *et al*, Physics Letters B375 (1996) 301

[378] J R Forshaw and G Poludniowski, hep-ph/0107068 (2001)

[379] A Donnachie and P V Landshoff, Physics Letters B185 (1987) 403

[380] C Adloff *et al*, H1 Collaboration, European Physical Journal C6 (1999) 587

[381] C Adloff *et al*, H1 Collaboration, Zeitschrift für Physik C76 (1997) 613

[382] J Breitweg *et al*, ZEUS Collaboration, European Physical Journal C6 (1999) 43

[383] B Kopeliovich, B Povh and I Potashnikova, Zeitschrift für Physik C73 (1996) 125

[384] M Glück, E Reya and A Vogt, Zeitschrift für Physik C53 (1992) 651

[385] M Glück, E Reya and A Vogt, Zeitschrift für Physik C67 (1992) 433

[386] M N Kapishin, H1 Collaboration, Nuclear Physics Proceedings Supplement 79 (1999) 321

[387] A Szczurek, N N Nikolaev and J Speth, Physics Letters B428 (1998) 383

[388] C Royon *et al*, Physical Review D63 (2000) 074004

[389] A Bialas, R Peschanski and C Royon, Physical Review D57 (1998) 6899

[390] S Munier, R Peschanski and C Royon, Nuclear Physics B534 (1998) 297

[391] J Bartels *et al*, European Physical Journal C7 (1999) 443

[392] A Brandt *et al*, UA8 Collaboration, Physics Letters B421 (1998) 395

[393] K Goulianos, Physics Letters B358 (1995) 379

[394] F Abe *et al*, CDF Collaboration, Physical Review Letters 78 (1997) 2698

[395] E Gotsman, E M Levin and U Maor, Physical Review D56 (1999) 5687

[396] E Gotsman, E M Levin and U Maor, Physical Review D60 (1999) 094011

[397] J D Bjorken and J Kogut, Physical Review D8 (1973) 1341

[398] G Bertsch *et al*, Physical Review Letters 47 (1981) 297

[399] C Adloff *et al*, H1 Collaboration, European Physical Journal C20 (2001) 29

[400] A Edin, G Ingelman and J Rathsman, Physics Letters B366 (1996) 371

[401] A Edin, G Ingelman and J Rathsman, Zeitschrift für Physik C75 (1996) 57

[402] J Rathsman, Physics Letters B452 (1999) 364

[403] K Golec-Biernat and M Wüsthoff, Physical Review D59 (1999) 014017

[404] K Golec-Biernat and M Wüsthoff, Physical Review D60 (1999) 114023

[405] J Bartels, H Jung and M Wüsthoff, European Physical Journal C11 (1999) 111

[406] J Bartels, H Jung and A Kyrieleis, hep-ph/0010300 (2000)

[407] J Bartels, Nuclear Physics B175 (1980) 365

[408] J Kwiecinski and M Praszalowicz, Physics Letters B94 (1980) 413

[409] R A Janik and J Wosiek, Physics Letters 82 (1999) 1092

[410] J Bartels, L N Lipatov and G P Vacca, Physics Letters B477 (2000) 178

[411] M A Braun, hep-ph/9805394 (1998)

[412] G P Vacca, in *Proceedings of the 9th International Workshop on Deep Inelastic Scattering and QCD (DIS 2001)* hep-ph/0106224 (2001)

[413] R Engel *et al*, European Physical Journal C4 (1998) 93

[414] J Bartels *et al*, European Physical Journal C20 (2001) 323

[415] E M Levin and L Frankfurt, Journal of Experimental and Theoretical Physics Letters 2 (1965) 65

[416] H J Lipkin, Physical Review Letters 16 (1966) 1015

[417] J J J Kokkedee and L Van Hove, Il Nuovo Cimento A42 (1966) 711

[418] D G Richards, Nuclear Physics B258 (1985) 267

[419] J F Gunion and D E Soper, Physical Review D15 (1977) 2617

[420] H Lipkin, Physics Letters 116B (1982) 175

[421] R F Streater, Il Nuovo Cimento 25 (1962) 274

[422] C Parrinello, Nuclear Physics Proceedings Supplement 53 (1997) 331

[423] O Nachtmann, Annals of Physics 209 (1991) 436

[424] B Andersson *et al*, Physics Reports C97 (1983) 31

[425] A Messiah, *Quantum Mechanics* volume I North-Holland (1961)

[426] P V Landshoff, A J F Metherell and W G Rees, *Essential Quantum Physics* Cambridge University Press (1997)

[427] W Buchmüller and A Hebecker, Nuclear Physics B476 (1996) 203

[428] H Verlinde and E Verlinde, hep-th/9302104 (1993)

[429] A Krämer and H G Dosch, Physics Letters B252 (1990) 669

[430] A Krämer and H G Dosch, Physics Letters B272 (1991) 114

[431] E R Berger and O Nachtmann, European Physical Journal C7 (1999) 459

[432] A Hebecker and P V Landshoff, Physics Letters B419 (1998) 393

[433] H G Dosch, T Gousset and H J Pirner, Physical Review D57 (1998) 1666

[434] P Ball *et al*, Nuclear Physics B529 (1998) 323

[435] H G Dosch, T Gousset, G Kulzinger and H J Pirner, Physical Review D55 (1997) 2602

[436] M Wirbel, B Stech and M Bauer, Zeitschrift für Physik C19 (1985) 637

[437] H G Dosch *et al*, European Physical Journal C21 (2001) 339

[438] S J Brodsky and G P Lepage, Physical Review D22 (1998) 2157

[439] S Munier, A M Stasto and A H Mueller, Nuclear Physics B603 (2001) 427

[440] M Rueter and H G Dosch, Physical Review D57 (1998) 4097

[441] G Kulzinger, H G Dosch and H J Pirner, European Physical Journal C7 (1999) 73

[442] M Rueter, European Physical Journal C7 (1999) 233

[443] E Ferreira and F Pereira, Physical Review D56 (1997) 179

[444] L Gerland *et al*, Physical Review Letters 81 (1998) 762

[445] J Hüfner and B Z Kopeliovich, Physics Letters B426 (1998) 154

[446] J Hüfner *et al*, Physical Review D62 (2000) 094022

[447] E Meggiolaro, European Physical Journal C4 (1998) 101

[448] M Arneodo *et al*, Nuclear Physics B429 (1994) 503

[449] S Anderson *et al*, CLEO Collaboration, Physical Review D61 (2001) 112002

[450] D Bisello *et al*, DM2 Collaboration, Physics Letters B220 (1989) 321

[451] D Aston *et al*, Physics Letters B92 (1980) 215

[452] A Donnachie and H Mirzaie, Zeitschrift für Physik C33 (1987) 407

[453] P Lebrun in *HADRON '97*, S Chung and H J Willetski eds, American Institute of Physics (1998)

[454] A Donnachie *et al*, European Physical Journal C9 (1999) 491

[455] M Rueter and H G Dosch, Physics Letters B380 (1996) 177

[456] E R Berger *et al*, European Physical Journal C14 (2000) 673

[457] T Golling, H1 Collaboration, in *International Europhysics Conference on High Energy Physics Budapest 2001*, D Horvath, P Levai and A Patkos eds, JHEP (http://jhep.sissa.it/) Proceedings Section, PrHEP-hep2001/034 (2001)

[458] A B Kaidalov and Y A Simonov, Physics Letters B477 (2000) 163

[459] L Brink and M Henneaux, *Principles of String Theory* Plenum Press (1988)

[460] D Kharzeev and E Levin, Nuclear Physics B578 (2000) 351

[461] V A Novikov *et al*, Nuclear Physics B191 (1981) 301

[462] A Ringwald, Nuclear Physics B330 (1990) 1

[463] O Espinosa, Nuclear Physics B343 (1990) 310

[464] A Ringwald and F Schrempp, Physics Letters B459 (1999) 249

[465] E Shuryak and I Zahed, Physical Review D62 (2000) 085014

[466] M A Nowak, E V Shuryak and I Zahed, Physical Review D64 (2001) 034008

[467] D E Kharzeev, Y V Kovchegov and E Levin, Nuclear Physics A690 (2000) 621

[468] P H Frampton, *Dual Resonance Models and Superstrings* World Scientific (1986)

[469] R A Janik and R Peschanski, Nuclear Physics B565 (2000) 193

[470] O Aharony *et al*, Physics Reports C323 (2000) 183

[471] R E Reedon *et al*, UA6 Collaboration, Physics Letters B147 (1989) 459

[472] M Bozzo *et al*, UA4 Collaboration, Physics Letters B147 (1984) 385

[473] E Meggiolaro, Zeitschrift für Physik C76 (1997) 523

[474] E Meggiolaro, Nuclear Physics B602 (2001) 261

[475] R A Janik and R Peschanski, Nuclear Physics B586 (2000) 163

[476] A Hebecker, E Meggiolaro and O Nachtmann, Nuclear Physics B571 (2000) 26

[477] E C Titchmarsh, *The Theory of Functions* Oxford University Press (1939)

[478] N Nikolaev and B G Zakharov, Zeitschrift für Physik C53 (1992) 331

[479] M Arneodo *et al*, Nuclear Physics B333 (1990) 1

[480] J C Collins, Physical Review D56 (1997) 2982

[481] M McDermott *et al*, European Physical Journal C16 (1999) 641

[482] J R Forshaw, G Kerley and G Shaw, Physical Review D60 (1999) 074012

[483] X Ji, Journal of Physics G24 (1998) 1181

[484] K Goeke, M V Polyakov and M Vanderhaeghen, Progress in Particle and Nuclear Physics 47 (2001) 401

[485] A V Radyushkin, *At the Frontier of Particle Physics/Handbook of QCD* World Scientific (2001)

[486] G Cvetic, D Schildknecht and A Shoshi, European Physical Journal C13 (2000) 301

[487] A H Mueller and B Patel, Nuclear Physics B425 (1994) 471

[488] N N Nikolaev, B G Zakharov and V R Zoller, Physics Letters B328 (1994) 486

[489] N N Nikolaev, J Speth and V R Zoller, Physics Letters B473 (2000) 157

[490] J Nemchik *et al*, Physics Letters B374 (1996) 199

[491] I P Ivanov and N N Nikolaev, Journal of Experimental and Theoretical Physics Letters 69 (1999) 294

[492] N N Nikolaev, A V Pronyaev and B G Zakharov, Journal of Experimental and Theoretical Physics Letters 68 (1998) 634

[493] A Donnachie and H G Dosch, Physical Review D65 (2002) 014019

[494] W Buchmüller, T Gehrmann and A Hebecker, Nuclear Physics B537 (1999) 477

[495] A Hebecker and H Weigert, Physics Letters B432 (1998) 215

[496] J Nemchik, N N Nikolaev and B G Zakharov, Physics Letters B341 (1994) 228

[497] L McLerran and R Venugopalan, Physical Review D49 (1994) 2233

[498] J Jalilian-Marian *et al*, Physical Review D55 (1997) 5414

[499] A H Mueller, hep-ph/9911289 (1999)

[500] H G Dosch *et al*, Nuclear Physics B568 (2000) 287

[501] A H Mueller, Nuclear Physics B558 (1999) 285

[502] E Gotsman *et al*, Journal of Physics G27 (2001) 2297

[503] A Capella *et al*, Physical Review D63 (2001) 054010

[504] A B Kaidalov *et al*, European Physical Journal C21 (2001) 521

[505] L McLerran, hep-ph/0104285 (2001)

[506] C Adloff *et al*, H1 Collaboration, Physics Letters B517 (2001) 97

[507] S Mandelstam, Annals of Physics 19 (1962) 254

[508] I S Gradshteyn and I M Ryzhik, *Table of Integrals, Series and Products* Academic Press (1980)

[509] R Cutler and D Sivers, Physical Review D17 (1978) 196

[510] S Weinberg, *The Quantum Theory of Fields* volumes I, II Cambridge University Press (1995 and 1996)

[511] G F Chew, M L Goldberger, F E Low and Y Nambu, Physical Review 106 (1960) 1337

[512] G Cohen-Tannoudji, A Morel and H Navelet, Annals of Physics 46 (1968) 239

[513] V Singh, Physical Review 129 (1963) 1889

[514] A D Lesquen *et al*, Physics Letters 40B (1972) 277

[515] G F Chew, M L Goldberger, F E Low and Y Nambu, Physical Review 106 (1960) 1345

Index